W9-AOM-908

# *Tectonic Geomorphology*

TITLES OF RELATED INTEREST

*Physical processes of sedimentation*
J. R. L. Allen

*The formation of soil material*
T. R. Paton

*Petrology of the sedimentary rocks*
J. T. Greensmith

*Elements of geographical hydrology*
B. J. Knapp

*Geology for civil engineers*
A. C. McLean & C. D. Gribble

*Thresholds in geomorphology*
D. R. Coates & J. D. Vitek (eds)

*Soil processes*
B. J. Knapp

*Fluvial geomorphology*
M. Morisawa (ed.)

*Glacial geomorphology*
D. R. Coates (ed.)

*Theories of landform development*
W. N. Melhorn & R. C. Flemal (eds)

*Geomorphology and engineering*
D. R. Coates (ed.)

*Geomorphology in arid regions*
D. O. Doehring (ed.)

*Tectonic processes*
D. Weyman

*Geomorphological techniques*
A. S. Goudie (ed.)

*Terrain analysis and remote sensing*
J. R. G. Townshend (ed.)

*British rivers*
J. Lewin (ed.)

*Soils and landforms*
A. J. Gerrard

*Soil survey and land evaluation*
D. Dent & A. Young

*Adjustments of the fluvial system*
D. D. Rhodes & G. P. Williams (eds)

*Applied geomorphology*
R. G. Craig & J. L. Craft (eds)

*Space and time in geomorphology*
C. E. Thorn (ed.)

*Pedology*
P. Duchaufour

*Sedimentology: process and product*
M. R. Leeder

*Sedimentary structures*
J. D. Collinson & D. B. Thompson

*Geomorphological field manual*
R. V. Dackombe & V. Gardiner

*Earth structures engineering*
R. J. Mitchell

*Statistical methods in geology*
R. F. Cheeney

*Groundwater as a geomorphic agent*
R. G. LaFleur (ed.)

*The climatic scene*
M. J. Tooley & G. S. Sheail (eds)

*Quaternary paleoclimatology*
R. S. Bradley

*Geomorphology and soils*
K. S. Richards, R. R. Arnett & S. Ellis (eds)

*Models in geomorphology*
M. Woldenberg (ed.)

*Quaternary environments*
J. T. Andrews (ed.)

*Introducing groundwater*
M. Price

*Geomorphology: pure and applied*
M. G. Hart

*Environmental chemistry*
P. O'Neill

*The dark side of the Earth*
R. Muir Wood

*Geology and mineral resources of West Africa*
J. B. Wright, et al.

*Principles of physical sedimentology*
J. R. L. Allen

*Experiments in physical sedimentology*
J. R. L. Allen

*Planetary landscapes*
R. Greeley

*Environmental magnetism*
F. Oldfield & R. Thompson

*Hillslope processes*
A. D. Abrahams (ed.)

# Tectonic Geomorphology

Edited by
M. Morisawa
*State University of New York at Binghamton*
and
J. T. Hack
*United States Geological Survey, Reston, Virginia*

*Proceedings of the 15th Annual Binghamton Geomorphology Symposium, September 1984*

Boston
ALLEN & UNWIN
London Sydney

This volume was prepared, proofed and passed for press by the Editors and Contributors.

**Allen & Unwin Inc.,**
**Fifty Cross Street, Winchester, Mass. 01890, USA**

George Allen & Unwin (Publishers) Ltd,
40 Museum Street, London WC1A 1LU, UK

George Allen & Unwin (Publishers) Ltd,
Park Lane, Hemel Hempstead, Herts HP2 4TE, UK

George Allen & Unwin Australia Pty Ltd,
8 Napier Street, North Sydney, NSW 2060, Australia



First published in 1985



Library of Congress Cataloging-in-Publication Data

"Binghamton" Geomorphology Symposium (15th : 1984 :
State University of New York At Binghamton)
Tectonic geomorphology.

Bibliography: p.
1. Geomorphology--Congresses. 2. Plate tectonics--Congresses. 3. Geology, Structural--Congresses. 4. Earthquake prediction--Congresses. I. Morisawa, Marie. II. Hack, John Tilton, 1913- . III. Title.
GB400.2.B56 1984 551.4 85-15095
ISBN 0-04-551098-9 (alk. paper)

---

**British Library Cataloguing in Publication Data**

Binghamton Geomorphology Symposium (*15th : 1984*)
Tectonic geomorphology.—— (Binghamton symposia in geomorphology. International series, ISSN 0261–3174; 15)
1. Geomorphology 2. Geology, Structural
I. Title II. Morisawa, Marie III. Hack, John
IV. Series
551.1'36 GB401.5
ISBN 0–04–551098–9

---

Printed in Great Britain by Mackays of Chatham

# Preface

Although accepting the paradigm of the influence of structure on the landscape, geomorphologists, with some exceptions, have not been generally concerned with the influence of active tectonism on the development of landforms. Two stimuli have given a recent impetus to change in the direction of geomorphic studies: (1) plate tectonic theory and (2) spaceborne remote sensing imagery. The plate tectonic revolution in the Earth Sciences has given geomorphologists a new perspective on morphology and processes on the Earth's surface. We are now looking at a world that is more dynamic than we had previously imagined. We must now examine continents that move apart and collide, with sea floors that rift, spread and are subducted beneath other sea floor plates or continents, with plates that move laterally against each other. Looked at in this light, we must now approach geomorphic studies with a regional - or even global perspective. According to Ollier (1981) we must learn to look at the forest rather than individual trees. We must turn our attention to the large, regional landscapes of continental and oceanic plates to determine the intimate relations between the exogenic and endogenic forces that shape them.

The new term "morphotectonics" has arisen in the literature. The erosion of large areal plate tectonic phenomena by subsequent subaerial processes produce characteristic landforms by which the tectonic events can be identified. Morphotectonics, hence, reflects the relationship between tectonics and surface features. According to Maroukian and Zamani, 1984,(p. 191) "In tectonically active areas the topography reflects the magnitude, intensity, frequency and duration of endogenic forces acting on them." An understanding of this relationship can allow the interpretation of tectonic events from the study of landforms. By integrating the study of plate tectonics with that of the surface features we can, thus, more easily unravel the geologic history of the earth.

But first we need to develop the principles and techniques of interpretation to create models combining global tectonics with the geomorphic processes of erosion and sedimentation. New tools have been provided by space imagery (Landsat, Seasat, SLAR, SIR and Shuttle imagery). These can be used by geomorphologists to observe the Earth on a regional scale. Remote sensing, combined with traditional methods of analysis, supply the techniques by which principles can be derived. It is up to us to adapt to the new global, tectonic, megageomorphology. A tremendous challenge lies before us.

This tectonic geomorphology symposium brings together some of the recent work done by geomorphologists relating plate tectonics and geomorphology.

Part I consists of papers which discuss the general relations of geomorphology to plate tectonics. Ollier describes morphologic characteristics of continental margins bounded by Great Escarpments. In discussing the mechanism of plateau uplift he points out that integration of geomorphic studies and tectonic principles is needed for a suitable model. Summerfield analyses South African tectonism in terms of the geomorphic processes which have affected the region and show how the landforms must be used to aid in our understanding of the tectonic events.

Geomorphic features such as scarps, straths, pediments and drainage changes are used as evidence of past tectonism in the Dead Sea Rift by Gerson et al. Hare and Gardner derive a model of neotectonic deformation in a subduction zone from various geomorphic forms. They utilize scarps, knickpoints, drainage characteristics, lineaments and erosion surfaces to delineate tectonic style as one plate is subducted beneath another.

Two studies relate geomorphology to tectonics as plates converge in New Zealand. Adams sees a dynamic balance occurring between uplift rate and erosion rate as evidenced by the landscape and invokes a dynamic cuesta model as characteristic of the geomorphology at a plate convergence. Bull, from a study of flights of marine terraces, derives a methodology for determining uplift rate at the convergence, allowing dates of the flights of terraces to be established.

The second group of papers are concerned with the use of geomorphology in studying the tectonism of smaller areas. Oberlander discusses drainage changes in orogenic fold belts, a number of folded mountain ranges as examples. Keller et al deal specifically with landforms generated by recent faulting in California. Petersen models alterations over time in features such as scarps created by faulting. According to his study, modification of the scarps by geomorphic processes can be used to date the time of origin of piedmont scarps along the Wasatch Front. Mayer and Young both deal with scarp degradation in the Colorado Plateau region. Mayer applies a diffusion model to scarp modification, as well as stream profile adjustment to determine the timing of tectonic events. Young uses paleontologic evidence to date tectonic events and relates the timing to rate of scarp retreat and scarp dissection.

A series of papers describe some of the current work by U.S. Geological Survey geologists on the neotectonics of the eastern United States. Markewich uses paleontology, isotopic data and geomorphology to determine the age and changing type of deformation in the Cape Fear region, N.C. Landforms such as scarps, terraces and river migration indicate deformation style, changing from arching to faulting to regional uplift over time. Erosion and distribution of sediments, as well as terrace formation, are controlled by deformation according to Newell's study in the Rappahannock estuary. Hence, one can infer tectonic style from erosional landscapes and the associated sediments. Pavich's study in the Appalachian piedmont, a relatively stable area, shows uplift from isostatic compensation is controlled by weathering and erosion rates. He finds there is an overall equilibrium between denudation rates and rates of uplift in the area for about 70 million years. However, weathering profiles indicate that in the late Cenozoic there was variation in uplift and erosion rates.

Part III contains two papers which explain the use of morphotectonics to predict earthquakes. In Japan surface features, especially marine terraces, have been related to earthquakes in many regions, according to Ota. Data obtained from strata exposed by trenching across known faults have enabled Japanese geomorphologists to determine recurrence intervals of seismic events. In China, Han has examined tectonic landforms such as faulted and deformed river terraces and related them to seismic events. Analysis of the morphology of terrace displacement gives an indication of style of tectonic deformation as well as pointing out specific locations for seismic events.

These types of geomorphic analysis has led to successful earthquake prediction. Thus the use of morphotectonics can have important human implications.

Many people helped make the 15th Geomorphology Symposium a success: the authors (speakers), the session leaders, the students who assisted in many ways. I would particularly like to thank my Co-Director, John Hack who spent a great deal of time and energy helping with the program and editing papers; Sanjeev Kalaswad graduate student who handled many of the logistics and Katherine Holley who word processed some of the manuscripts. My appreciation also for the financial support of the SUNY Foundation Conversations in the Disciplines and to the National Science Foundation for their travel support which enabled the foreign speakers to attend.

Marie Morisawa
SUNY Binghamton
March, 1985

REFERENCES

Ollier, C.D., 1981, Tectonics and landforms: London, Longman.

Maroukian, H. and Zamani, A., 1984, Morphotectonic observations in the drainage basin of the Sperkhios River, Central Greece, in Mishev, K. and Vaptsarov, I., Problems of morphotectonics, Bulgarian Acad. of Science, p. 191-203.

# Contents

# Contributors

John Adams - Division of Seismology and Geomagnetism, Earth Physics Branch, Energy, Mines and Resources, Ottawa K1A OY3, Canada

Dan Bowman - Department of Geography, Ben-Gurion University, Beer Sheva 84105, Israel

William B. Bull - Department of Geosciences, University of Arizona, Tucson, Arizona 85721

Thomas W. Gardner - Geosciences Department, Penn State University, University Park, Pennsylvania 16802

Ran Gerson - Institute of Earth Sciences, The Hebrew University of Jerusalem, Jerusalem 91904, Israel

Sari Grossman - Institute of Earth Sciences, The Hebrew University of Jerusalem, Jerusalem 91904, Israel

Mukang Han - Department of Geography, Peking University, Beijing, China

Paul W. Hare - Dunn Geosciences, 5 Northway Lane N, Latham, New York 12110

Donald L. Johnson - Department of Geography, University of Illinois, Urbana, Illinois 61801

Edward A. Keller - Department of Geological Sciences,University of California, Santa Barbara, California 93106

Helaine W. Markewich - U. S. Geological Survey, National Center, Reston, Virginia 22092

Larry Mayer - Department of Geology, Miami University , Oxford, Ohio 45056

Wayne Newell - U. S. Geological Survey, National Center, Reston, Virginia 22092

Theodore M. Oberlander - Department of Geography, University of California, Berkeley, California 94720

Cliff D. Ollier - Bureau of Mineral Resources, G.P.O. Box 378, Canberra A.C.T. 2601

Yoko Ota - Department of Geography, Yokohama National University, Yokohama, Japan

Mario Panizza - Chair, Professor of Geomorphology, Institute of Geology, Modena University, Modena, Italy

Milan J. Pavich - U. S. Geological Survey, National Center, MS 926, Reston, Virginia 22092

James F. Petersen - Department of Geography and Planning, Southwest Texas State University, San Marcos, Texas 78666

Thomas K. Rockwell - Department of Geological Sciences, San Diego State University, San Diego, California 92182

Michael A. Summerfield - Department of Geography, University of Edinburgh, Scotland

Richard A. Young - Department of Geological Sciences, State University of New York, Geneseo, New York 14454

# PART I

# 1 Morphotectonics of continental margins with great escarpments

*Clifford D. Ollier*

ABSTRACT

A generalized continent has a low in the middle, and a rise towards the margins where plateaus are found, bounded on the outer side by Great Escarpments beyond which are coastal lowlands. Several large land masses demonstrate this morphology. Consideration of 20 postulated mechanisms of plateau uplift shows that most ignore the geomorphic setting, highlighting the need to integrate tectonic and geomorphic studies. If the Great Escarpment model is accepted, several models of continental erosion and isostatic response need revision. A favoured model relates marginal uplift to the break-up of Pangaea, followed by scarp retreat from newly created continental edges.

## INTRODUCTION

The landscape feature here termed a Great Escarpment is a very major landform present in many continents, which has generally been overlooked!

Great Escarpments tend to run parallel to the coast, and they separate a high plateau from a coastal plain. When whole continents are considered further generalizations are apparent. It seems that a generalized continent has a low in the middle, and a rise towards the margins, bounded symmetrically by two Great Escarpments. If this pattern is real, some major tectonic and geomorphic generalizations are required to account for the tectonic uplift of the continental rims, and the mechanism of

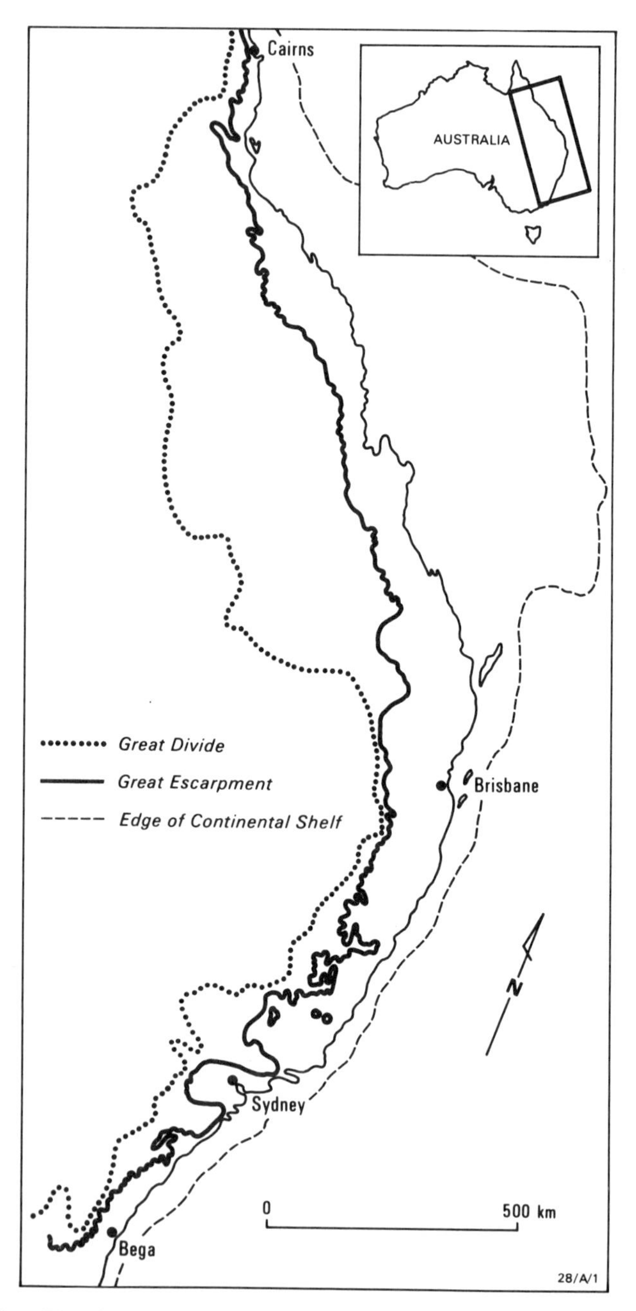


Figure 1. The Great Escarpment and associated features of eastern Australia. Mainly based on analysis of topographic maps at 1:100,000 (after Ollier, 1982).

formation of the escarpments. Since many scarps occur on continental margins they may relate to continental edges formed at the inception of modern continents, or 'plates' by the fragmentation of supercontinents such as Gondwanaland, Laurasia, or perhaps Pangaea.

REGIONAL DESCRIPTION

Australia. Most maps of Australia mark a Great Dividing Range running inland of the eastern coast, perpetuating a cartographic myth. The Great Divide is a watershed which for much of its length crosses wide and gently undulating plains: lakes and airstrips are commonly located right on the divide. Much more spectacular is the Great Escarpment (Figure 1). This is a landform on the grand scale, thousands of kilometres long and often approaching a thousand metres high. The escarpment, when seen from the coastal lowlands, looks like a mountain range, so the idea that there was a Great Dividing Range probably derives from the views of early settlers along the Australian coast. The many different names given to the "ranges" was probably responsible for the failure to realise the unity of the Great Escarpment as a single, continuous escarpment rather than many separate features. The Great Escarpment separates two regions of vastly different geomorphology: the tablelands with many palaeoforms, low relief and slow process rates, and the coastal zone with few palaeoforms, moderate relief and rapid process rates.

The Great Escarpment is very irregular in plan and is clearly an erosional feature, with embayments and large waterfalls at the knickpoints where rivers cross the escarpment. It is particularly well-marked on flat-lying, hard rocks such as basalts and sandstones, but is often sharp on other rocks. Erosional processes are still active on the escarpment and it is presumably still retreating. It is a diachronic landform, so if we ask, "when was it formed", we usually mean "when was it initiated". A few clues to its age are available. In one place a lava flow runs over the escarpment, so the escarpment is older than the lava flow, but as the flow is only three million years old this is not much help. Elsewhere the escarpment has cut back to intersect dated, near-horizontal lava flows, where the lava and underlying sediments are about 20 million years old.

The age of uplift of the eastern highlands, a prerequisite for the formation of the escarpment, remains controversial. Wellman (1980) emphasised the close connection evident between the Great Divide and volcanic eruptions in eastern Australia which have been continuous since Cretaceous times. He considered that uplift had been continuous over the past 90 million years. Ollier (1982), adopting a rift valley model for the uplift, concluded that the time of uplift was most likely to be about 80 million years ago. Jones and Veevers (1982) noted three cycles of sedimentation in associated basins, and three phases of volcanic activity, and concluded that there have been three phases of uplift alternating with periods of tectonic settlement.

In Western Australia there is no Great Escarpment, but there is a very interesting small escarpment which appears to be analogous to the Great Escarpment of the east, but lacks its size. The southern part has been termed the Meckering Line (Mulcahy, 1967). Inland of the Meckering Line deep zones of weathering are as common beneath the floors of the broad valleys as under the divides. Deep profiles are stripped on the coastward side. Inland are broad flat-floored valleys with salt lakes; the Meckering Line marks the downstream limit beyond which such valleys are not found as a connected system. The Meckering Line has been extended further north by Horwitz and Hudson (1977, p. 126), as the "Line separating areas with rapid drainage to the coast from sluggish and/or inland drainage areas (includes Meckering Line of Mulcahy (1967)". It is apparent that it is a continental feature.

In northern Australia the northern edge of the Barkly Tableland is a similar escarpment, separating the Tableland from the lowlands that merge into the Gulf of Carpentaria. Ollier and D'Addario (1983) put all Australian "Great Escarpments", including the extended Meckering Line and the Barkly escarpment onto one map. It is intriguing to note that the escarpments seem to be absent on the soft rocks of the inter-cratonic basins, and across south Australia. A fuller account of the Great Escarpments of Australia will be provided by Pain (1985).

South Africa. The Great Escarpment has been recognized for a long time in southern Africa. W.M. Davis (1906) describes a visit to part of it: he was aware of its significance, and complained that his party were taken to climb some hill instead of examining the escarpment. Rogers (1920) gave a presidential address on the subject of the Great Escarpment, provided a map, and discussed many of the topics that now concern us again. He worked before continental drift became respectable, even in South Africa, when modern ideas of plate tectonics and continental margin tectonics were far in the future. Knetch (1940) suggested a probable Oligocene age for marginal uplift of the western side of southern Africa, and formation of the Great Escarpment in the Oligocene and Miocene. Obst and Kayser (1949) provided a detailed description of the Great Escarpment (Randstufe) of the east side of southern Africa, and discussed the age of uplift (an early Tertiary marginal uplift and scarp formation, followed by a later epeirogenic uplift). Wellington (1955) wrote a physical geography of South Africa in which the Great Escarpment was described in detail, and many useful geomorphic ideas were presented. King continued the work, but his emphasis on several erosion surfaces and escarpments took attention away from the different scale and uniqueness of the Great Escarpment. Nevertheless his work on the Drakensberg in Natal (King, 1982) provides a good detailed study of one sector of the escarpment.

As shown in Figure 2 the Great Escarpment makes a great horse-shoe shaped curve, roughly parallel to the coast and about 200 km inland. Within the horseshoe is a rim of high plateau country (Highveld), and the Kalahari depression lies at the centre. The escarpment itself is best preserved where horizontal rocks add structural control, but it is also quite clear on other rocks. Outside the escarpment the coastal and intermediate zone is a complex erosional area. The major rivers on the upland - the Orange-Vaal and the Limpopo - appear unrelated to the form of Africa, and are probably antecedent rivers dating back to an origin on the palaeoplain before fragmentation of Gondwanaland created the modern continental margins and perhaps the raised rim of southern Africa.

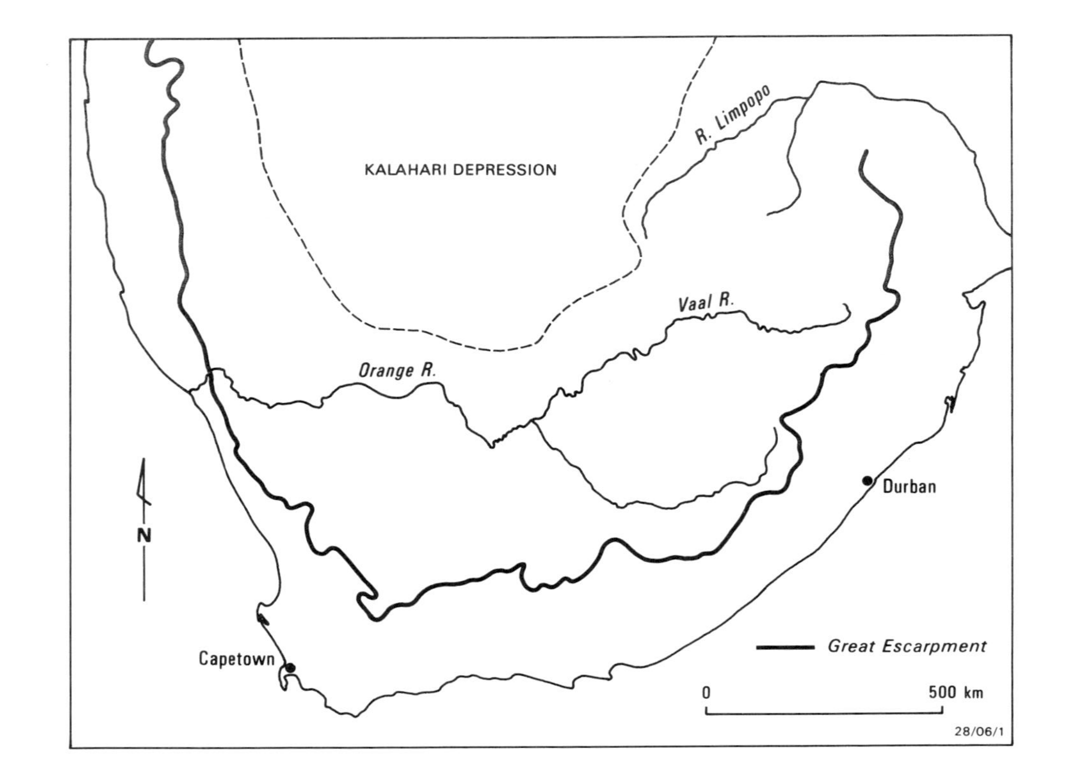


Figure 2. The Great Escarpment of southern Africa, based on analysis of maps at 1:250,000 by M.E. Marker.

The escarpment crest often coincides with a watershed divide, but there are some areas where river capture and antecedent drainage cause complications. There are no associated volcanics except one volcano in Namibia.

Several morphotectonic ideas related to the Great Escarpment will be mentioned later in this paper. A fuller account will be given by Marker and Ollier (1985), who conclude that the Great Escarpment was initiated by erosion following uplift associated with the formation of new continental edges in Jurassic times.

India. Peninsular India displays some features of an escarpment-bounded continent, with the Western Ghats in particular providing a splendid example of a Great Escarpment. The escarpment of the Western Ghats (Fig. 3) has a total length of over 1500 km, and the distance between the escarpment and the coast is seldom more than 60 km. The topography is complicated by isolated hills both above and below the escarpment, but the escarpment is a large, continuous and distinct feature. The escarpment is at its clearest where the basalts of the Deccan Traps help preserve a steep cliff face. To the south on Precambrian gneisses the escarpment is more varied by differential erosion, structural control and river capture.

South of Coimbatore is the remarkable east-west Palghat Gap, which makes a break in the Great Escarpment. The gap has been interpreted in various ways (rift valley, palaeoriver) but remains a morphotectonic enigma. South of Palghat Gap is a dissected triangular plateau, bounded in the west by a section of the Western Ghats. Here the main divide is about 50 km east of the front of the escarpment, which is very dissected by west-flowing rivers.

The Eastern Ghats are more complex, and not simply a great escarpment. (The word 'ghat' simply means an ascent, and can refer to any hill or escarpment). Only in the southern section, south of the latitude of Madras, does the Eastern Ghats refer to a Great Escarpment, which is here the eastern side of the Mysore Plateau. North of Madras the eastern Ghats refers to various structural ridges (such as the Nalamala Ranges), or broad regions of uplift.

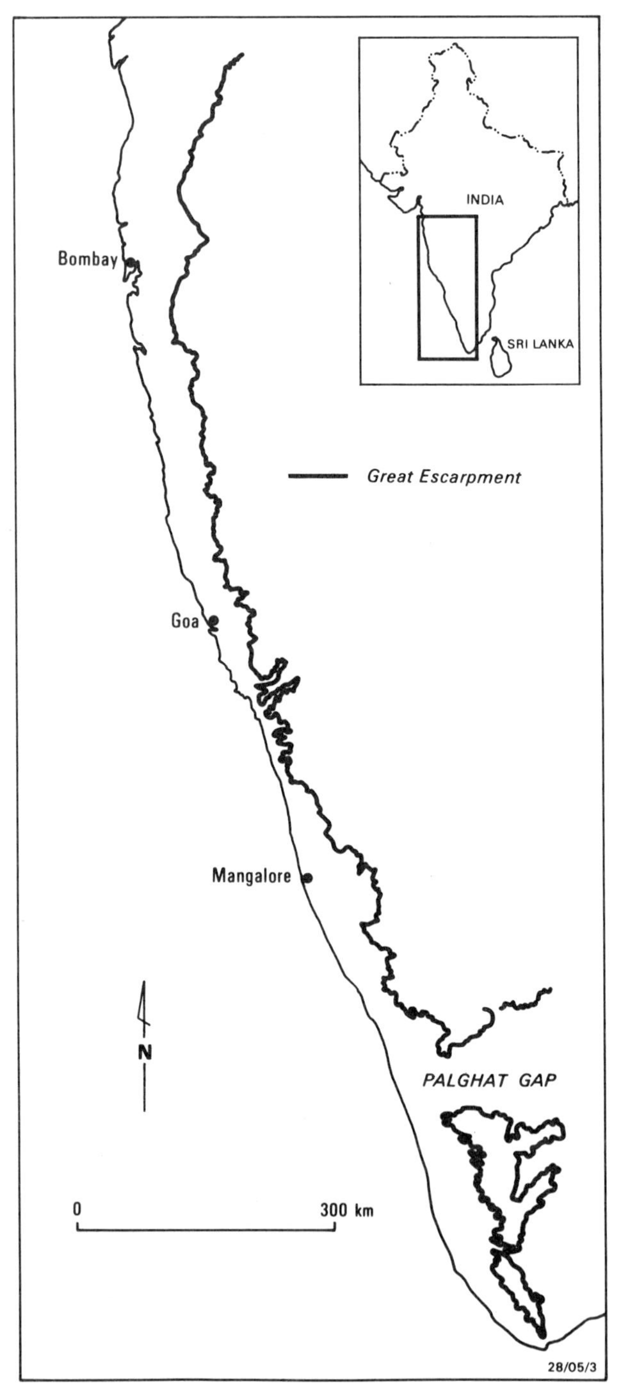


Figure 3. The Western Ghats escarpment of India, based on analysis of maps at 1:63,360 and 1:250,000 and satellite images, by C.D. Ollier.

The drainage pattern of peninsular India is dendritic from a watershed in the Western Ghats, but major drainages cross the eastern Ghats in gorges, regardless of structure. These antecedent gorges indicate that the drainage pattern was initiated after the uplift of the Western Ghats. Hence the uplift of the Eastern Ghats took place later than the uplift of the Western Ghats. A fuller account of the morphotectonics of Peninsular India will be given by Ollier and Powar (1985).

South America. South America consists of three cratons separated by the Amazon-Plata lowlands, and each of the cratons has an escarpment. In east Brazil there is a very well marked escarpment, the crest of which corresponds to the watershed. The morphology of this has been described in part by Maack (1969). It seems to resemble the Western Ghats of India. A further account will be given by Christofoletti (1985).

In the northern part of South America similar topography is found, with the Guyana plateau bounded to the north by a large escarpment. Some of the worlds largest waterfalls occur here. The Guyana Highlands are largely undulating, but scattered remnants of flat-lying sandstone and conglomerate of the resistant Roraima Formation (Proterozoic) form the mesas and plateaus of the Pakaraima Mountains in Guyana, Brazil and Venezuela. There is a major escarpment but it exists only where there is strong structural control, so it may not have the same genesis as Great Escarpments elsewhere. McConnell (1968) records five planation surfaces, similar to those described by King (1962) for east Brazil, and if the morphotectonic history of the two areas is really so similar it may still be proper to regard the Roraima escarpment as a morphotectonic Great Escarpment. A further account of the morphotectonics of part of this area will be given by Zonneveld (1985).

Another major escarpment has been described from Chile (Paskoff, 1978). This cliff in northern, arid Chile is about 700 m high and 800 km long. Paskoff describes it as a marine cliff, and indeed it is still an active cliff under wave action north of Iquique. Elsewhere it is an abandoned cliff with a wave-cut platform at its foot. Paskoff believes the cliff may have been initiated as a Miocene fault scarp, which has retreated during the Pleistocene. This escarpment is different in many ways from those considered elsewhere in this paper, and its location on a Pacific type coast suggests a different morphotectonic setting. Nevertheless in sheer size this is undoubtedly a "Great Escarpment", and may have lessons for the study of those in other parts of the world.

Greenland. Greenland has a mountainous rim, much dissected on the outer side by erosion. Some of this erosion may have been originally fluvial, but evidence is generally obliterated by glaciation. The central part is occupied by an ice cap, and indeed the weight of the ice cap is commonly assumed to have caused the sag of central Greenland and the associated rise of the marginal mountains. However, the parallelism of gross morphology with continents that do not have an ice-cap suggests that the origin of this form may warrant further investigation.

Brooks (1979, 1980) has presented evidence for tectonic uplift in east Greenland. In the late Mesozoic the area was a plain, and low relief persisted through a long volcanic episode. Later came massive domal upwarping, which has been a dominant feature of the landforms up to the present day. Brooks studied the Kangerdlugssuaq dome, which is elliptical with a major axis of 300 km and a height of 6.5 km. Updoming occurred about 50 million years ago. This dome was rapidly eroded by a radial drainage system, relics of which can still be found.

About 35 million years ago the entire area underwent epeirogenic uplift. At present the plateau is undergoing dissection from the seaward side, but considerable areas are preserved under thin, horizontal ice-caps. The age of the plateau basalts in the area is about 55 million years, so the volcanicity is somewhat older than the uplift. Further data on the Greenland morphotectonics will be presented by Brooks (1985).

Antarctica. Antarctica is rather like Greenland, in having a mountainous rim and a depression in the centre, under a huge ice-cap. The mountains often have the form of plateaus, and erosion is mainly on the seaward side, so there is some chance that tectonic uplift may have affected the uplift of the rim, and the rise is not only an isostatic response to the ice-cap.

North America. There seems to be two possible contenders for the role of Great Escarpment in the United States. In the west a very significant escarpment has been described by Peirce, Damon and Shafiqullah (1979) from Arizona. This escarpment marks the southern termination of Permian cliff-making strata, and runs diagonally across Arizona for a distance of over 500 km. It is commonly known as the Mogollon Rim.

In the past the origin of the escarpment was generally associated with a late Cenozoic "plateau uplift" and faulting associated with the late Cenozoic Basin and Range tectonism. However Peirce et al. show that an escarpment had evolved by mid-Miocene time, prior to the onset of Basin and Range faulting and the development of the modern Grand Canyon.

On the plateau above the escarpment about 150 m of Cretaceous rocks are overlain by Miocene gravels associated with a northerly drainage. This north-sloping plateau is cut off abruptly by the south-facing escarpment, often referred to as a drainage reversal. Peirce et al. suggest that the scarp formation dates to Oligocene times, and is certainly younger than oldest Eocene. The modern canyon cutting cycle dates from about 10 million years.

In the eastern U.S. the fall line east of the Blue Ridge appears to be the main contender for the role of Great Escarpment. Davis described this in 1903 p. 214, in his own exciting, anthropomorphic style. "The divide between the eastern and western streams is known as the Blue Ridge; but in southern Virginia and North Carolina it is not a ridge, with a crestline and well-defined slopes on either side; it is an escarpment, descending from the hilly and mountainous upland of the western drainage area to the rolling and hilly lower land of the eastern streams." On p. 240 he writes: "The escarpment itself is by no means a straight and simple wall. The ruins of the upland often form a labyrinth of hills and spurs at the back of the piedmont coves. Even where the form is relatively simple, the escarpment is not retreating in good order but is falling back irregularly. Here a hardy resistance or a less active attack determines a salient; there a vigorous attack or an enfeebled resistance causes a reentrant". The height of the escarpment is commonly 300-500 m, but occasionally 1000 m. Davis offers a very simple explanation, p. 214: "The cause of this markedly unsymmetrical form is that the eastern streams have worn down their basins to a much lower level than the western streams have. The chief consequence of the unsymmetrical form is that the eastern streams are rapidly ... pushing the escarpment westward. The divide at the crest of the escarpment is therefore, a migrating divide ..." Davis (page 222) quotes a Mr. Campbell who offers a more explicit statement: "I think that exceptional conditions have prevailed in this region which permitted the formation of this scarp; the exceptional conditions ... are those of unsymmetrical uplift or warping, the unsymmetrical uplift having its axis east of and presumably near the present front of the Blue Ridge with its longer and more gentle slope to the west and its steeper slope to the east."

The Blue Ridge escarpment has been described more recently by Hack (1982 p. 1). "The Southern Blue Ridge province is bounded on its southeast margin by a bold escarpment that corresponds to the major drainage divide. ... at least the northeastern part of this escarpment has retreated toward the northwest. The origin of the escarpment is probably a monoclinal flexure or a movement along a series of undetected faults as much as 30 km or more southeast of the present position of the escarpment."

Hack reviewed the hypotheses of four workers:

1. Hayes and Campbell (1894). A monoclinal flexure that was later dissected.

2. Davis (1903). A topographic discontinuity between a pair of peneplains of the same age but on opposite sides of the major divide between streams taking a short route to the Atlantic and those taking the longer route to the Gulf of Mexico.

3. White (1950). The scarp was caused by movement on a series of en echlon faults. This hypothesis is readily disproved.

4. Thornbury (1965). The scarp may have been initiated by faulting, but the faults are now some distance from the scarp. The discontinuity in the topography produced by faulting was maintained as the divide slowly shifted northwest.

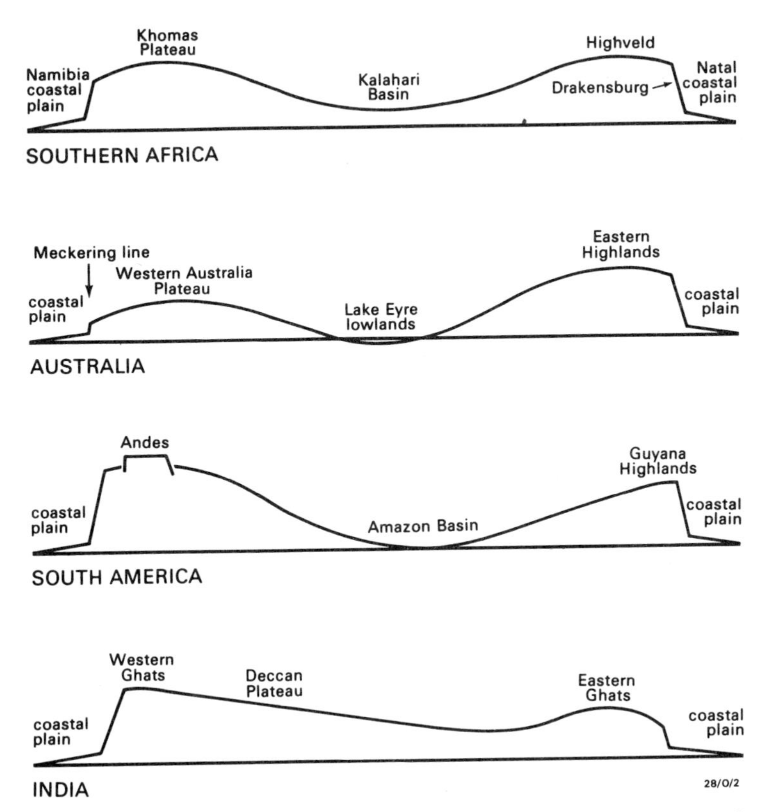


Figure 4. Diagram to show possible common features of Great Escarpments on different continents.

Hack (1982, p. 36) considers hypotheses 1 and 4 to be "close to a satisfactory explanation". He is keen to minimise the amount of divide migration: "Divide migration on a large scale need not have taken place" (1982, p. 45). "... the major drainage divide within the southern Appalachian highlands corresponds closely to a topographic high and a negative gravity anomaly ... This coincidence suggests a long period of stability for the divide and suggests that a hypothesis of escarpment evolution should be considered that does not involve substantial migration of the divide" (1982, p. 44).

The time scale involved is apparently shorter than in most Great Escarpments, though Hack evidently considers it long when he writes (1982, p. 45): "The general form of the topography was determined at a much higher erosional level, possibly as long ago as the Middle Mesozoic".

Other areas. Several other regions display continental margins that seem to show marginal uplift and possible development of Great Escarpments. Northwest Europe is one such margin, with uplift of Scandinavia and the northwest of the British Isles (Linton, 1951). Madagascar and Sri Lanka may be other contenders.

## A GENERALISED MODEL FOR CONTINENTAL MARGINS

The descriptions given from different regions suggest a basic common form (Ollier, 1984). There is a central sag in the continents, bounded by gentle rises to high plateaus. The plateaus are bounded by abrupt, out-facing escarpments, beyond which are coastal plains. Figure 4 shows some of these diagrammatically.

A generalised morphotectonic model for continental margins with Great Escarpments may be expressed simply, as in Figure 5.

1. An original continental slab is isolated from the original supercontinent (e.g. Gondwanaland) by chasmic faults and new sea floor.

2. Erosion and scarp retreat from the new continental margins creates the coastal lowlands backed by a major escarpment.

Then two routes are possible:

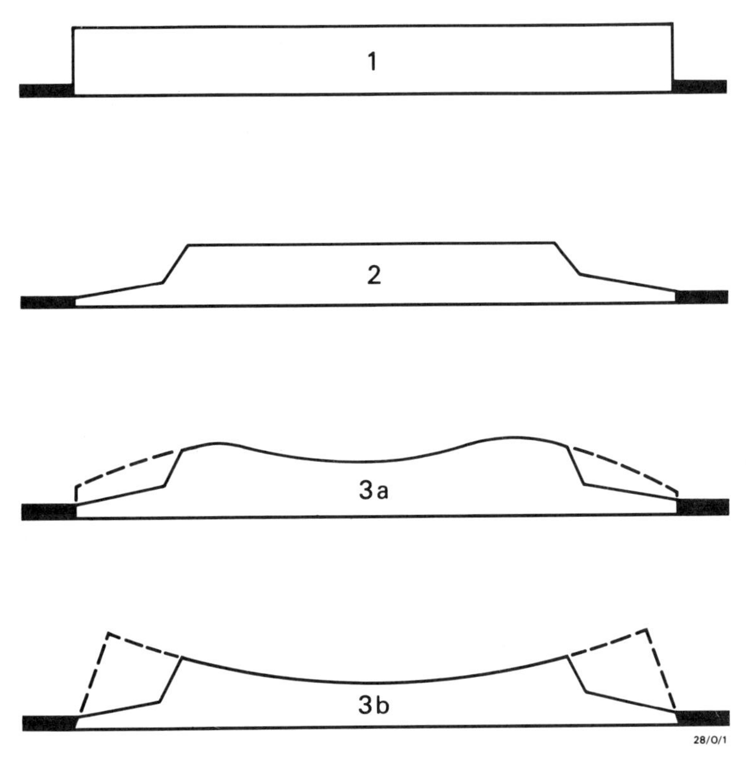


Figure 5. A possible sequence to form Great Escarpments and associated plateau uplift and mid-continental sags. For explanation see text.

3a. The continental plate tends to sag in the middle and rise towards the edges and then down again to the coast.
or

3b. The continental plate tends to sag in the middle and rise to the edges, so the top of the escarpment is the watershed.
The two variants, 3a and 3b, should have different associated landforms as well as different tectonic history and causes.

Type 3a. The broad swell changes the gradient on the plateau, reversing the direction of drainage on one side of the swell. A modern analogue is the Western Rift Valley in Uganda, where the plateau is warped along an axis about 80 km east of the Lake Albert fault scarp. The old east-to-west drainage is disrupted, with reversal of rivers to the east of the warp axis and rejuvenation to the west. The watershed (Great Divide) does not correspond to the top of the Great Escarpment, and examples of reversed drainage are found on one side of the watershed. However, in time the escarpment might retreat to a position beyond the watershed, when a cross section like that of 3b will ensue.

Type 3b. A simple tilt block is formed, bounded by a fault scarp, and the watershed corresponds to the edge of the block. Modern examples are the Red Sea and the Ethiopian Rift. The fault scarp will be eroded and the erosional scarp will retreat, beheading the long rivers on the top of the tilt block. Apart from occasional capture this will not cause significant reversal or disruption of the drainage pattern. The East Brazil highlands, and the Western Ghats, appear to be nearest to this type of development.

At least two further complications should be borne in mind in assessing this model. Firstly, the model assumes that uplift occurs after formation of the first escarpment. It is perhaps more probable that doming precedes the formation of a new continental margin and erosion of the Great Escarpment. Secondly, although the continental sag is lower than the surrounding plateaus, it may nevertheless have been uplifted, though less than the plateaus. The Kalahari Basin is uplifted; the Lake Eyre lowlands are below sea level.

THE FORMATION OF GREAT ESCARPMENTS

As envisaged by most workers in the field who have considered great escarpments, two problems arise; what is the mechanism that creates a plateau that can be eroded into an escarpment, and what is the mechanism of erosion that creates an escarpment. We shall consider these in turn.

Mechanisms of Plateau formation.

I. Constructional plateaus.

In areas where the rocks are simply deposited, as in the basalt plateaus of the Deccan, or the Drakensberg, there may be no need to postulate uplift. However, if escarpments run continuously from such basalt areas to normal bedrock areas that were uplifted, it is probable that the basalt plateaus have undergone some uplift too.

II. Tectonic uplift of plains.

Assuming that a plateau gained its low relief by long continued erosion at relatively low altitude, some sort of vertical uplift is required to turn the original plain into a plateau. Many suggested mechanism are available. McGetchin et al. (1980) list 14 mechanisms, which may be summarised as follows:

1. Thermal expansion due to a mantle plume or hot spot. Hawaii is a example, but the method could work inland.

2. Thermal expansion due to overriding and subduction of a hot midocean ridge or spreading centre.

3. Thermal expansion due to shear heating along lithosphere-asthenosphere interface.

4. Expansion accompanying partial melting. The volume of fusion for basaltic magma is about 8%.

5. Hydration reactions such as serpentinization, which could produce about 10% volumetric expansion. This has been suggested for the Colorado Plateau.

6. Introduction of volatiles due to deep-seated dehydration of hydrous minerals. King (1983) proposes volatiles as the prime force in vertical tectonics.

7. Expansion due to depletion of mantle fertile in garnet and iron resulting from basalt genesis.

8. Crustal thickening due to horizontal transfer of mass in the lower crust.

9. Deep-seated solid state reactions such as eclogite-basalt transformation.

10. Subduction at very shallow angle, perhaps horizontal.

11. Simple subduction, or continent-continent subduction. The altiplano of South America and the Tibet Plateau are given as examples.

12. Cessation of subduction and resulting thermal equilibration of static slab.

13. Isolation of plateau by listric normal faulting in surrounding areas. This has been proposed for the Colorado Plateau.

14. Cooling lithosphere detaches from the crust, and is replaced by a counterflow of asthenosphere, which warms the crustal rocks and causes uplift and volcanism.

To this list a few other mechanisms can be added:

15. Intrusion of magma into the lower crust. McKenzie (1984) suggested this as a new mechanism of epeirogenic uplift, possibly associated with volcanism, but it was previously listed by Hobbs, Means and Williams (1976).

16. Intrusion of sills. This seems a plausible mechanism, especially in those areas where surface volcanic activity is prevalent. It has been considered for eastern Australia, but the widespread regional gravity low suggests it is an inadequate explanation.

17. Isostatic uplift after scarp retreat. This will not initiate a plateau, but if, for example, there is erosion around the edge of a continent, the continental margin will eventually be isostatically uplifted, taking the remaining plateau up with it. As a hypothesis related to scarp retreat this mechanism was first put forward by King (1955).

18. Erosion-isostatic rebound. This hypothesis is proposed by Stephenson and Lambeck (1984), and applied to southeastern Australia. As in mechanism 16, uplift is first driven by some other (tectonic) mechanism, but after a first uplift erosional forces unload the crust and isostatic rebound occurs such that the earlier base level of the terrain is regionally uplifted.

This model does not use scarp retreat, but assumes erosion of mountains is regional, and is proportional to their height.

19. Underplating. Wellman (1979) suggested this rather unspecified mechanism for the uplift of southeast Australia. "The underplated material is the 20 km of abnormally high velocity material that forms the base of the crust in this area."

20. Isochemical phase change with volume increase in the lower crust of upper mantle (Hobbs, Means and Williams, 1976).

Mechanisms of escarpment formation. The mechanism of parallel slope retreat has been championed by King for many years. The widespread occurrence of Great Escarpments suggest that the mechanism occurs in a wide variety of settings, and is not controlled by either structure or climate. The amount of scarp retreat is extremely variable; Hack wishes to minimise it in the case of the Blue Ridge escarpments; King claims the Drakensberg escarpment has retreated over 150 km.

General textbooks of geomorphology commonly deal with the principles of slope development and with parallel slope retreat. One recent study of mechanisms on Great Escarpments is by Moon and Selby (1983). They demonstrate a widespread occurrence of equilibrium slopes indicating the continuing adjustment of slope form to rock mass strength. They say "rock faces retreat with angles which are in conformity with the mass strength of their rocks."

## WHY WERE GREAT ESCARPMENTS NEGLECTED BY GEOMORPHOLOGISTS?

If the Great Escarpments of the world are as extensive and important as I am trying to make out, the question arises: "Why have they been neglected for so long?"

Some possible explanations are:

1. I am overstating the case, and in fact Great Escarpments have had as much treatment as they deserve.

2. Great Escarpments are not very easy to see in Europe and North America, where the greatest amount of geomorphic research is carried out.

3. In earlier times the emphasis in geomorphic work was on

erosion surfaces, not the steeper slopes bounding them. Davis (1903) was more concerned with the peneplains and the "stream contest" than with the Blue Ridge escarpment: the same author in South Africa noted the Great Escarpment of Natal, complained of lack of time to examine it, and then went on to consider if it is possible to create an erosion surface at a high level. King (1962) certainly noted the Great Escarpment many times but most of his writings are more concerned with his hypothesis of pediplanation, as opposed to peneplanation.

4. Several geomorphologists have been concerned in recording successions of erosion surfaces, separated from each other by steeper country. King (1962) is the most persistent in this, and much of his work has concerned the recognition and decription of several pediplains which he considers to be of world-wide importance. The different pediplains are all separated by steeper terrain, and the record of many scarps hides the fact that there is one Great Escarpment which is very much bigger, more obvious, and probably more significant than the others.

5. Since the advent of plate tectonics, there has been more concern with subduction and the morphotectonics of collision sites than with trailing edges and continental interiors. There has also been a greater interest in "mountains", usually meaning fold belts, than with "plateaus". We now know that most mountains are dissected plateaus, and there are no fold mountains! (Ollier, 1981). Plateaus and vertical tectonics have only recently become a topic of major interest amongst tectonicists (e.g. McGetchin et al., 1980; McKenzie, 1984).

DISCUSSION

The understanding of the morphotectonics of coastal regions calls for a bringing together of geomorphic, tectonic and geophsyical information. With cross checks it should be possible to reduce some of the speculation on tectonic processes, and also on geomorphic models.

For example, Smith (1982) proposed that during the last 100 million years Africa moved over a heat source and "those areas affected by plateau uplift now overlie or have passed across the former positions of the oceanic ridges". His model is based on a perceived asymmetry of southern Africa, with the east being uplifted while the west remains low. But in fact there is a remarkable symmetry. It is necessary to account for uplift in both east and west, and for that matter in the south. There is a complete horse-shoe of uplift, and Africa cannot have been drifting in all directions at once over conveniently-located heat sources.

Several models of landscape development relate isostasy to regional erosion, perhaps in some relation to elevation (e.g. Ahnert; 1970: Hack, 1982; and Stephenson and Lambeck, 1984). If, in the face of new data of mapped escarpments, these models have to be abandoned, new models based on isostatic response to scarp retreat must be developed. King (1955) provided an early attempt at such a model, but the topic needs to be up-dated.

Most models assume tectonic uplift followed by erosion and perhaps further tectonic uplift. Even this can be questioned, as by Jones and Veevers (1982) who proposed alternating tectonic uplift and tectonic subsidence.

It now seems that many explanations of individual regions have "special case" explanation of uplift and escarpment formation. Some have icecaps, but others do not: some have volcanoes, others do not. It seems to me that the only common feature is the relationship between Great Escarpments and continental margins, and so I favour a model that relates marginal uplift to the creation of new continental edges, as for example by the break-up of Gondwanaland. Whatever the details may be, we should be looking for a general explanation that can account for all continental margins and not merely a sub-sample. We are not yet able to say what the basic mechanism is, but the morphotectonic facts now put firm limits on the type of explanation that is acceptable.

This paper is published with the permission of the Director, Bureau of Mineral Resources, Geology and Geophysics, Canberra.

REFERENCES

Ahnert, F., 1970, Functional relationships between denudation, relief and uplift in large mid-latitude drainage basins: American Journal of Science, v. 268, p. 243-263.

Brooks, C.K., 1979, Geomorphological observations at Kangerdlugssuaq, East Greenland: Greenland Geoscience, v. 1, p. 3-21.

__________, 1980, Episodic volcanism, epeirogenesis and the formation of the North Atlantic Ocean: Palaeogeog., Palaeoclim., Palaeoecol., v. 30, p. 229-242.

__________, 1985, Vertical crustal movement in the Tertiary of central East Greenland: Zeitschrift fur Geomorph., in press.

Christofoletti, A., 1985, Geomorphology of the marginal escarpments of southeast and northeast Brazil: Zeitchrift fur Geomorph., in press.

Davis, W.M., 1903, The stream contest along the Blue Ridge: Bull. Geog. Soc. Philadelphia, v. 3, p. 213-244.

__________, 1906, Observations in South Africa: Bull. Geol. Soc. Amer., v. 17, p. 433.

Hayes, C.W. and Campbell, M.R., 1894, Geomorphology of the Southern Appalachians: Nat. Geog. Mag., v. 6, p. 63-126.

Hack, J.T., 1982, Physiographic divisions and differential uplift in the Piedmont and Blue Ridge: U.S.G.S. Prof. Paper 1265.

Hobbs, B.E., Means, W.D. and Williams, P.F., 1976, An outline of structural geology: Wiley, New York.

Horwitz, R.C. and Hudson, D.R., 1977, Australites from northern Western Australia: Jour. Roy. Soc. Western Australia, v. 59, p. 125-128.

Jones, J.G. and Veevers, J.J., 1982, A Cainozoic history of Australia's southeast highlands: Geol. Soc. Aust. Jour., v. 29, p. 1-12.

King, L.C., 1955, Pediplanation and isostasy: an example from Southern Africa: Quart. Jour. Geol. Soc. Lond., v. 111, p. 353-359.

__________, 1962, The morphology of the earth: Oliver and Boyd, Edinburgh.

__________, 1982, The Natal monocline: University of Natal Press, Pietermaritzburg.

__________, 1983, Wandering continents, and spreading sea floors on an expanding earth: Wiley, New York.

Knetch, G., 1940, Zur Frage der Kustenbildung und der Entwicklung des Oranjetals in Sudwestafrikas. Sonderveroffentlichung III d. Georg. Ges., Hannover.

Linton, D.L., 1951, Problems of Scottish scenery: Scot. Geog. Mag., v. 67, p. 65-85.

Maack, R., 1969, Die Serra do Mar im Staate Parana: Die Erde, v. 100, p. 327-347.

McConnell, R.B., 1968, Planation surfaces in Guyana: Geogr. Jour., v. 134, p. 506-520.

McGetchin, T.R., Burke, K.C., Thompson, G.A. and Young, R.A., 1980, Mode and mechanisms of plateau uplifts: p. 99-110 in Bally, A.W., Bender, P.L., McGetchin, T.R., and Walcott, T.I. (eds.), 1980, Dynamics of Plate Interiors: Geodynamic Series, 1., Am. Geophys. Union, Washington, D.C.

McKenzie, D., 1984, A possible mechanism for epeirogenic uplift: Nature, v. 307, p. 616-618.

Marker, M.E. and Ollier, C.D., 1985, The Great Escarpment of southern Africa: Zeitschrift fur Geomorph., in press.

Moon, B.P. and Selby, M.J., 1983, Rock mass strength and scarp forms in southern Africa: Geografiska Annaler, v. 65A, p. 135-145.

Mulcahy, M.J., 1967, Landscapes, laterites and soils in south-western Australia: p. 211-230 in Jennings, J.N. and Mabbutt, J.A., Landform studies from Australia and New Guinea: Aust. Nat. Univ. Press, Canberra.
Obst, E. and Kayser, K., 1949, Die Grotse Randstufe auf der Ostseite Sudafrikas und ihr Vorland: Geogr. Ges., Hannover.
Ollier, C.D., 1981, Tectonics and landforms: Longman, London.
___________, 1982, The Great Escarpment of eastern Australia: tectonic and geomorphic significance: Geol. Soc. Aust. J., v. 39, p. 431-435.
___________ and D'Addario, G.W., 1983, Problems of regolith distribution in Mainland Australia: Bureau of Mineral Resources Record 1983/27, p. 101-107, Canberra.
___________, 1984, Drainage pattern analysis and morphotectonics: the southern continents: Proc. Morphotectonics Working Group Meeting, Sofia, 1983, Bulgarian Acad. Sci., Sofia, in press.
___________ and Powar, K.B., 1985, The Western Ghats and the morphotectonics of peninsular India: Zeitshrift fur Geomorph., in press.
Pain, C.F., 1985, Morphotectonics of Australian continental margins: Zeitschrift fur Geomorph., in press.
Paskoff, R.P., 1978, Sur l'evolution geomorphologique du Grand Escarpment cotier du desert Chillien: Geogr. Physique et Quaternaire, v. 32, p. 351-360.
Peirce, H.W., Damon, P.E. and Shafiqullah, M., 1979, An Oligocene (?) Colorado Plateau edge in Arizona: Tectonophysics, v. 61, p. 1-24.
Rogers, A.W., 1920, Geological survey and its aims: and a discussion on the origin of the Great Escarpment: Trans. Geol. Soc. S. Africa (Anniversay Address) xix-xxxiii.
Smith, A.G., 1982, Late Cenozoic uplift of stable continents in a reference frame fixed to South America: Nature, v. 296, p. 400-404.
Stephenson, R.S. and Lambeck, K., 1984, Erosion-isostatic rebound models for uplift: An application to southeastern Australia: Jour. Geophys. Res., in press.
Thornbury, W.D., 1965, Regional geomorphology of the United States: Wiley, New York.
Wellington, J.H., 1955, Southern Africa: Cambridge University Press, Cambridge.
Wellman, P., 1979, On the isostatic compensation of Australian topography: Bureau of Mineral Resources Jour. Aust. Geol. Geophys., v. 4, p. 373-382.
Wellman, P., 1980, On the Cainozoic uplift of the southeastern Australian highland: Geol. Soc. Aust. Jour., v. 26, p. 1-9.
White, W.A., 1950, Blue Ridge front - A fault scarp: Bull. Geol. Soc. Amer., v. 61, 1309-1346.
Zonneveld, J.E.S., 1985, Geomorphological notes on the continental border in the Guyanas, in press.

# 2
# Plate tectonics and landscape development on the African continent

*Michael A. Summerfield*

ABSTRACT

Major discrepancies exist between the widely accepted landscape chronology of Africa proposed by King and the record of denudational events represented by continental margin stratigraphy constructed from seismic and borehole data. These are attributable both to problems in dating of landscape cycles and complex geomorphic response of passive margins to base level changes. The latter is complicated by factors including changes in rate of sea level movement in relation to rate of margin subsidence and variations in gradient of the shelf zone over which the shore migrates. Most significantly, a relative fall in sea level will not invariably promote landscape rejuvenation. Models of continental rifting and passive margin evolution predict an initial zone of thermal uplift along a nascent margin following continental rupture. This is subsequently replaced by a coastal uplift induced by flexure and rotation of the margin associated with post-rifting thermal subsidence, sediment loading off-shore and denudational unloading inland. This model appears to accord broadly with both stratigraphic history and present gross morphology of the African continental margin. The presence of a marginal upwarp has promoted development of a dual system of drainage, one part supplying sediment to inland basins, the other eroding back into the seaward flank of the marginal upwarp and providing the off-shore sediment record. Most of the major present-day African drainage systems probably developed long after formation of the continental margin through capture of large interior networks by local breaching of the coastal upwarp. Continent-wide erosion surfaces are unlikely to have formed under such a dual system of drainage and even after development of extensive exterior drainage continuing uplift along the margin would have prevented the direct transmission of coastal landscape rejuvenation inland. Moreover, likely mechanisms of epeirogeny, such as sub-lithospheric hot spots and phase changes in the mantle, also appear incapable of generating the pulses of extensive contemporaneous uplift required to promote the development of continent-wide erosion surfaces on the African continent.

## INTRODUCTION

Africa has been the subject of numerous studies of long-term, large-scale landscape development. Investigations from the 1930s to the 1960s were concerned primarily with establishing a landscape chronology related to the continent's geological history. Of the various interpretations proposed those of King (1962, 1972) have been the most widely applied. Although his scheme of globally synchronous denudational cycles has recently been incorporated into a coherent, if speculative, tectonic model (King, 1983), earlier attempts to explain the evolution of the African landscape lacked a comprehensive tectonic framework. Over the past decade there has been a growing appreciation of the significance of plate tectonics in the interpretation of macroscale geomorphic features (Melhorn and Edgar, 1975; Morisawa, 1975; Ollier, 1979, 1981, in press). Although this model has been applied with most success in the continental context to the geological structures of plate margins, it is over the much more extensive areas remote from contemporary plate boundaries where large-scale tectonic activity is likely to have its most pervasive, if indirect, effects on the long-term evolution of landscapes.

My aim here is to highlight some of the more important

aspects of the relationship between large-scale tectonics and long-term landscape development on the African continent. Attention is focused on the continental margin since this zone has not only experienced the most active recent uplift but also provides the crucial link between phases of landscape development inland and the depositional record off-shore. Three main issues are addressed: 1) correlation between the stratigraphic record of the continental margin and the denudational cycles thought to have affected the continent; 2) relationship between continental margin tectonics and landscape development; and 3) relative importance of tectonic and eustatic controls in initiating new denudational cycles.

## MORPHOLOGIC AND STRUCTURAL BACKGROUND

Although largely unaffected by orogeny during the Mesozoic and Cenozoic, Africa has experienced significant epeirogenic uplift throughout this period generating a gross morphology of broad upwarps separating extensive basins (Holmes, 1965) (Fig. 1). These swells, locally surmounted by smaller domal uplifts, attain heights of about 1000m in the north and west but in excess of 2000m in the south and east. An almost continuous line of swells and domes extends through eastern Africa from the Afar region to the Zambezi, the uplifts being cut by rifts (Fig. 1) which in several cases link triple junctions located on dome crests (East African Rift System). Other rift structures occur both in the interior and along the continental margin (Burke and Whiteman, 1973). Some of these are currently active Cenozoic features while others are inactive or recently reactivated structures originating in the Mesozoic or earlier. Developed on these upwarps and basins are a succession of surfaces separated, especially in the south and east, by steep escarpments. Although primarily erosional, these surfaces are extensively mantled by duricrusts and other weathering deposits. Along the coastal margin and around the edge of interior basins they merge into depositional surfaces.

The continental margins of Africa have evolved from the

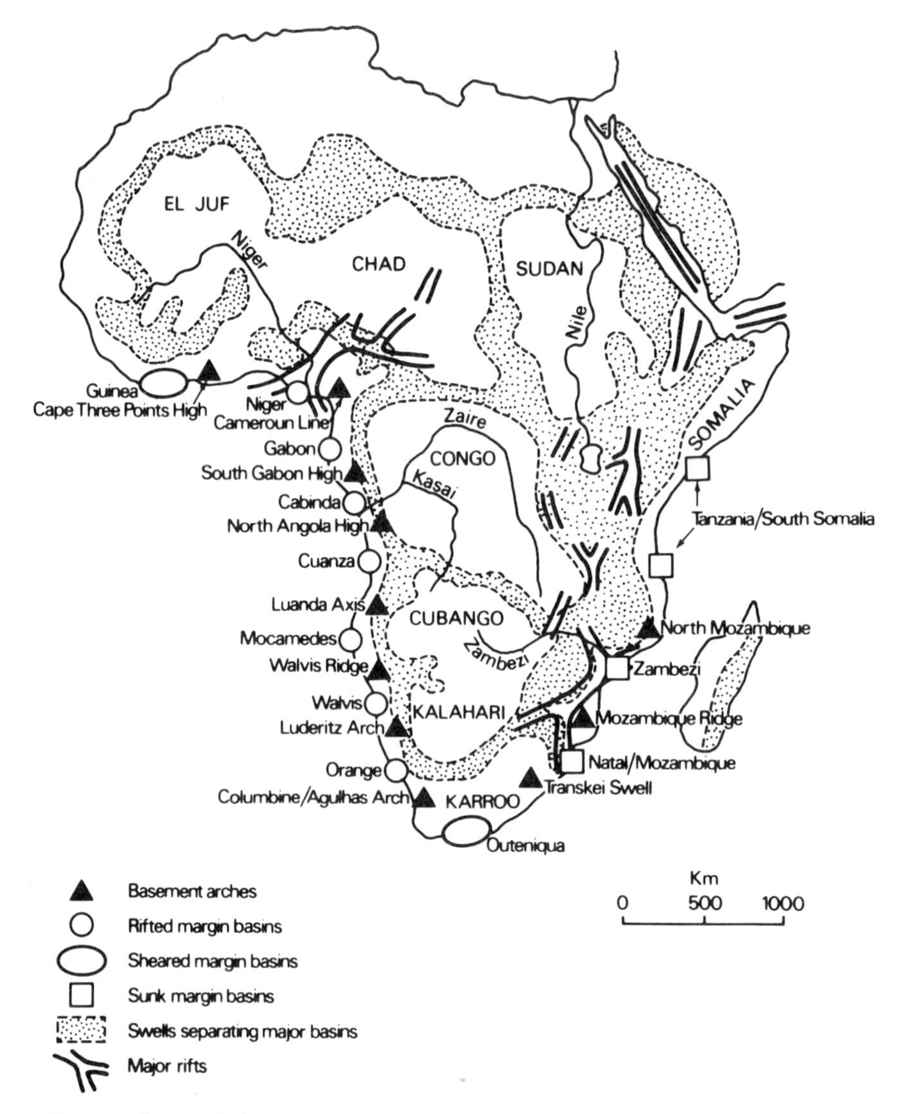


Figure 1. Africa: Major morphologic and structural features. Interior basin nomenclature and delimitation from Holmes (1965, Figure 763, p. 1054). Location and categorization of continental margin basins and basement arches from Dingle (1982).

lithospheric rupture associated with the break-up of Gondwanaland (200 - 95Ma B.P.). Three types of margin can be distinguished related to distinct tectonic styles (Dingle, 1982) (Fig. 1). Rifted margins, associated with normal tension, occur along the west coast from the Orange to the Niger basins between the two large off-sets of the Equatorial and Agulhas fracture zones. These off-sets, characterized by shear tension, from sheared margins along the east-west trending sections of the continental margin. Large parts of the east coast from Natal to Somalia have experienced prolonged vertical movement (Kent, 1974) and are thereofre categorized as sunk margins. Both the western and eastern margins consist of basins of deposition separated by less extensive regions of relative uplift. Subsidence has clearly predominated since basins account for nearly 70% of the length of the continental margins of the east and west coasts (Dingle, 1982). Paralleling the contrasts in tectonic style between the two coasts are differences in the spacing of basins and intervening upwarps. Along the western margin the crests of upwarps are typically 400-500km apart whereas on the eastern margin the mean distance is around 1000km with basins being proportionately further apart.

## LANDSCAPE CHRONOLOGY

Several detailed regional and continent-wide investigations have attempted to unravel the chronology of landscape development in Africa (see De Swardt and Trendall, 1969, for an excellent critical summary). Although dateable fossiliferous covering sediments or volcanic rocks can provide a reliable minimum local age in the few restricted localities where they are present, dating of most erosion surfaces rests on the assumption that they can be traced seaward into unconformities in marine sediments along downwarped continental margins. Correlation of erosion surfaces in the continental interior relies heavily, at least in King's scheme (King, 1962, 1976, 1983), on visual recognition of features considered to characterize surfaces of a particular age. While King has confidently maintained that "all major problems of dating and correlation seem to be resolved" (King, 1962, p. 301) others have emphasized the considerable uncertainties involved (Bishop, 1966; De Swardt and Trendall, 1969; De Swardt and Bennet, 1974). In addition to doubts about erosion surface correlation over large areas on the basis of morphologic and altimetric data, dating of marine sediments along the continental margin, especially those of the late Cenozoic, has yet to attain general agreement (Dingle *et al.* 1983). Further doubts as to the validity of King's global chronology have recently been raised by K-Ar dating of landscapes in southeastern Australia. Here deep canyons near the coast ascribed by King (1962) to the effects of Quaternary uplift are now thought to have begun to form in the early Cenozoic (Young, 1983).

There are clearly major uncertainties in the chronology of landscape development proposed for Africa by King (1962) (Table 1) but it provides a useful starting point for the discussion presented here. Three aspects of the model are worth underscoring. First, denudational cycles are considered to be initiated at the coast by continental uplift and to extend along the drainage system by scarp retreat and pediplanation. Secondly, uplift and denudation are held to have been essentially synchronous over the continent (and indeed globally), although the probability of considerable local differences in timing, especially during prolonged cycles such as the 'Moorland' planation, is acknowledged. Thirdly, emphasis is placed on the flexure of the continental margin. Subsidence seaward of an axis parallel to the margin is complemented by uplift inland.

## TECTONIC EVOLUTION of PASSIVE MARGINS

Before considering the relationship between the marine stratigraphic record of passive margins and denudational events inland it is necessary to examine briefly the margin evolution. Two phases can be identified - a rifting stage, and a subsidence stage. Several of the earlier attempts to model continental rifting emphasized the importance of doming prior to rifting (Sleep, 1971; Falvey, 1974; Kinsman, 1975) and recently Veevers (1981) has used the East African Rift System, the Red Sea, and the southern and western continental margins of Australia as an analogue of the temporal sequence thought to characterize the evolution of passive margins. While it is tempting to regard the East African Rift System as an example of incipient continental rupture, particularly in view of the continuity of seismicity and fault structures between the oceanic ridge system of the Indian Ocean and estern Africa via the Gulf of Aden, recent work has highlighted the variation in rift structure and evolution (Bermingham *et al*. 1983; Brown and Fairhead, 1983). For example, although the Afro-Arabian Rift System has uplifted rift flanks, sedimentary evidence of marine transgressions in the Jurassic and Cretaceous shows that it developed initially as a subsiding basin.

It is now considered that the sub-aerial erosion of thermal uplifts associated with continental rifting was overestimated in earlier models (Sleep, 1984) and most of the lithospheric thinning associated with rifting is now widely ascribed to crustal stretching (McKenzie, 1978; Cochran, 1983). After rifting, the nascent passive margin lying on attenuated lithosphere experiences subsidence. Initially this results from thermal contraction but gradually sediment loading becomes predominant (Steckler and Watts, 1982). The response of passive margins to sediment loading is affected by the way in which the load is isostatically compensated. Seismic reflection profiles indicate that the early stages of rifting are characterized by active faulting with local (Airy) compensation probably dominant, but subsequently flexure becomes significant (Watts, 1982). Along the African continental margin significant differences are found in the subsidence behaviour of rifted, sheared and sunk margins (Dingle, 1982). Only the rifted margins appear to broadly match the 'standard' subsidence history of passive margins to be expected from the combined effects of thermal contraction and normal sediment loading.

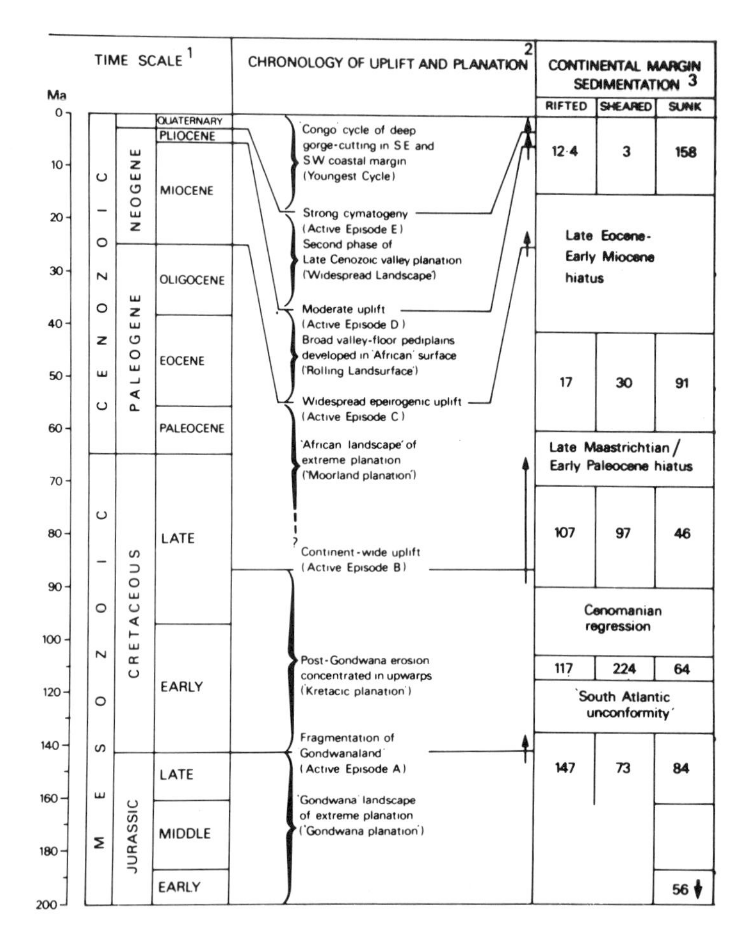


Table 1: Chronology of uplift and planation, continental margin sedimentation, and sea level change for Africa during the late Mesozoic and Cenozoic.

1) Time scale and nomenclature based on Harland et al. (1982) - note change of scale at 100Ma.

2) From King (1962) (Table VII, p. 242 and Table VIII) primarily for central and southern Africa with later modified terminology (King, 1976) in parentheses. Vertical arrows emphasize phases of uplift. Dating based on King (1976, 1983) where this differs from King (1962).

3) Subsidence/sedimentation rates for western, southern and eastern continental margins according to Dingle (1982). Figures refer to mean sediment accumulation rates ($mMa^{-1}$) for rifted margin basins (Gabon, Cabinda, Cuanza, Walvis and Orange), sheared margin basins (Guinea and Outeniqua) and sunk margin basins (Zambezi, Tanzania, Kenya and Somalia) (see Figure 1 for locations). Approximate age and duration of major hiatuses also indicated.

4) From seismic sequence analysis of Vail (1977).

5) Estimated eustatic changes of sea level based on curves in Vail et al. (1977) and Vail and Todd (1981).

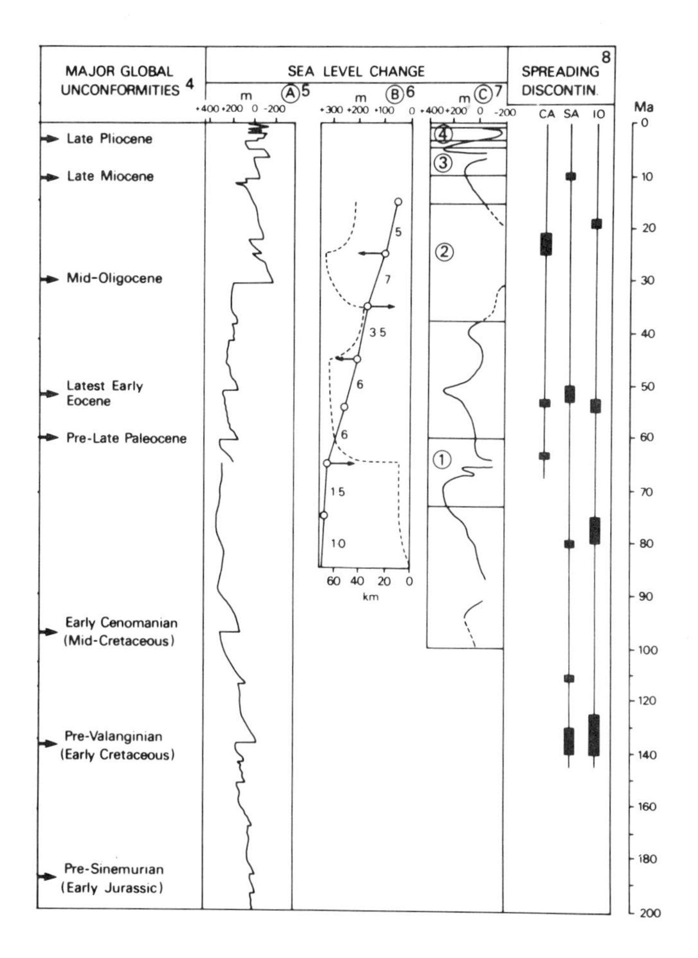


6) Change in sea level due to change in oceanic ridge volume as estimated by Pitman (1978) (solid line). 0–15Ma not shown as glacio-eustasy assumed to be predominant during this period. Figures refer to approximate mean rates of sea level fall ($mMa^{-1}$). Arrows indicate major changes in rate of fall (arrows pointing to right indicate increase in rate, those pointing to left indicate decrease in rate). Dashed line shows distance (km – bottom scale) of shoreline from hinge line as a function of rate of sea level change.

7) Late Cretaceous and Cenozoic sea level movement for southern Africa according to Dingle et al. (1983) and incorporating data from Hendey (1981). Approximate duration of hiatuses in shelf sediments shown to emphasize major regressions: 1) Late Cretaceous (Maastrichtian) – early Paleocene; 2) Oligocene – early Miocene; 3) Late Miocene – early Pliocene; 4) Late Pliocene – early Pleistocene (from Siesser and Dingle, 1981).

8) Discontinuities in sea-floor spreading and/or change in plate motion according to data compilation of Schwan (1980): CA – Central Atlantic; SA – South Atlantic; IO – Indian Ocean.

STRATIGRAPHIC RECORD OF THE CONTINENTAL MARGIN

The recent availability of seismic and borehole records of submarine continental margin sediments has provided a body of data of potential value in monitoring phases of uplift and denudation during the Cenozoic and late Mesozoic (Chappell, 1983; King, 1983). Major regressions are represented on such records by unconformities which can be traced across the continental shelf. Several studies have been published of the seismic stratigraphy of the African continental margin (Emery _et al._, 1975; Siesser and Dingle, 1981; Austin and Uchupi, 1982; Gerrard and Smith 1982; Jansen _et al._, in press). From a compilation of data from an extensive section of the African coast Dingle (1982) has identified four major unconformities, though not all are ubiquitous (Table 1). As these hiatuses can be dated with some precision it is interesting to compare them with the King (1962) chronology of uplift and planation for the adjacent continental hinterland. It is evident from Table 1 that the degree of correspondence is rather poor. Uplift phases in the continental chronology coincide with both active periods of sedimentation off-shore, as in the late Cretaceous, and depositional breaks, as in the case of the late Eocene - early Miocene hiatus.

Examination of the records for specific areas suggests that this is not an artifact of the aggregation of data from widely dispersed regions. Along the margin of southern Africa, for example, Siesser and Dingle (1981) have identified four major unconformities since the late Cretaceous (on sea level curve C in Table 1). The two most recent (3 and 4) are of late Pliocene - early Pleistocene and late Miocene - early Pliocene age; they correlate well with the two late Cenozoic uplifts postulated in the continental chronology. Although the later of these was widespread, there is no evidence of the earlier regression on the east coast or Agulhas Arch (Siesser and Dingle, 1981). The Oligocene - early Miocene hiatus (2) is the most widespread of the four and the most prolonged. It coincides in its later phases with the extensive late Oligocene - early

Miocene epeirogeny proposed by King (1962, 1976, 1983) but any correspondence during its initial stages is conjectural because of uncertainty in the timing of the start of uplift. The clearest lack of correlation is found in the late Cretaceous - early Paleocene unconformity (1) which coincides with termination of the uplift preceding the 'African' denudational cycle.

Data from the Congo - southern Gabon continental shelf (Jansen *et al.*, in press) suggest that the late Eocene - early Miocene and late Miocene - early Pliocene hiatuses were widespread. During the late Eocene - Oligocene regression identified by Jansen *et al.* (in press) the continental shelf in this area was subject to deep fluvial erosion. The subsequent mid- to late Miocene unconformity is also recorded on the Congo mainland and the continental shelf of Cameroun.

Various factors serve to complicate relationships to be expected between denudational and tectonic events on land and the marine stratigraphic record of the adjacent continental margin (Chappell, 1983). One arises from continental margin subsidence and its interrelationship with sea level (Pitman, 1978; Pitman and Golovchenko, 1983). During a period of falling eustatic sea level the shoreline will be located on the subsiding continental margin at that point where rate of sea level fall is balanced by rate of subsidence minus rate of deposition. Conversely, when eustatic sea level is rising the shoreline will migrate to a position where rate of rise is equal to deposition rate minus rate of subsidence of the margin. Along gently subsiding passive margins, therefore, regressions and transgressions are not related directly to rises or falls in eustatic sea level but to changes in rates of sea level movement. When sea level is falling, a decrease in rate of fall will cause the shoreline to move landward and an increase will cause it to migrate seaward. Such a falling sea level regime seems to have prevailed since the late Cretaceous due to a decreasing oceanic ridge volume (Pitman, 1978) (Table 1). The significance of these relationships lies in the fact that changes in base level cannot be interpreted simply as a result of eustatic sea level fall or continental uplift. Consequently, care must be taken in relating regressive phases recorded in marine sediments to active uplift of the adjacent landmass or eustatic sea level fall, since such regressions can result either from an increase in rate of sea level fall or a decrease in rate of continental margin subsidence. Changes in rate of sediment supply can also cause local variations.

A second complication involves changes in gradient across the shelf zone over which the shoreline migrates. It is generally assumed that a relative fall in sea level will promote incision of river systems and rejuvenation of the landscape. Whether this, in fact, occurs will depend, at least in part, on the gradient of the newly exposed land in relation to existing channel gradients at the coast. Figure 2 illustrates schematically three simple situations. In the first (Fig. 2A) a regression occurs over a continental margin of constant gradient. Seaward of the new shoreline ($S_2$) deposition will continue, but on its landward side, as far as the original shoreline ($S_1$), deposition will cease, or be greatly reduced, as rivers will tend to transport terrestrial debris across this zone. However, as the newly extended channels of these rivers do not have a steeper gradient we would not expect any significant incision in zone $S_1$ to $S_2$ and consequently no marked rejuvenation of the continental margin landward of the original shoreline. Some channel incision might, of course, occur if the newly exposed coastal sediments were markedly more susceptible to erosion than the previously exposed sediments inland. If sea level returns to $L_1$ a lack of deposition will be recorded by an unconformity between $S_1$ and $S_2$ but there will be no significant associated landscape rejuvenation.

In the second case (Fig. 2B) there is an increase in gradient between the initial ($S_1$) and subsequent ($S_2$) shoreline positions. Here, we would expect the increase in erosional energy of river systems below the slope break to induce a regrading of the shoreline (dashed line) first to the initial shoreline ($S_1$) and ultimately beyond. After a subsequent transgression returns sea level to $L_1$, net erosion

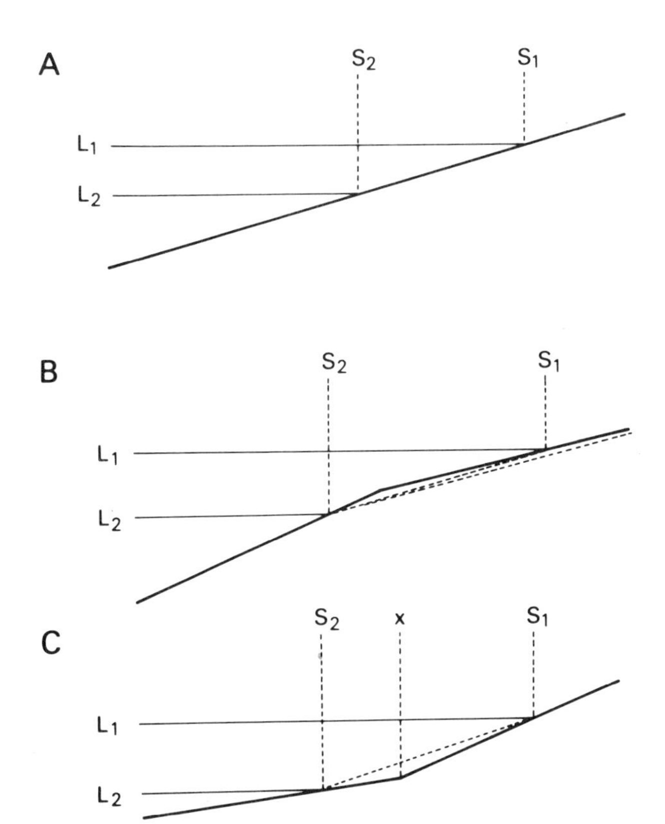


Figure 2. Possible effects of sea level change along continental margins with various configurations.

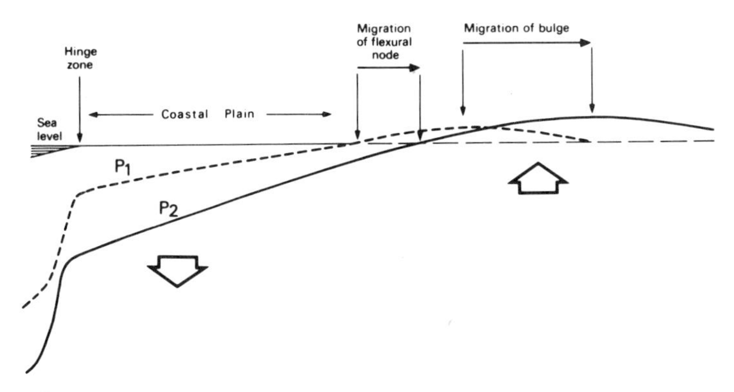


Figure 3. Inland migration of flexural bulge on a passive margin. $P_1$ represents an early and $P_2$ a later position of the basement. Ongoing sedimentation maintains the coastal plain above sea level as the basement subsides.

causes a hiatus in the marine sedimentary sequence between $S_1$ and $S_2$ and a new landscape cycle is initiated on the adjacent continent. In the third example (Fig. 2C) there is a decrease in gradient over the zone of shoreline regression. In this case there would be a period of little or no net erosion or deposition between the initial shoreline ($S_1$) and the slope break (x) since no change of channel gradient is involved. However, deposition would be expected where the gradient is lower below the slope break. Eventually this could cause a regrading of the new coastal plain with aggradation possibly extending landward of the original shoreline ($S_1$). In the marine sedimentary sequence this would be recorded as a phase of low deposition rates between $S_1$ and $S_2$ reaching a maximum at point x. Such a regression would, therefore, not be recorded by an unconformity and would be difficult to detect stratigraphically.

The possible effect of a variation in gradient along the continental margin on the consequences of a fall in base level can be assessed with reference to southern Africa (Dingle et al., 1983, p. 313). Assuming a major regression across the present submarine topography the gradients of river channels extending over the newly exposed broad coastal plains of the western and southern margins would in general be less than the mean gradients along the present-day coastal zone. Only along a section of the east coast, where there is a very steep off-shore slope, would there be a significant and widespread increase in seaward channel gradients. This situation probably obtained during the major Oligocene - early Miocene lowering of sea level by 200 to 300m. Interestingly, during this period the numerous upper slope and outer shelf submarine canyons of the eastern margin of southern Africa were probably cut by newly extending river channels (Dingle et al., 1983, p. 314).

The foregoing discussion suggests that there is not a simple or direct relationship between unconformities in the marine sedimentary record and continental planations, as implied by King (1983). Indeed unconformities marking regressions on the continental margin may be quite unrelated to the initiation of new landscape cycles inland. Clearly, attempts to relate in detail the marine sedimentary record to denudational cycles must incorporate not only the factors considered above but also isostatic adjustments of the margin to sea level change (Chappell, 1983), the source region for sediment and the zone of deposition off-shore, sediment redistribution by submarine currents and variations in erosion rates associated with climatic change. For instance, the general trend of declining Cenozoic sedimentation rates along much of the continental margin of Africa (Table 1) may be partly a response to an overall decrease in precipitation. By the end of the Cenozoic, for example, deposition rates on the shelf of southern Africa had fallen in most localities to between 20 and 60% of the rates typical of the late Cretaceous (Dingle *et al.*, 1983, p. 299-300).

## CONTINENTAL MARGIN TECTONICS AND LANDSCAPE DEVELOPMENT

Continental margins experience extensive warming during the rifting phase as heat flows laterally from the adjacent rift (Sleep, 1971; McKenzie, 1978). This causes expansion and uplift along the continental margin during the early stages of rifting (Steckler and Watts, 1982). Models developed with respect to the eastern seaboard of North America suggest that the uplift produced by this thermal bulge reaches a maximum value of about 500m about 15-20Ma after rifting has begun. By 60-70MA after the initiation of rifting the amplitude of the thermal bulge is greatly reduced. In these later rifting stages, flexure of the continental margin in response to off-shore sediment loading becomes important. The behaviour of the plate along the continental margin is affected by its flexural strength, which is low seaward of the hinge zone but increases on the landward side. Variations in flexural strength across the margin are large when it is young but decrease with time as the margin cools and becomes more rigid. Sediments progressively loading the margin eventually create a flexural depression which overcomes the thermal bulge. The lithosphere is sufficiently rigid by this stage for the flexural depression to promote a bulge landward of the hinge zone. As the margin becomes more rigid this flexural bulge migrates inland (Fig. 3). The associated landward movement of the flexural node leads to a widening of the coastal plain and a migration inland of the axis of maximum uplift. Evidence for the rotation of the continental margin slope about the hinge zone is provided in Africa by the relationships of coastal plain and off-shore sedimentary sequences (Dingle et al., 1983; King, 1983). More specifically, Siesser and Dingle (1981) found a correlation between shelf width and the maximum level of erosion of off-shore sediments indicating subsidence of the shelves through a similar angle since the formation of the late Pliocene - Pleistocene unconformity. This rotation implies uplift along a common axis some 80-120km inland of the present coastline, approximately coinciding with the line of the Great Escarpment.

In addition to uplift associated with the flexural bulge, erosion along the continental margin will promote isostatic uplift. Because of the flexural rigidity of the lithosphere the pattern of isostatic compensation will depend on the area subject to denudational unloading. The lithosphere is infinitely rigid with respect to short wavelength topography whereas for very long wavelength topography it has no rigidity and is fully compensated. Making reasonable assumptions about the physical properties of the lithosphere and mantle it can be estimated that topography should be 50% compensated at a wavelength of around 420km (Turcotte and Schubert, 1982). At wavelengths shorter than 420km the topography is largely supported by the rigidity of the lithosphere whereas at progressively longer wavelengths it is increasingly compensated.

This idea has been applied previously to the mode of isostatic uplift associated with landscape development through escarpment retreat and pediplanation by King (1955) and Pugh (1955). They argue that isostatic adjustment to denudational unloading would be delayed until a threshold distance scarp retreat had been achieved. They both employ the estimate by Gunn (1949) of 480km for the critical crustal width over which isostatic compensation becomes significant. Once escarpment retreat from the coast associated with a particular denudational cycle exceeds this threshold distance, isostatic re-adjustment is thought to begin. The resulting uplift, in turn, initiates a new denudational cycle which extends inland. It is notable that a significant proportion of high land in Africa, not obviously related to hot spots or major rifting, is located within about 400 to 500km of the present coastline. Residuals from earlier landscape cycles could eventually be elevated to considerable heights by isostatic adjustments occurring in response to denudational unloading over the landscape at lower levels. A further point is that denudational unloading results from the effects of both chemical and mechanical denudation, whereas off-shore loading results from sediment accumulation only.

An important additional element in the relationship between continental margin tectonics and the nature of denudation is the effect of a more or less constantly maintained coastal upwarp on river systems draining the continent. During formation of the continental margins of Africa following the fragmentation of Gondwanaland, two distinct drainage systems developed, one draining the steep seaward flank of the newly rifted margin and the other orientated towards interior basins. This interpretation is supported by the extensive continental deposits of the inland basins which extend back to the late Jurassic, the age of the oldest off-shore sediments (De Swardt and Bennet, 1974). It is also suggested by the present-day drainage pattern of the continent which still retains extensive areas of endoreic drainage (the Kalahari and Chad Basins, for example) and several major tributaries in externally drained catchments which are oriented away from the coast (for instance, the Kasai in the Zaire Basin). Details of the anomalous form of some of the major drainage systems, such as the Zambezi, Orange and Niger, suggest that they may have developed at a late stage by capture of inland drainage (De Swardt and Bennet, 1974). In view of this separation of drainage, which may have prevailed up to the end of the Mesozoic, it is difficult to envisage how coastal base level changes during this period can be correlated with erosion surfaces inland developed with respect to local base levels in interior basins (De Swardt and Bennet, 1974). Moreover, the persistance of a marginal upwarp, experiencing periodic uplift related to off-shore sediment loading, which eventually replaced the initial thermal bulge, would also have interrupted the transmission of new denudational cycles inland along the newly integrated continental drainage systems. It seems, therefore, unlikely that continent-wide erosion surfaces, as envisaged by King (1962) and numerous other workers, can exist over Africa. A direct link between coastal base level changes and denudational cycles is more probable along the seaward limb of the continental margin upwarp, and this is where the succession of landscape cycles in Africa has been most clearly documented (King, 1972). Even here, though, considerable doubts remain concerning the number and timing of denudational cycles (De Swardt and Bennet, 1974).

## EUSTATIC AND TECTONIC CONTROLS OF LANDSCAPE DEVELOPMENT

Although King (1962) favoured widespread epeirogenic uplift or cymatogeny as the mechanism initiating new landscape cycles over the African continent, recent evidence, as indicated above, tends to support a more complex interpretation. If landscape rejuvenation is to be continent-wide it must be generated by a fall in base level at the coast (although, as discussed above, a base level fall does not necessarily cause rejuvenation). This can be accomplished either by an extensive uplift of the entire continent or by a eustatic fall in sea level.

The possible role of eustatic sea level change can be evaluated by comparing the curve (A) derived by Vail *et al.* (1977) from seismic stratigraphy with the chronology of King (1962) (Table 1). This curve indicates gradual transgressions punctuated by rapid regressions, most notably in the Oligocene. The major regressions, which Vail *et* al.(1977) suggest are global in extent, are labelled in the adjacent column. Although elements of this curve may be tectonic in origin since the pattern of onlap used to identify eustatic changes is a characteristic feature of subsiding passive margins (Watts, 1982), the major regressions are probably largely eustatic. The rapidity of sea level change indicated by these regressions is, however, controversial as there is no accepted non-glacial eustatic mechanism to explain them.

Using the curve as a basis for anticipating when landscape rejuvenation might be initiated it is clear that there are major discrepancies (Table 1). The uplift episodes D and E of King (1976) (Table 1) appear to correspond approximately to the complex late Cenozoic fluctuations. This is confirmed to a certain extent by the regional curve for southern Africa (Table 1(D)) but the correlation for earlier events is poor. Another approach to analyzing eustatic fluctuations is to use a curve based on the predicted effect of the decrease in oceanic ridge volume throughout the Cenozoic associated with reduced rates of sea-floor spreading (Pitman, 1978) (Table 1(B)). This curve indicates a gradual decline in eustatic sea level since the late Cretaceous, with transgressive and regressive events being related to changes in the rate of sea level movement (and locally the rate of passive margin subsidence). Again there is little correspondence between periods of increased rates of sea level fall and the initiation of denudational cycles. Nor indeed is there a good correlation between this hypothetical curve and the empirical curve based on seismic stratigraphy, indicating that regional and local effects in most cases override the eustatic control of sea-floor spreading rates (Chappell, 1983). The lack of correspondence between the regressive events indicated on sea level curves and proposed landscape cycles in Africa (Table 1) could be due both to errors in the landscape chronology and the fact that eustatic sea level changes have had little or no effect in promoting continent-wide landscape cycles.

The net relative movements of continental surfaces with

respect to sea level can be further analyzed by calculating percentages of continental flooding over specific time intervals and plotting these percentages on the hypsometric curves for each continent (Bond, 1978). Such calculations indicate that, of all the continents, Africa has experienced the most substantial net uplift during the late Cenozoic (Bond, 1979). However, as the steep lower slope of the hypsometric curve of Africa illustrates, this uplift was probably concentrated in the lowland and shelf areas (Bond, 1979) corresponding to the upwarped continental margin. This approach to estimating net uplift has received some criticism (Harrison et al., 1983) and it provides information on the continental margin rather than the continent as a whole.

Further evaluation of the role of large-scale uplift necessitates consideration of possible epeirogenic mechanisms. Several workers have examined the likely effects of localised sub-lithospheric thermal anomalies (hot spots) which are thought to have generated the regions of domal uplift characteristic of large areas of Africa (Gass et al., 1978; Thiessen et al.1979, Summerfield, in press). It has been suggested that the phase of uplift proposed for Africa around 25Ma B.P. coincided with the coming to rest of the continent with respect to hot spots which then induced localised volcanism and uplift on the overlying lithosphere (Burke and Wilson, 1972). Although this coincides with King's (1962) postulated late Oligocene - early Miocene uplift episode, this mechanism would have promoted only localised areas of landscape rejuvenation which, in any case, are not likely to have been exactly temporally coincident given the variations in thickness of the African lithosphere (Gass et al., 1978). Moreover, there is no direct evidence of hot spot activity in southern Africa, nor can this model easily account for the recurrent pulses of uplift proposed for Africa throughout the late Mesozoic and Cenozoic. It might be argued that deep mantle processes could engender periodic activation of sub-lithospheric thermal perturbations but this would presumably also affect sea-floor spreading rates. There is little evidence, however, for a link between discontinuities in sea-floor spreading on the oceanic ridges surrounding Africa and the timing of proposed uplift events on the continent itself (Summerfield, in press) (Table 1). Although hot spot activity has contributed to domal uplift, especially in northern Africa, it could not alone have precipitated continental scale landscape cycles.

Another mechanism recently proposed to account for epeirogeny in Africa invokes uplift in response to a phase change in the mantle associated with the passage of the continent across the former site of an oceanic spreading ridge (Smith, 1982). This model predicts a wave of uplift migrating across southern Africa from east to west following, with a delay of 50-60Ma, the relative movement of the sub-lithospheric heating promoting the phase change and density decrease in the overlying continental lithosphere. Although this could account for the gross altitudinal contrast between high (south and east) and low (north and west) Africa, it fails to explain the marked line of continental margin uplift which, while varying in elevation, can be traced around much of the coastal hinterland of southern and western Africa. More significantly, in terms of the notion of continent-wide denudational cycles, such a mechanism could only produce diachronous uplift at the regional scale.

## CONCLUSIONS

This has been very much a preliminary, qualitative exploration of some of the issues involved in relating large-scale tectonic models and continental margin stratigraphy to the problem of long-term landscape development. Four main conclusions arise from this survey. First, there appear to be significant discrepancies between the chronology of landscape cycles widely accepted for Africa and the stratigraphic evidence of the continental margin. Secondly, these discrepancies may be due to both errors in the landscape chronology and the complex response of passive margins to base level changes. Thirdly, the generation of distinct denudational cycles by base level changes at the coast is likely to be confined to the vicinity of continental margin upwarps. Finally, the most probable mechanisms of epeirogeny would not lead to the kind of extensive uplift sufficient to produce continent-wide landscape cycles. Clearly, detailed quantitative analyses of the relationships considered here will be required in order to improve significantly our understanding of the long-term evolution of landscapes in plate interiors.

## REFERENCES CITED

Austin, J.A. and Uchupi, E., 1982, Continental-oceanic crustal transition off southwest Africa: Geological Society of America Bulletin, v. 66, p. 1328-1347.

Bermingham, P.M., Fairhead, F.D. and Stuart, G.W., 1983, Gravity study of the Central African Rift System: A model of continental disruption 2. The Darfur domal uplift and associated Cainozoic volcanism: Tectonophysics, v. 94, p. 205-222.

Bishop, W.W., 1966, Stratigraphical geomorphology: A review of some East African landforms, in Dury, G.H., ed., Essays in Geomorphology: London, Heinemann, p. 139-176.

Bond, G., 1978, Evidence for Late Tertiary uplift of Africa relative to North America, South America, Australia and Europe: Journal of Geology, v. 86, p. 47-65.

________, 1979, Evidence for some uplifts of large magnitude in continental platforms: Tectonophysics, v. 61, p. 285-305.

Brice, S.E., Cochran, M.D., Pardo, G. and Edwards, A.D., 1982, Tectonics and sedimentation of the South Atlantic rift sequence: Cabinda, Angola: in Watkins, J.S. and Drake, C.L. eds., Studies in continental margin geology: Association of American Petroleum Geologists Memoir, no. 34, p. 5-18.

Browne, S.E. and Fairhead, J.D., 1983, Gravity study of the Central African Rift System: A model of continental disruption. 1. The Ngaoundere and Abu Gabra rifts: Tectonophysics, v. 94, p. 187-203.

Burke, K. and Whiteman, A.J., 1973, Uplift, rifting and the break-up of Africa: in Tarling, D.H. and Runcorn, S.K. eds., Implications of continental drift for the earth sciences: London, Academic Press, p. 735-755.

__________ and Wilson, J.T., 1972, Is the African plate stationary?: Nature, v. 239, p. 387-389.

Chappell, J., 1983, Aspects of sea levels, tectonics, and isostasy since the Createceous: in Gardner, R. and Scoging, H. eds., Mega-Geomorphology: Oxford, Clarendon Press, p. 56-72.

Cochran, J.R., 1983, Effects of finite rifting times on the development of sedimentary basins: Earth and Planetary Science Letters, v. 66, p. 289-302.

De Swardt, A.M.J. and Bennet, G., 1974, Structural and physiographic development of Natal since the late Jurassic: Transactions of the Geological Society of South Africa, v. 77, 309-322.

_______________ and Trendall, A.F., 1969, The physiographic development of Uganda: Overseas Geology and Mineral Resources, v. 10, p. 241-288.

Dingle, R.V., 1982, Continental margin subsidence: A comparison between the east and west coasts of Africa, in Scrutton, R.A. ed., Dynamics of Passive Margins, Geodynamic Series, v. 6: Washington, American Geophsyical Union: Boulder, Geological Society of America, p. 59-71.

___________, Siesser, W.G. and Newton, A.R. eds., 1983, Mesozoic and Tertiary geology of southern Africa: Rotterdam, A.A. Balkema.

Emery, K.O., Uchupi, E., Bowin, C.O., Phillips, J. and Simpson, E.S.W., 1975, Continental margin off western Africa: Cape St. Francis (South Africa) to Walvis Ridge (south west Africa): American Association of Petroleum Geologists Bulletin, v. 59, p. 3-59.

Falvey, D.A., 1974, The development of continental margins in plate tectonic theory: Australian Petroleum Exploration Association Journal, v. 14, p. 95-106.

Gass, I.G., Chapman, D.S., Pollack, H.N. and Thorpe, R.S., 1978, Geological and geophysical parameters of mid-plate volcanism: Philosophical Transactions of the Royal Society London A., v. 288, p. 581-596.

Gerrard, I. and Smith, G.C., 1982, Post-Paleozoic succession and structure of the southwestern African continental margin: in Watkins, J.S. and Drake, C.L. eds., Studies in continental margin geology: Association of American Petroleum Geologists Memoir, no. 34, p. 49-74.

Gunn, R., 1949, Isostasy - extended: Journal of Geology, v. 57, p. 263-279.

Harland, W.B., Cox, A.V., Llewellyn, P.G., Pickton, C.A.G., Smith, A.G. and Walters, R., 1983, A geological time scale: Cambridge, Cambridge University Press.

Harrison, C.G.A., Miskell, K.J., Brass, G.W., Saltzman, E.S. and Sloan, J.L., 1983, Continental hypsography: Tectonics, v. 2, p. 357-377.

Hendey, Q.B., 1981, Palaeoecology fo the late Tertiary fossil occurrences in 'E' quarry, Langebaanweg, South Africa, and a reinterpretation of their geological context: Annals of the South African Museum, v. 84, p. 1-104.

Holmes, A., 1965, Principles of physical geology: London, Nelson, 2nd edn.

Jansen, J.H.F., Giresse, P. and Moguedet, G. in press, Structural and sedimentary geology of the Congo and southern Gabon continental shelf, a seismic and acoustic reflection survey: Netherlands Journal of Sea Research.

Kent, P.E., 1974, Continental margin of East Africa - a region of vertical tectonics, in Burke, C.A. and Drake, C.L., eds., Geology of continental margins: Berlin, Springer, p. 313-320.

King, L.C., 1955, Pediplanation and isostasy; an example from South Africa: Quarterly Journal of the Geological Society London, v. 111, p. 353-359.

King, L.C., 1962, The morphology of the earth: Edinburgh, Oliver and Boyd, 2nd edn. 1967.

__________, 1972, The Natal monocline: Explaining the origin and scenery of Natal, S. Africa: Durban, University of Natal Press, 2nd edn. 1982.

__________, 1976, Planation remnants upon high lands: Zeitschrift fur Geomorphologie, v. 20, p. 133-148.

__________, 1983, Wandering continents and spreading sea floors on an expanding earth: Chichester, Wiley.

Kinsman, D.J.J., 1975, Rift valley basins and sedimentary history of trailing continental margins, in Fischer, A.G. and Judson, S., eds., Petroleum and global tectonics: New Jersey, Princeton University Press, p. 83-126.

McKenzie, D., 1978, Some remarks on the development of sedimentary basins: Earth and Planetary Science Letters, v. 40, p. 25-32.

Melhorn, W.N. and Edgar, D.E., 1975, The case for episodic, continental-scale erosion surfaces: a tentative geodynamic model, in Melhorn, W.N. and Flemal, R.C. eds., Theories of landform development: Binghamton, State University of New York, p. 243-276.

Morisawa, M., 1975, Tectonics and geomorphic models, in Melhorn, W.N. and Flemal, R.C. eds., Theories of landform development: Binghamton, State University of New York, p. 199-216.

Ollier, C.D., 1979, Evolutionary geomorphology of Australia and Papua - New Guinea: Transactions of the Institute of British Geographers, New Series, v. 4, p. 516-539.

__________, 1981, Tectonics and landforms: London, Longman.

Pitman, W.C., 1978, Relationship between eustacy and stratigraphic sequences of passive margins: Geological Society of America Bulletin, v. 89, p. 1389-1403.

__________ and Golovchenko, X., 1983, The effect of sea-level change on the shelfedge and slope of passive margins: Society of Economic Paleontologists and Mineralogists Special Publication no. 33, p. 41-58.

Pugh, J.C., 1955, Isostatic readjustment and the theory of pediplanation: Quarterly Journal of the Geological Society London, v. 111, p. 361-369.

Schwan, W., 1980, Geodynamic peaks in Alpinotype orogenesis and changes in ocean-floor spreading during late Jurassic - late Tertiary time: American Association of Petroleum Geologists Bulletin, v. 64, p. 359-373.

Siesser, W.G. and Dingle, R.V., 1981, Tertiary sea-level movements around southern Africa: Journal of Geology, v. 89, p. 83-96.

Sleep, N.H., 1971, Thermal effects of the formation of Atlantic continental margins by continental break up: Geophysical Journal of the Royal Astronomical Society, v. 24, p. 325-350.

__________, 1984, Contraction and stretching in basin formation: Nature, v. 308, p. 771.

Smith, A.G., 1982, Late Cenozoic uplift of stable continents in a reference frame fixed to South America: Nature, v. 296, p. 400-404.

Steckler, M.S. and Watts, A.B., 1982, Subsidence history and tectonic evolution of Atlantic-type continental margins, in Scrutton, R.A., ed., Dynamics of passive margins: Geodynamics Series, v. 6: Washington, D.C., American Geophysical Union: Boulder, Geological Society of America, p. 184-196.

Summerfield, M.A., 1981, Macroscale geomorphology: Area, v. 13, p. 3-8.

________________, in press, Tectonic background to long-term landform development in tropical Africa, in Douglas, I. and Spencer, T., eds., Tropical geomorphology: London, Allen and Unwin.

Thiessen, R. Burke, K. and Kidd, W.S.F., 1979, African hotspots and their relation to the underlying mantle: Geology, v. 7, p. 263-266.

Turcotte, D.L. and Schubert, G., 1982, Geodynamics: Applications of continuum physics to geological problems: New York, Wiley.

Vail, P.R., 1977, Seismic recognition of depositional facies on slopes and rises: Association of American Petroleum Geologists Continuing Education Course Note Series, no. 5, Geology of continental margins, p. F1-9.

__________, Mitchum, R.M. and Thompson, S., 1977, Seismic stratigraphy and global changes of sea-level, Part 4: Global cycles of relative changes of sea-level: American Association of Petroleum Geologists Memoir no. 26, p. 83-97.

__________, and Todd, R.G., 1981, Northern North Sea Jurassic unconformities, chronstratigraphy and sea-level changes form seismic stratigraphy, in Cling, J.V. and Hobson, C.D., eds., Petroleum geology of continental shelf of north-west Europe: London, Institute of Petroleum, p. 216-235.

Veevers, J.J., 1981, Morphotectonics of rifted continental margins in embryo (East Africa), youth (Africa-Arabia), and maturity (Australia): Journal of Geology, v. 89, p. 57-82.

Watts, A.B., 1982, Tectonic subsidence, flexure and global changes of sea level: Nature, v. 297, p. 469-474.

Young, R.W., 1983, The tempo of geomorphological change: evidence from south-eastern Australia: Journal of Geology, v. 91, p. 221-230.

# 3

# Stages in the creation of a large rift valley – geomorphic evolution along the southern Dead Sea Rift

*Ran Gerson, Sari Grossman, and Dan Bowman*

ABSTRACT

The southern Dead Sea Rift is a complex feature composed of hundreds of different structures, each having a very different geomorphic evolution. Phases of rift bottom lowering were studied by using scarps, straths and pedimented surfaces, to document past periods of tectonism and quiescence. Ten such phases were identified in the Timna Valley in the southern Negev. This is a minimal number, because periods of continuous activity, degradation of multiple scarps and lack of preservation of scarps due to lithologic and topographic conditions, preclude the possibility of identifying additional phases.

Faults and lineaments in alluvium are related to the later stages of rift activity. The fault scarps were degraded rapidly. The distance of their recession is short compared with fault scarps in hard bedrock. The short distance of recession is also related to the lower initial relief. Most of the normal fault scarps were formed during individual (single) events. Normal faulting along the Gulf of Elat has cut alluvial surfaces of late Holocene and earlier ages, whereas faults along the Arava Valley appear to be of early Holocene and older ages. Is the Arava Rift a seismic gap at present?

Changes in drainage along the rift shoulders accompany rifting. In the Grofit Pass area reversals of flow, piracies and impeded drainage in the divide belt are frequent. Rift formation also affected the nature of fluvial deposits: Fine deposits are now found in many major stream channels, that transported mostly gravel prior to rift formation.

## THE DEAD SEA RIFT - A GENERAL SKETCH

The Dead Sea Rift is a transform feature between the spreading Red Sea and the collision zone of the Taurus - Zagros Mountains, (Freund, 1965). It is a feature some 1000km long between the Sinai-Levant microplate and the Arabian plate (fig. 1; Freund & Garfunkel, 1976).

Several structural characteristics of the Dead Sea Rift may be emphasized:
(1) A major left-lateral offset of some 105-110 km has occurred along a belt of strike slip faults across the rift (Quennell, 1959; Freund, 1965). In the Gulf of Elat (G.E. in fig. 1) some of the offset has taken place along shear several km wide (Eyal et al., 1981). Most of the movement was over by the early Quaternary (Freund et al., 1968; Zak & Freund, 1981). Yet, in the middle Miocene, much of the area now in Jordan drained to the Mediterranean Sea, across the rift belt which now drains to the Dead Sea and the Gulf of Elat (Neev, 1960; Garfunkel & Horowitz, 1966; Garfunkel, 1970).
(2) Rhomb-shaped grabens, that appear as depressions in the rift bottom, were formed by strike slip faults, many of which are arranged en-echelon (Quennell, 1959; Freund et al., 1968; Garfunkel, 1970).
(3) Normal faulting along the rift margins has led to a relief of several hundreds to a few thousands m high.

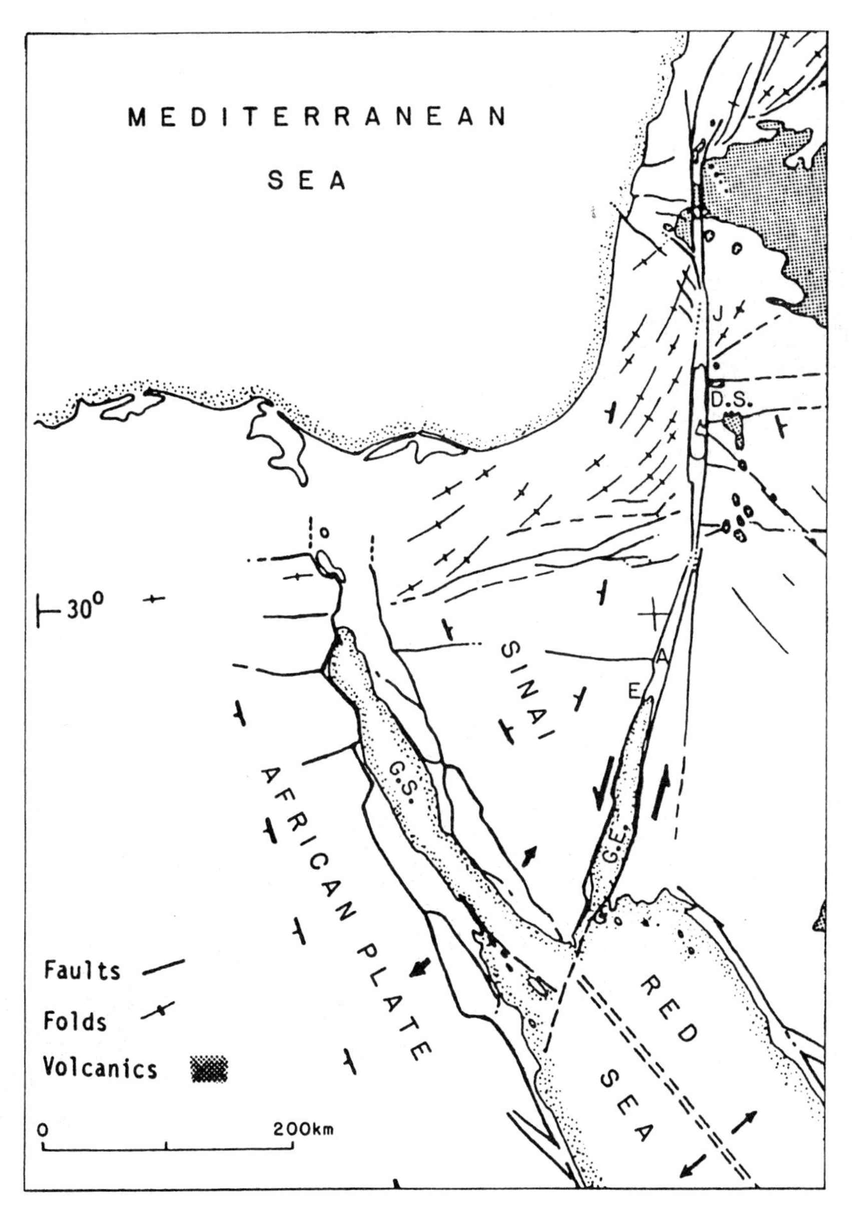


FIGURE 1  The southern Dead Sea Rift - general setting. A = Arava Valley; D.S. = Dead Sea; E = Elat; G.E. = Gulf of Elat; G.S. = Gulf of Suez; J = Jordan Valley; N = Negev (after Garfunkel & Fruend, 1976).

(4) Uplifted rift shoulders are apparent along large stretches of the rift. However, downfaulted stepped blocks form the rift margins in many areas (fig. 2).
(5) The rift margins are subdivided into separate structures, usually several km long and wide, striking at different angles to the general NNE trend of the rift. Most faults and flexures are not parallel to the rift but trend up to 45° to it, thus forming structural embayments, as is typical to the eastern Arava Valley margins (fig. 2. Bender, 1974). Along segments of the rift, such as the central Arava Valley and the central Jordan Valley in their western margins, one finds flexures forming the rift boundaries.

The bottom of the Dead Sea Rift is floored by Neogene and Quaternary sediments, up to several km thick. They are faulted in places, and together with morphotectonic features related to the rift margins, allow detailed definition of faulting stages of the rift.

The portion of the Dead Sea Rift considered here includes the Gulf of Elat and the Arava Valley - the southern rift (fig. 1,2). The study does not cover the whole length of this 350km long stretch, but draws conclusions from several typical features and case studies.

## A GENERAL STATEMENT, OBJECTIVES AND SCOPE

It is widely accepted that tectonism has a geomorphic expression in the regions where it occurs and sometimes around them. Such a geomorphic expression in taphrogenic regions is a major source of evidence relating to tectonic phase separation, to the definition of repeated faulting, to relative quiescence and to stages and degrees of ground tilting.

Along the margins of the southern Dead Sea Rift as well as at its bottom one finds various types of geomorphic features, which appear to be the only evidence that allows the separation of different stages of rift development and especially the assessment of

FIGURE 2 The Arava Rift Valley. Note the strike slip fault traces along the Arava Valley, crossing diagonally from Elat towards the east side of the Arava. Structural and morphologic segmentation of the eastern Rift shoulder is seen (Landsat imagery, NASA). E = Elat; G = Grofit area; G.E. = Gulf of Elat. T = Timna.

relatively young (Quaternary) activity.

The segmentation and separation of the rift margins and bottom into hundreds of different blocks (fig. 2) precludes the possibility of assigning a detailed morphotectonic evolution of the rift as a whole. Adjacent blocks show different geomorphic age. Yet, we feel it is possible to draw generalizations concerning the development of several morphotectonic elements: (1) Fault escarpments; (2) Geomorphic surfaces, both rock-cut and alluvial; (3) Faults and lineaments in alluvium; (4) Drainage systems.

The objectives of the present article are:
(1) To assess and differentiate some types of geomorphic expressions of rift movement.
(2) To enumerate different phases of rift activity.
(3) To set a time frame to the latter phases of rift activity and implications regarding late quaternary seismicity.
(4) To point at possible age differences in activity in the various parts of the southern Dead Sea Rift.

## FAULTING, RIFT LOWERING, SCARP RETREAT AND GEOMORPHIC SURFACE EVOLUTION

The evolution of erosion surfaces initiated during periods of tectonic quiescence has been discussed by many authors since Powell (1875) and Davis (1895, 1899). Comprehensive models, such as Penck's (1953) and King's (1962), assist significantly in understanding the existence of broad rock-cut surfaces separated by escarpments, that truncate different structural and lithologic terrains. The evolution of pediments at the base of retreating tectonically-generated escarpments is well illustrated in the Basin and Range Province in SW United States (King, 1968; Bull & McFadden, 1977). Such features should be well documented in cases where for a while "Rate of Uplift > Rate of Incision and Denudation" of elevated rock blocks (Kennedy, 1962).

As demonstrated by Quennell (1958, 1959) and Garfunkel (1970), there is a regional and high erosion surface, truncating old structures in upper Cretaceous sedimentary rocks in southern Jordan and the southern Negev. Garfunkel (1970) suggests that this erosion surface was formed during the upper Miocene or lower Pliocene. A later stage was marked by valley incision of several dozen meters and deposition of gravel (Arava Conglomerate?) on valley floors. During both stages, prior to the evolution of the southern Dead Sea Rift as a regional drainage attracting feature, streams draining southern Jordan flowed west across the present-day rift belt. In watersheds now draining to the rift, there are remnants of conglomerates along valley sides (the reversal of drainage is discussed in the last chapter).

It is generally accepted that the rift has developed in two major phases of sinistral movement. (Quennell, 1958, 1959; Freund, 1965; Freund et al., 1968; Freund et al., 1970), but many other problems exists. For example, how many stages of rift lowering by vertical (normal) movement can be diagnosed? What, in that respect, is a "stage"? How sensitive are the stage indicators? How long can the evidence of discrete stages be preserved? How may changing palaeo-climates have affected the landscape during rift formation? These problems were addressed while studying one of the case histories - the Timna and Avroa Valleys along the margins of the southern Arava.

Timna Valley (T in fig. 2) is suitable for the study of rift margins and their reaction to faulting for several reasons:
(1) The valley has a diversity of rock types that may have different reaction to erosion, scarp retreat and drainage net evolution. The generalized columnar section includes, from bottom to top, hard plutonic igneous rocks, various sandstones, hard limestones and dolomites, chalks and marls (fig. 4).
(2) A domal structure around Mt. Timna, cut in half by the rift, and faults in the structure, allows rapid erosion sandstones, thus transmitting the effect of base level lowering into the mountainous terrain.
(3) A broad mountain front exits at the rift margins (Mount Timna), in-between large incised valleys. Scarp retreat may be examined there.

(4) Pediments and rock-cut straths may be correlated.

Figures 3, 4 present some of the evolution of Timna Valley. Several points are emphasized:

(1) The old erosion surface encircles the Timna Valley; the two are separated by an escarpment some 300m high. Along the margins of the erosion surface there are ten saddles, representing beheaded valleys, draining areas now in the Timna Valley and Arava rift belt.

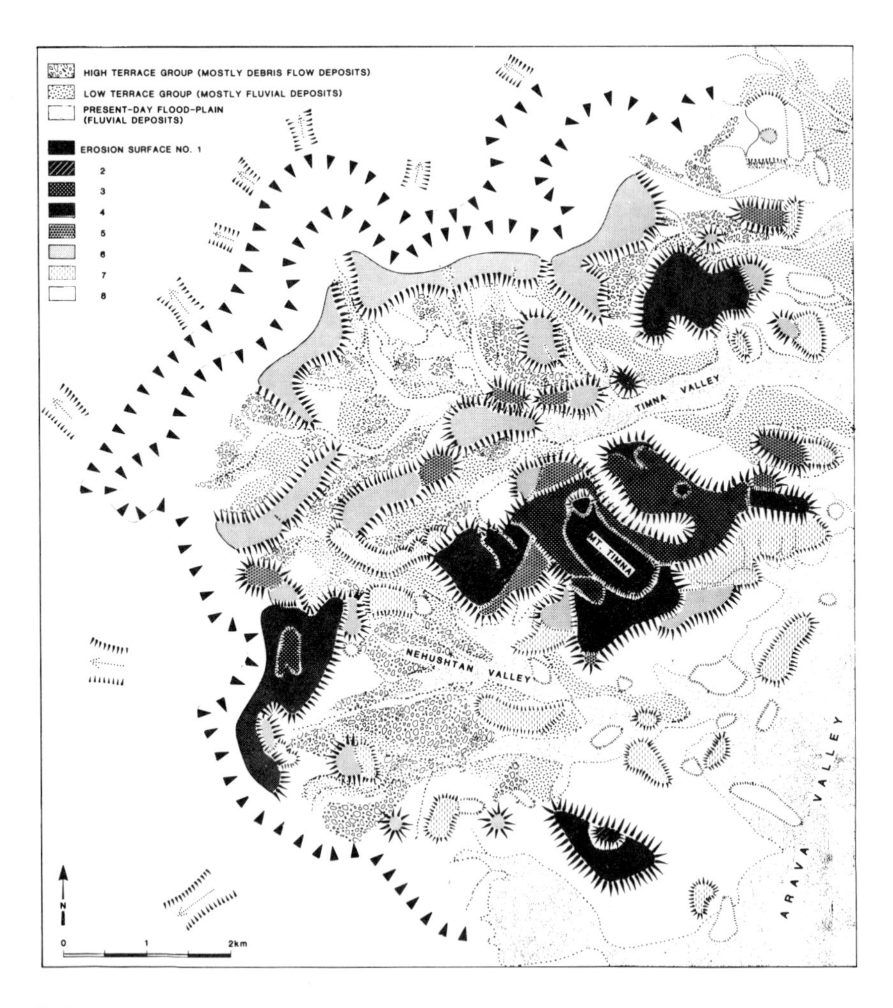


FIGURE 3 Map of rock-cut geomorphic surfaces in Mt. Timna and Timna Valley, draining to the Arava Rift. Above the encircling escarpment there are ten saddles - remnants of former valleys flowing from east to west and northwest.

(2) Eight high rock-cut geomorphic surfaces were identified in the Timna Valley. Their correlation and reconstruction were made possible using detailed topographic maps (scales - 1:5000, 1:10000; contour interval - 5m, 10m). The surfaces smoothly truncate fault structures as well as different rock types of highly varied strength (such as granite, syenite, mudstones, sandstones). They slope towards the rift at a gradient similar to present-day stream channels.

(3) Scarps and elevation differences separating successive surfaces are 20-40m high, whereas the width of rock-cut surfaces is usually much larger than 150m. Scarp gradients are 20°-30° and their longitudinal profile is mostly rectilinear. In many cases, where scarps are generated by incision of major tributaries, several scarps coalesce into one and elevation difference between relicts of erosion surfaces may reach 80-100m (fig. 3).

(4) Fault generated scarps 20-40m high may be preserved, even though segmented, to a distance of more than 2 km from their initial position. Similar horizontal distances were measured for scarps generated by stream or flood plain incision, where straths were widened by pedimentation through scarp retreat. Since scarp retreat is affected by factors such as scarp relief as well as bedrock erodibility and water and sediment yield, one should not at this time, use horizontal distance of retreat from the generating plane as a measure of age.

(5) The geomorphic surfaces discussed are not covered by diagnostic deposits, such as fluvial sediments or clearly defined soils. Although weathered, these surfaces are preserved mostly because of lack of sufficient power to erode them. There are, then, no deposits to correlate the rock-cut geomorphic surfaces with any rift bottom sediments.

(6) Evidence of later stages and climatic changes in the Timna Valley has been assembled and interpreted elsewhere (Gerson, 1981, 1982 a,b). The lowermost surfaces in the valley bottom and the high-relief encircling escarpment bare evidence of major climatic fluctuations: Talus aprons veneering the high escarpments and debris flow deposits are typical of both hillslopes and flood plains during wet (pluvial or only moderately arid) climatic modes. Erosion of talus aprons, incision of valley bottoms and fluvial

sediments are characteristic of dry (arid, extremely arid) climatic modes. Alternating major climatic fluctuations are indicated by talus relicts and valley bottom sediments of the middle and late quaternary. It was observed that such alternating terrains are related to escarpments having an available relief of 40m and higher. Scarps of lower relief have not developed the above mentioned sequences. This may explain why we do not observe the effects of climatic change in the older terrain, in addition to possible stripping of deposits during longer periods of time.

(7) Two (or possibly three) stages of faulting along the Rift boundary have affected the Quaternary alluvial fans along the Timna mountain front. As in the Elat region, this faulting occurred prior to the latest Quaternary. Evidence of this faulting is observed close to the rift boundaries (fig. 4) and by reconstruction of the longitudinal profiles of the highest terraces in the Timna Valley.

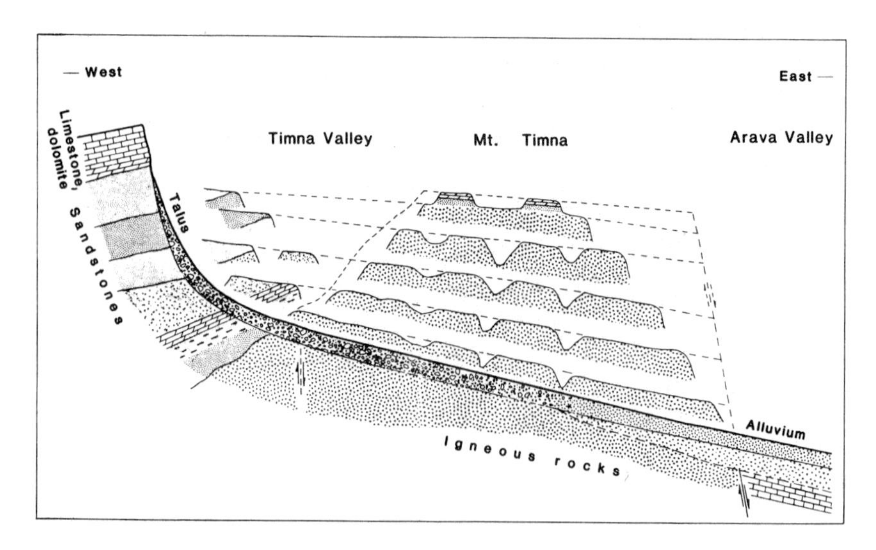


FIGURE 4 A schematic composite cross section illustrating geomorphic surface formation in the Timna Valley. Geomorphic surfaces in Mt. Timna are projected and separated.

Several general conclusions may be drawn from these observations:
(1) About ten phases of faulting are evident in the rift segment of Timna Valley. The evidence should be interpreted as phases of intensive and frequent faulting/earthquakes with long periods of tectonic quiescence between them. The number of stages is probably a minimal figure, since the type of evidence assembled and the method of analysis are bound to cluster tectonic events. It has to be noted that multiple scarps, representing several frequent faulting events may not have been detected. They may have formed single scarp or a gently sloping ground through degradation. Adjacent structures along the southern Arava margins show fewer stages; five to six stages were identified in the Mt. Amram area and a lesser number in the Grofit area (G in fig. 2).
(2) How do we estimate rates of scarp retreat, especially along valley bottoms governed by changing erosional and depositional regimes? In areas similar to the Timna Valley, where no absolute dating methods are applicable to the fluvial sediments or to the gypsic and salic soils (rather than calcic soils), rates have to be based on speculative models. A further complication results from the fact that different climatic regimes cause different flood plain width. Narrow linear flood plains are formed during dry regimes (like the one prevailing at present) and wide flood plains are formed during wet regimes (Gerson, 1982a). At present we do not have data for a quantitative estimate of valley-side scarp retreat. General considerations, based on the age of the rift as a whole lead us to the conclusion that valley-side and mountain-front scarps having initial relief of more than 20m may be preserved for $10^5$-$10^6$ years, whereas scarps in coarse alluvium may be degraded in a period of $10^4$-$10^5$ years.
(3) While the total relief of the Timna Valley is subdivided into about ten scarps, there are many segments of rift margins which have escarpments hundred of meters high with no appreciable break of slope, implying intensive continuous activity during a relatively short period of time.

## FAULTS IN ALLUVIUM - NATURE, DISTRIBUTION AND SOME IMPLICATIONS

Faulted alluvium is found along the bottom of the southern Dead Sea Rift both in the center and extensively along the margins. Faults cut alluvium of different ages - Pliocene and quaternary. These faults may be subdivided into two groups: (1) Faults having long traces (several or tens km) occur mostly in the Arava Valley center, somewhat diagonal to its general trend (Shaw, 1947; Quennell, 1958, 1959; Bentor et al., 1965; Zak & Freund, 1966; Garfunkel, 1970; Garfunkel et al., 1981). These are strike slip faults of different ages; (2) Faults having relatively short traces (several hundred to several thousand m) are found along the margins of the rift bottom. One of the few exceptions is the fault along the northwestern segment of the Gulf of Elat (fig. 7). Most of these faults are normal.

Several features characterize the faulted surfaces, as described by previous authors:
(1) Rhomb shaped grabens and pressure ridges at the end of strike slip faults (Zak & Freund, 1966; Garfunkel, et al., 1981).
(2) Displaced drainage patterns, ridges built of alluvial deposits and fans detached from the streams that formerly fed them (Garfunkel, et al., 1981).
(3) Old alluvial fill (Garof conglomerate, of Pliocene (?) age) faulted along the Arava Valley margins, and "very young to recent" fill is faulted by the strike slip faults at the center of the Arava Valley (Garfunkel, et al., 1981).

Further inspection of faults in alluvium discloses some additional characteristics:
(1) Normal faults along the Gulf of Elat cut Pleistocene and/or Holocene alluvial surfaces (fig. 5). Normal faults along the western margins of the southern Arava Valley do not cut latest Pleistocene and Holocene deposits (fig. 6). The youngest faults (several hundred to few thousand years old) along the Gulf of Elat are normal faults.
(2) Morphological features of scarps of many of the normal faults in alluvium and colluvium along the Gulf of Eilat bare evidence to

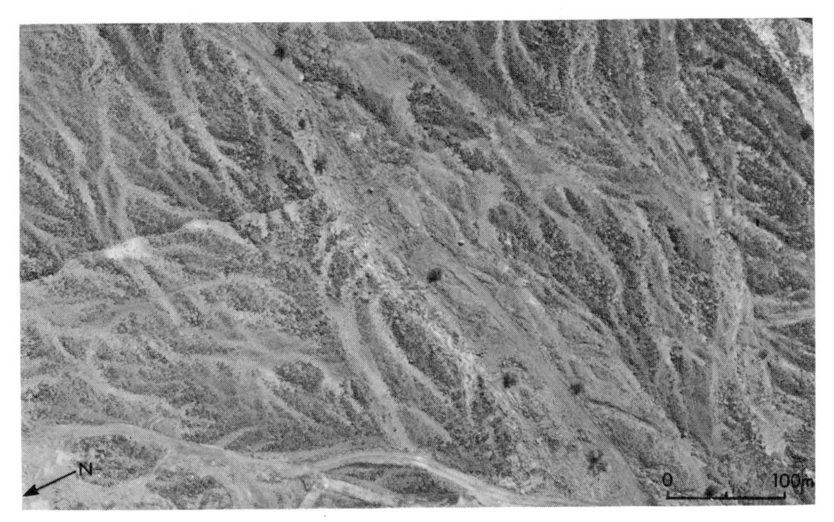


FIGURE 5 A fault cutting an early to Middle Holocene alluvial fan (Wadi Mukeibila, eastern Sinai).

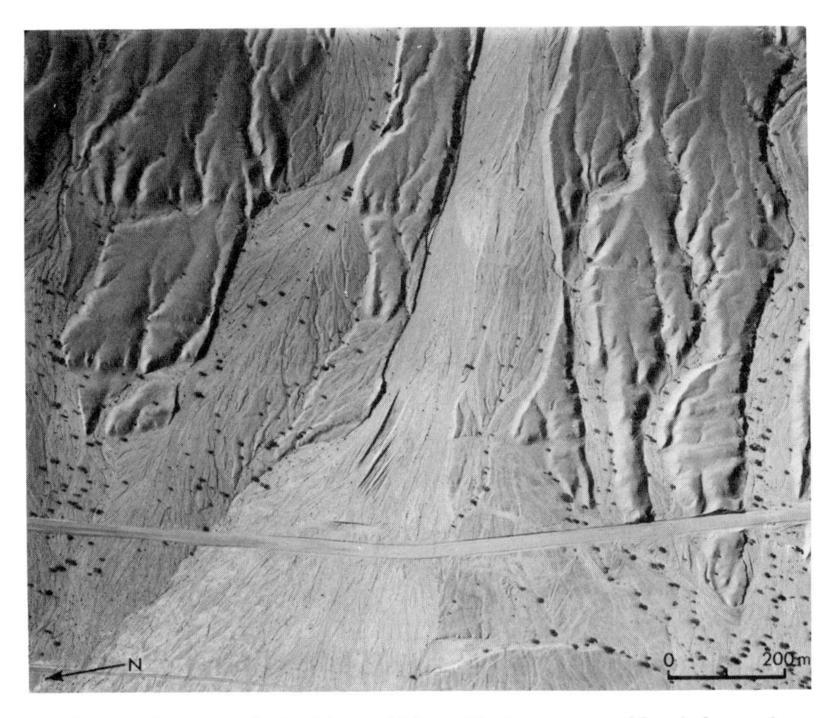


FIGURE 6 Traces of faults cutting Pleistocene alluvial surfaces. (southern Arava Valley, near Elat).

their recency: fresh and steep (40°-60°) free faces; sharp unrounded crests; short debris slopes; none or only slight dissection; faulted, well-preserved gravel bars; steep scarps still well preserved on talus mantles; silt and fine sand in small ponds along scarps facing upfan; furrows at the base of scarps. Such features were rarely observed along fault scarps in the Arava Valley.
(3) Strike slip faults along the southern Arava Rift cut alluvial surfaces considered to be of early Holocene age.
(4) Examination of gradients of fault scarps related to alluvial surfaces of generally known age (Bowman & Gerson, in prep.) leads to several conclusions: Gradients of most fault scarps range between 10° and 35°. Only very young scarps, especially those carved in cohesive debris flow deposits, are definitely steeper, 40° to 60°. Scarps of faults cutting Holocene coarse-alluvial surfaces are steeper than 20° whereas scarps of faults cutting definitely older alluvial surfaces, but not adjoining Holocene surfaces, are 10° to 20° in gradient. Since more precise ages for most surfaces are still wanting we have not been able to correlate slope and age in a more accurate way.

Many of the normal faults at the margins of the southern Arava Valley strike as much as 15°-45° from the strike slip fault traces at the Arava Valley Rift bottom (fig. 7). As these faults have small throws (0.5-5.0m) and short traces (usually less than 3km), they may be considered as satellite faults or perhaps slumps related to major earthquakes generated along the larger faults. This point is of special significance since we are unable to observe the major strike slip faults masked by the water of the Gulf of Elat. There we inspect only the satellite faults along the rift margins. The density of these faults (figs. 5,7) may mean that they form as one fracture plane at depth, and branch into separate fractures in the upper alluvial fill. Most of these faults do not show any multiple activity - they are single event scarps.

A general conclusion, then, is that the rift in the Gulf of Elat segment has been more active, tectonically and seismically, in the very recent past, than the Arava Valley segment. This conclusion

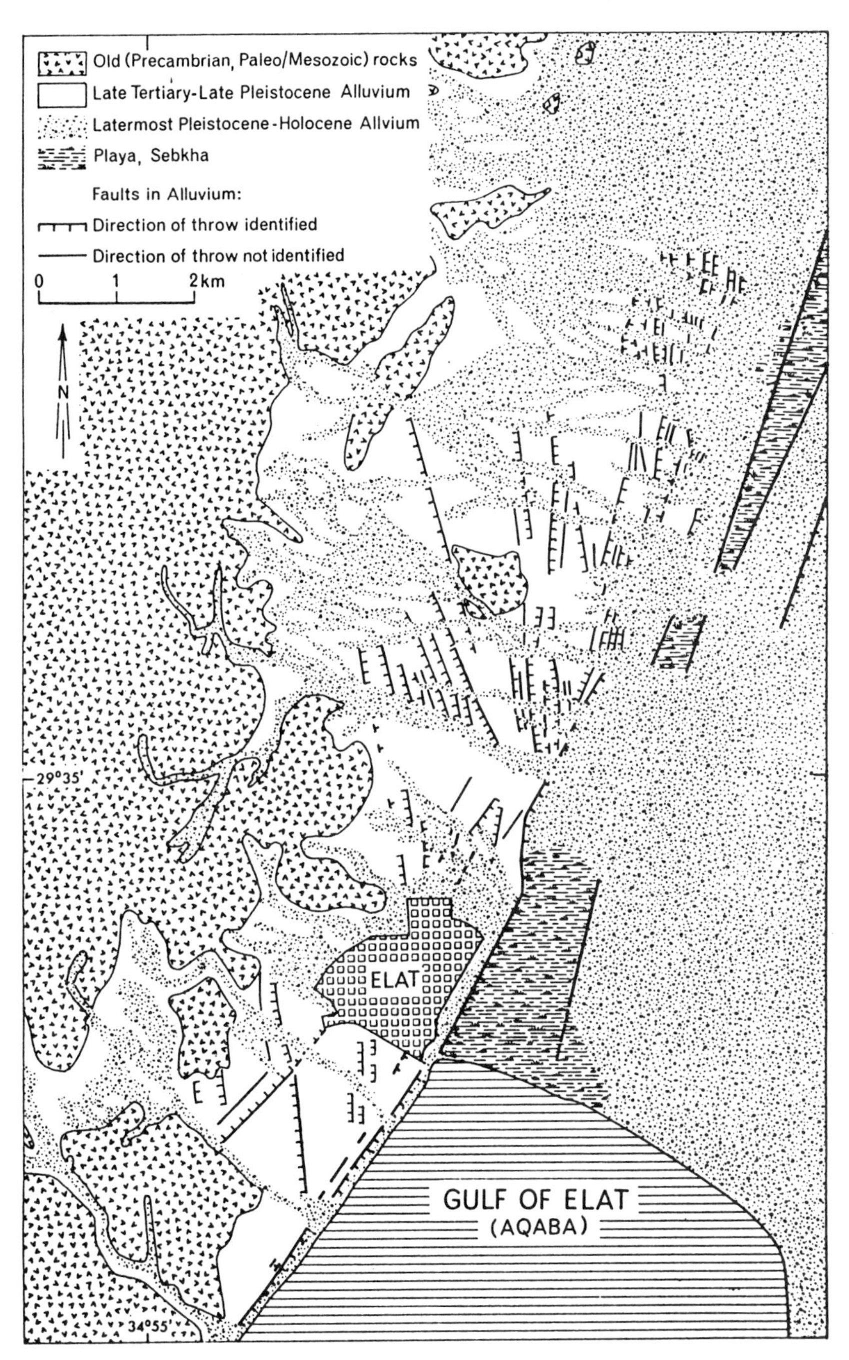


FIGURE 7 Faults in alluvium in the southern Arava Valley near Elat. The slightly thicker lines in the eastern portion of the area represent faults mapped by Garfunkel (1970).

coincides with earthquake foci distribution in recent years (Ben-Menahem, 1981; Ben-Menahem & Abodi, 1981). It appears that the Arava Valley has been a "seismic gap" in the last few thousand years.

A possible explanation for the discrepancy in age between the old normal faults and the younger strike-slip faults may be a change in mode of operation from normal faulting and extension to horizontal movement.

It is rather rare, along the southern rift, that recent faulting has occurred right at the bottom of a large (high and long) rock cut fault escarpment. Most faulted alluvium extends further into the rift than the precise base of fault escarpments. There is some sort of gentle debris slope (a portion of either a talus or a fan) between the rock-cut escarpment and the rift bottom. Generally it implies that the fault escarpments as such have not been active for a while.

## CHANGES AND REVERSALS OF DRAINAGE-NET PATTERNS

In the last 25 years several authors have discussed major changes in drainage-net patterns due to the Dead Sea Rift formation (Neev, 1960; Garfunkel & Horowitz, 1966; Garfunkel, 1970; Zak & Freund, 1981). A repeated theme in these discussions is the changes and reversals in drainage courses during the Pliocene and the Quaternary.

We have recently studied several drainage systems along the western Arava rift margins. One example is the Grofit Pass area (figures 8 A, B). The area now includes two sets of faults systems: (1) An old fault system, forming blocks that were truncated by the high erosion surface but now does not have any prominent morphologic expression. Its age is Miocene-early Pliocene (?); (2) A later fault system, of Pliocene-Quaternary age, forming stepped fault blocks separated by high fault scarps.

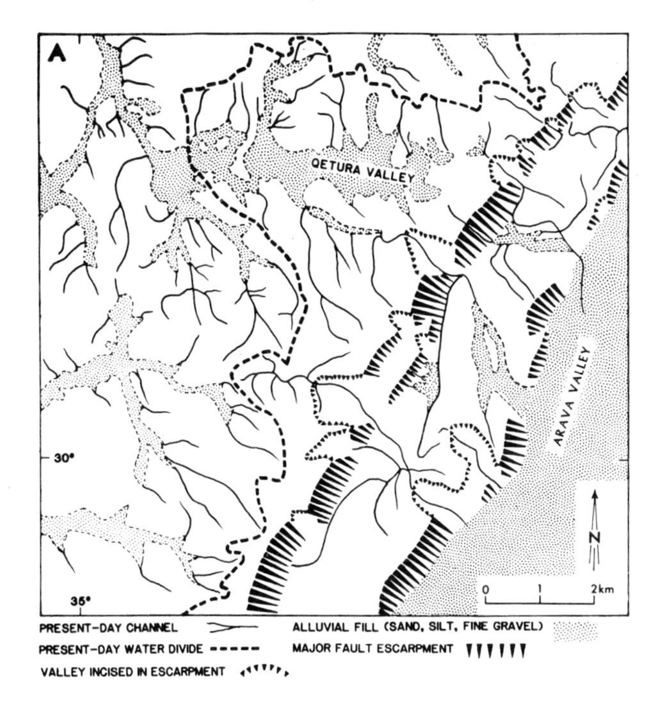


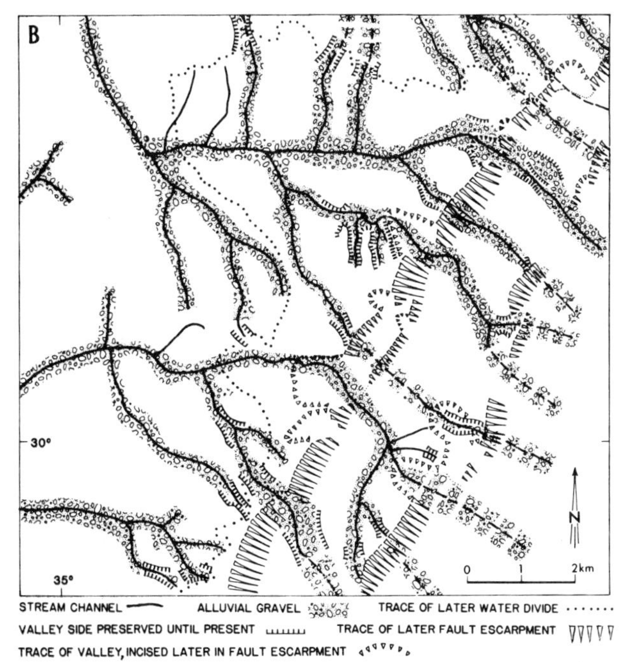


FIGURE 8 Stages in the evolution of drainage net patterns in the Grofit Pass area (western margin of the southern Arava valley). A. At present; B. Before rifting. See text for explanation.

The situation at present is illustrated in Figure 8A: (1) Fault escarpments form total relief of 400m; (2) The dividing line between streams draining directly to the rift and west or northwest to the Hyyon Valley system, is 4-7km west to the Arava Valley; (3) Deeply incised streams flow eastward in valleys formerly accommodating streams flowing westward; (4) Portions of former valleys are left as wind-gaps; (5) Most present-day stream channels are narrow and have small catchment areas; (6) Streams utilize former valleys flowing in reversed direction; (7) Many incised new first order streams have been added; (8) A playa and impeded drainage are evident in the northern end of the area presented in Figure 9A.

Figure 8B shows a suggested reconstruction of the drainage pattern as it appears to have been prior to step faulting. The reconstruction is based on several geomorphic elements: (1) Wide saddles, often filled with fluvial gravel across present-day divide belts; (2) Presence of valley fill high above present flood plains; (3) Drainage nets presenting anomalous patterns, such as tributary junctions opposite to flow directions, streams flowing west and turning sharply east, playas and impeded drainage in divide zones and further down stream; (4) Overfit valleys draining streams of small catchment areas and narrow channels. In the past these valleys have drained large watersheds now buried under the rift bottom fill and possibly in the plateau east of the Rift.

A special feature is the change in type of deposits in valleys not yet affected by incision due to later morphotectonic relief. Such valleys, used to carry larger streams in the past, still have gentle gradients. After being beheaded from their large catchments they have not been able to carry the sediment supplied to them by their small catchment tributaries, alluvial fans formed at their valley bottom sides and segmentation into small basins of endoreic drainage and playas took place. This sequence of events may have been enhanced by possible slight tilting of the area to the east. Similar evolution was observed in several other instances a few km to the south and north of the area here presented.

## CONCLUSIONS

The southern Dead Sea Rift margins is built of hundreds of separate structures - fault blocks, flexures and folds. Adjacent structures have different structural and geomorphic histories. Mountain fronts display different degrees and ages of activity as analysed by geomorphic methods (Wallace, 1978; Bull, 1978).

Strike slip faulting along the bottom of the Rift as studied directly at the Rift bottom, and by satellite normal faulting at the Rift margins, was active during the late Quaternary, but at different rates at different rift segments. The evidence here presented points at more intensive Holocene faulting (and seismic) activity along the Gulf of Elat than along the Arava Valley. Very young normal faults are present along the Gulf's margins; faults scarps along the southern Arava are somewhat older. Is it possible that the Arava Valley is a seismic gap between the more active Gulf of Elat and Dead Sea segments?

The Valley of Timna was selected for a detailed study of older stages of rift bottom lowering. Ten stages of lowering were identified, using rock-cut straths and pediments of various sizes, receding scarps of considerable relief and faulted alluvial terraces. Ten is considered to be a minimal number of stages, since long continuous faulting events or fault scarps of low relief may not be preserved as geomorphic evidence for long periods of time (1 million years or more); only initial scarps of 20m or higher appear to be preserved while receding. Most of the faulting occurred during the Pliocene. During the Quaternary, faulting affected fluvial gravel covered surfaces of alternating wet and dry climatic regimes, as reflected in talus and fluvial landforms and deposits. The retreat of high escarpments appears to be significantly dominated by major climatic changes and cannot be used directly as a morphotectonic indicators.

Changes in drainage patterns along the right margins due to rift formation are abundant: stream captures, reversal of stream direction in existing valleys and impeded drainage, accompanied by change of type of the sediments deposited and playa formation. These are demonstrated in the Grofit Pass area along the southern Arava Rift margins. Such changes vary from structure to structure along the southern Dead Sea Rift.

## REFERENCES CITED

Bender, F., 1968, Geologie von Jordanien: Borntraeger, Berlin, 230 p.

Bender, F., 1974, Explanatory notes on the geological map of the Wadi Araba, Jordan, scale 1:100,000, 3 sheets: Geologishes Jahrbuch Reihe B. Heft 10, 62 p.

Ben-Menahem, A., 1981, Variation of slip and creep along the Levant Rift over the past 4,800 years: Tectonophysics, v. 80, p. 183-197.

Ben-Menahem, A., and Abodi, E., 1981, Micro- and macroseismicity of the Dead Sea Rift and off-coast eastern Mediterranean: Tectonophysics, v. 80, p. 199-233.

Bentor, Y.K., Vroman, A., and Zak, I., 1965, Geological map of Israel, scale 1:250,000, southern sheet: Government printer, Jerusalem.

Bowman, D., and Gerson, R., 1984 (in preparation), Late Quaternary faulting in the Gulf of Elat Region, Eastern Sinai.

Bucknam, R.C., and Anderson, R.F., 1979, Estimation of fault-scarp ages from a scarp-height-slope-angle relationship: Geology, v. 7, p. 11-14.

Bull, W.B., 1978, Geomorphic tectonic activity classes of the south front of the San Gabriel Mountains, California: U.S. Geological Survey, 59 p.

Bull, W.B., and McFadden, L.D., 1977, Tectonic geomorphology north and south of the Garlock Fault, California, in Doehring, D.O., ed., Geomorphology in arid regions: Proceedings, 8th, Geomorphology Symposium, Binghamton, p. 115-138.

Davis, W.M., 1895, The Geographical cycle: Geographical Journal, v. 14, p. 481-504.

Davis, W.M., 1899, The rivers and valleys of Pennsylvania: National Geographic Magazine 1, p. 81-110.

Eyal, M., Eyal, Y., Bartov, Y., and Steinitz, G., 1981, The tectonic development of the western margin of the Gulf of Elat (Aqaba) Rift: Tectonophysics, v. 80, p. 39-66.

Freund, R., 1965, A model of the structural development of Israel and adjacent areas since upper Cretaceous times: Geological Magazine, v. 102, p. 189-205.

Freund, R., and Garfunkel, Z., 1976, Guidebook to excursion along the Dead Sea Rift: Mimeograph, Department of Geology, The Hebrew University, Jerusalem, 27 p.

Freund, R., Garfunkel, Z., Zak, I., Goldberg, M., Derin, B., and Weissbrod, T., 1970, The shear along the Dead Sea Rift: Philosophical Transactions, Royal Society of London, Series A., 276, p. 107-130.

Freund, R., Zak, I., and Garfunkel, Z., 1968, Age and rate of the sinistral movement along the Dead Sea Rift: Nature, v. 220, p. 253-255.

Garfunkel, Z., 1970, The tectonics of the western margins of the southern Arava (Ph.D. thesis, in Hebrew): Jerusalem, The Hebrew University, 204 p.

Garfunkel, Z., 1978, The Negev, regional synthesis of the sedimentary basins: Guidebook to excursion A2, Tenth International Congress on Sedimentology, Jerusalem, p. 35-110.

Garfunkel, Z., Zak, I., and Freund, R., 1981, Active faulting in the Dead Sea Rift: Tectonophysics, v. 80, p. 1-26.

Gerson, R., 1981, Geomorphic aspects of the Elat Mountains, in Dan, J., and others, eds. Aridic Soils of Israel: International Conference on Aridic Soils, Israel, p. 279-293.

Gerson, R., 1982a, Talus relicts in deserts - a key to major climatic fluctuations: Israel Journal of Earth-Sciences, v. 31, p. 123-132.

Gerson, R., 1982b, The Middle East: Landforms of a planetary desert through environmental changes: Striae, v. 17 (The Geological Story of World's Deserts, T.L. Smiley, ed.), p. 52-78.

Kennedy, W.Q., 1962, Some theoretical factors in geomorphological analysis: Geological Magazine, v. 99, p. 304-312.

King, L.C., 1962, Morphology of the earth: Oliver and Boyd, Edinburgh, 699 p.

Nash, D.B., 1981, Morphologic dating of degraded normal fault scarps: Journal of Geology, v. 88, p. 353-360.

Penck, W., 1953, Morphological analysis of landforms: Macmillan, London, 429 p.

Quennell, A.M., 1958, The structure and geomorphic evolution of the Dead Sea Rift with discussion: Quarterly Journal of the Geological Society of London, v. 114, p. 1-24.

Quennell, A.M., 1959, Tectonics of the Dead Sea. Intern. Geol. Cong., 20th, Mexico, Assoc. Serv. Geol. Afr., p. 385-405.

Shaw, S.H., 1947, Southern Palestine, Geological map, scale 1:250,000, with explanatory notes: Government Printer, Jerusalem, 42 p.

Wallace, R.E., 1977, Profiles and ages of young fault scarps, north central Nevada: Geological Society of America Bulletin, v. 88, p. 1267-1281.

Wallace, R.E., 1978, Geometry and rates of fault-generated mountain fronts, north central Nevada: U.S. Geological Survey Journal of Research, v. 6, p. 637-650.

Wallace, R.E., 1980, Degradation of the Hagben Lake fault scarps of 1959: Geology, v. 8, p. 225-229.

Zak, I., and Freund, R., 1966, Recent strike-slip movements along the Dead-Sea Rift: Israel Journal of Earth-Sciences, v. 15, p. 33-37.

# 4
# Geomorphic indicators of vertical neotectonism along converging plate margins, Nicoya Peninsula, Costa Rica

*Paul W. Hare and Thomas W. Gardner*

ABSTRACT

Four large-scale geomorphic surfaces on the Nicoya Peninsula of northwestern Costa Rica are used to infer neotectonic deformation. The deeply weathered Cerro Azul Surface, bounded by a continous topographic scarp and fluvial knickpoints with 150-250 m of relief, represents an uplifted and deformed erosion surface. Elevation data gathered from this high-level surface are used to statistically model deformation using regression and trend surface analyses. In conjunction with LANDSAT lineament analysis, a "faulted half-dome" model is proposed for neotectonic deformation. The model is further supported by stratigraphic information, drainage basin asymmetry, floodplain development, terrace height, and the degree of erosional consumption of the Cerro Azul Surface.

The Nicoya Peninsula-Tempisque Valley-Gulf of Nicoya region is divided into four geomorphic provinces based on drainage network and topographic characteristics and the distribution of geomorphic surfaces. The two southern provinces have undergone positive intermittent uplift whereas the two northern provinces have experienced stability or subsidence in the recent past. The different geomorphic provinces are the result of differing neotectonic movements related to subduction of the Cocos Plate at the Middle American Trench.

## INTRODUCTION

The use of landscape features as indicators of vertical tectonism began in earnest nearly a century ago with the Geographical Cycle (Davis, 1899). Initiated by uplift, the "ideal cycle" allows for systematic changes in the landscape through geologic time with the development of the penultimate landscape feature, a peneplain. Davis realized, however, that "old age" may never be achieved due to tectonic interruptions. Though such an interruption marks the beginning of a new cycle in some respects, Davis (1899, pg. 499) suggests that the landscape "can only be understood by considering what had been accomplished in the preceeding cycle previous to its interruption." Though other variables such as climatic change must be considered, the landscape can remember its tectonic history.

Studies of large-scale geomorphic surfaces that invoke erosional processes ranging from Appalachian peneplanation (Davis, 1889; Johnson, 1931; Campbell, 1933; Ashley, 1935) to African pediplanation (Dixey, 1942; King, 1947; King, 1951; Fair and King, 1954) offer some useful and occasionally controversial examples of geomorphic surfaces as tectonic indicators. Furthermore, systematic variation in surface elevation and drainage adjustments are used as indicators of tectonic deformation. Notable examples come from the Dead Sea Rift Zone (Quennell, 1956), the East African Rift Zone (Doornkamp and Temple, 1966), and the Mesozoic rift along the Appalachians (Judson, 1975).

The Nicoya Peninsula of northwestern Costa Rica (Figure 1), an area of rapid and pronounced vertical tectonism, provides an excellent location to examine the interaction between landscape evolution and vertical tectonism. Elevation trends of several large-scale (20-100 kilometers) geomorphic surfaces are used to construct a model for vertical neotectonic deformation of the Nicoya Peninsula. The deformation model is supported by drainage basin morphology, drainage network morphometry, and LANDSAT lineament analysis.

The Nicoya Peninsula (Figure 1) is part of an outer arc province extending along the Pacific Coast of southern Central

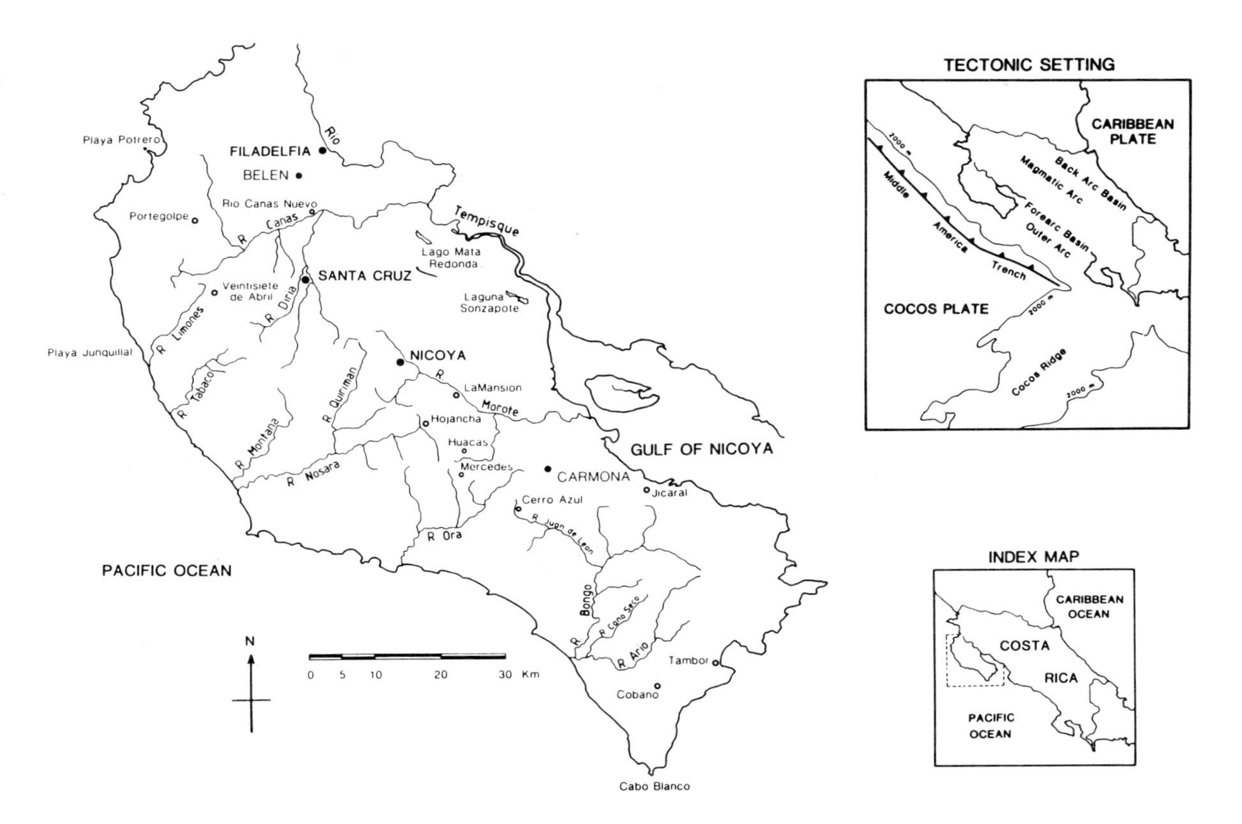


Figure 1. Location map of the Nicoya Peninsula with insert on the general tectonic setting.

America (Dengo, 1962; Case, 1974; Kuijpers, 1980). The bedrock geology is largely composed of the Nicoya Complex, a Jurassic-Cretaceous basic igneous complex of pillowed basalts, diabase dikes, small gabbroic intrusions, and intercalated pelagic sediments. Cretaceous to Tertiary sedimentary rocks unconformably overlie the Nicoya Complex locally along the margin of the peninsula (figure 2).

The current plate tectonic setting (Figure 1) is marked by northeast subduction of the Cocos Plate beneath the Caribbean Plate at the Middle America Trench (Larson and Chase, 1970; Herron, 1972; Aubouin et al., 1981). Rates of subduction ranging from 10 cm/yr (Aubouin et al. 1981; Larson and Chase, 1970) to as low as 2-3 cm/yr (Holden and Dietz, 1972; Molnar and Sykes, 1969) have been proposed for this section of the Middle America trench. Although tectonic models differ significantly in their treatment of the Nicoya Complex, all invoke Cretaceous initiation of subduction at the Middle America Trench (deBoer, 1979; Galli-Olivier, 1979; Schmidt-Effing, 1979; Kuijpers, 1980; Lundberg, 1981). More importantly, however, all tectonic models for the Nicoya Peninsula recognize the importance of spatial and temporal variability in rates of vertical neotectonism.

Structural and strarigraphic analyses suggest that emergent conditions have apparently existed since Oligocene or Miocene time for most of the peninsula (deBoer, 1979; Kuijpers, 1980) with subsequent deformation continuing to the present. Evidence of Quaternary tectonism is abundant. Recent vertical tectonism on a regional scale is indicated by relevelling surveys (Miyamura, 1975), wave-cut coastal benches (Fischer, 1980), and coastal terraces (Alt et al., 1980). Estuarine areas adjacent to the Gulf of Nicoya represent drowned topography due to subsidence or eustacy. Uplift is indicated on the southern tip of the peninsula where several marine benches are displayed at elevations up to 200 m; the lowest marine terrace at 7 m has a radiocarbon date of 6600 y.b.p. (Bergoeing, 1983).

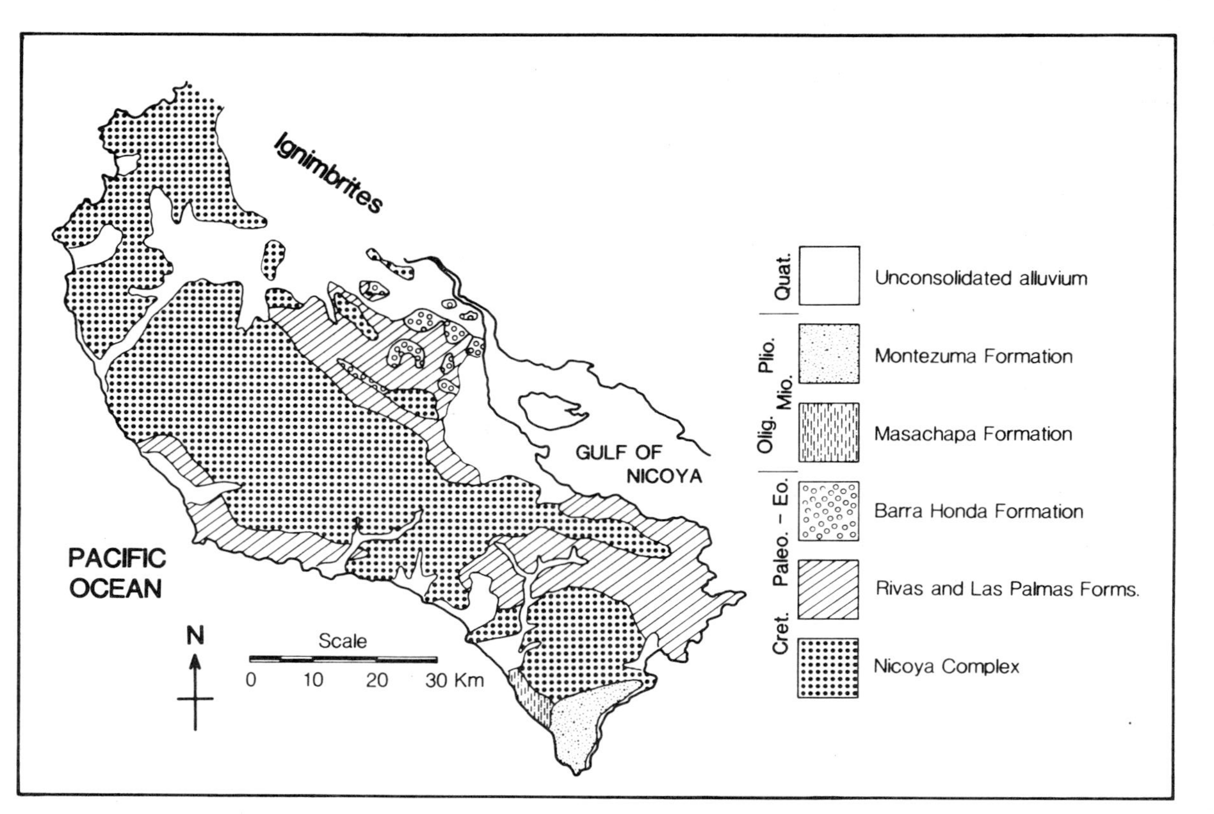


Figure 2. Generalized geologic map of the Nicoya Peninsula region. Modified from the Mapa Geologico de Costa Rica (1982).

## GEOMORPHIC SURFACES OF THE NICOYA PENINSULA

Four distinct geomorphic surfaces have been delineated on the Nicoya Peninsula: the Santa Cruz, Cobano, La Mansion, and Cerro Azul surfaces (Figures 3 and 4). All surfaces provide information on the deformational history of the peninsula, but the Cerro Azul Surface, the highest and oldest, will be addressed in the most detail.

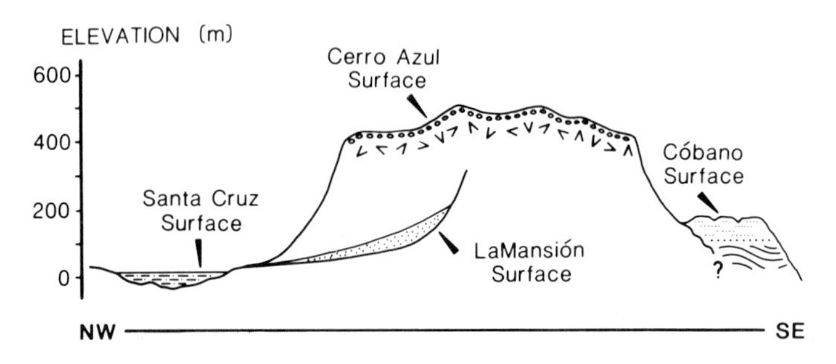


Figure 3. General topographic relationships between the four geomorphic surfaces of the Nicoya Peninsula. Illustration is a composite section drawn along a NW-SE line through the peninsula.

The Santa Cruz Surface with an average elevation of 20 m is an aggradation surface of low relief constructed by active floodplain sedimentation of Rio Canas and its tributaries (Figure 4). Ponded drainage, marshes, and seasonal lagoons with occasional hills rising abruptly from the floodplains are characteristics of this surface (Figure 5a). The drainage from this large, basin-like portion of the northern part of the Nicoya Peninsula is funnelled through a relatively small gap to the aggradational plains of Rio Tempisque which separate the peninsula from the mainland.

The Cobano Surface, marked by strongly accordant summits (Figure 5b), is largely coincident with gently-dipping exposures of the Miocene to Pliocene Montezuma Formation (figure 2) (Dengo, 1962; deBoer, 1979), a shallow-water, transgressive marine sequence

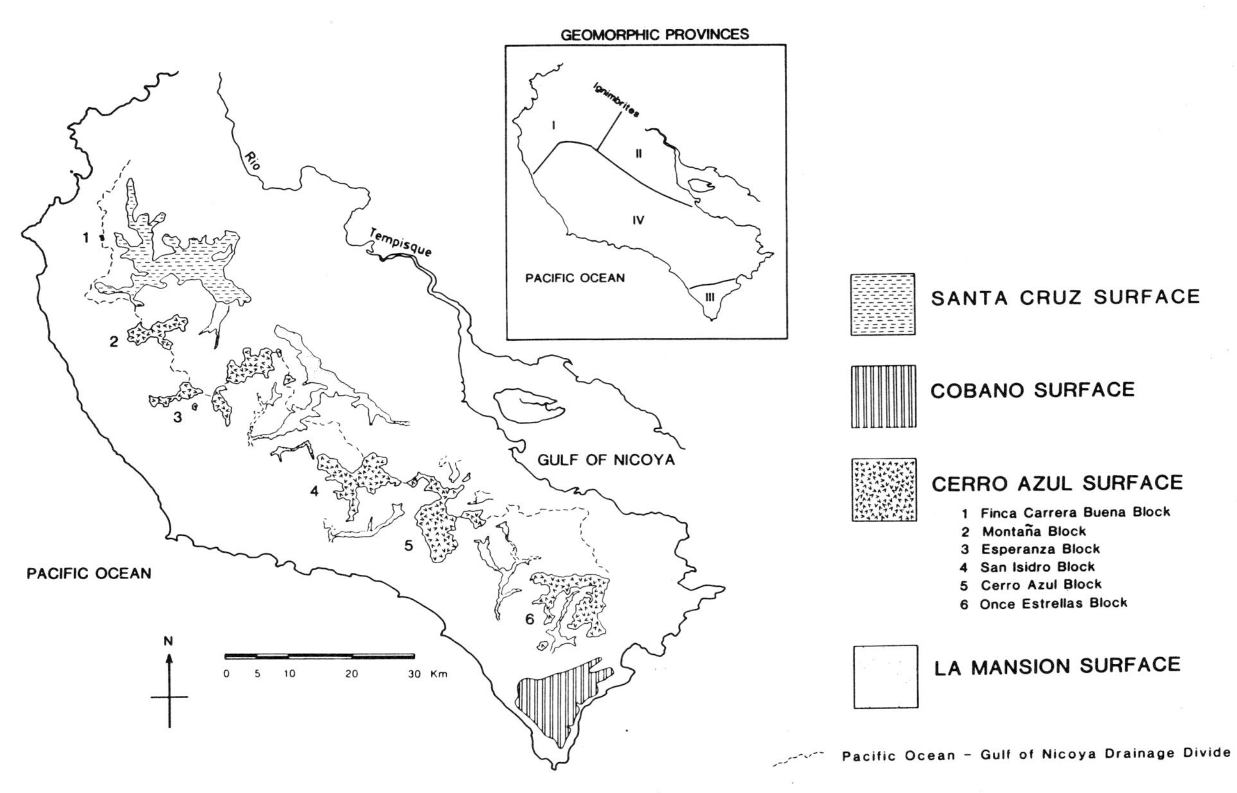


Figure 4. Map of the geomorphic surfaces of the Nicoya Peninsula and the Pacific Ocean-Gulf of Nicoya drainage divide. Insert shows the subdivision of the region into four geomorphic provinces which are described later.

Figure 5. Photographs of (a) the low aggradational plains of the Santa Cruz Surface (Rio Diria), (b) the accordant summits of the Cobano Surface (foreground).

Figure 5 (continued). (c) the incised valley fill of the La Mansion Surface between Nicoya and La Mansion, and (d) a portion of the Cerro Azul Surface west of Carmona as seen in the upper portion of figure 6 near Rio Ora.

(Lundberg, 1981). The Cobano Surface is an uplifted marine bench that is presently being dissected by joint-controlled rectangular drainage. Flat-topped summits remain and indicate that erosion has been insufficient to appreciably lower the elevation of the Cobano Surface. The flat-topped residuals, at elevations of 160-200 m, suggest similar amounts of tectonic uplift since formation of the surface.

The La Mansion Surface, a terrace of wide-spread occurrence, is developed along major river valleys in the mountains of the central and southern portion of the Nicoya Peninsula. The surface is characterized by wide, open valleys, and a distinct break in slope at the base of flanking hills (Figure 5c). In cross-section the surface is broadly concave with slopes reaching 5°-10° near valley sides. Apophyses of the La Mansion Surface extend into tributary valleys where maximum elevations near 200 m are reached. The La Mansion Surface is largely composed of stratified fluvial sands and gravels. Clasts are dominantly basalt and chert derived from the Nicoya Complex. Incision by the modern drainage has created a depositional terrace typically 4-10 m above stream level. Rivers in the central part of the peninsula have incised to bedrock and do not as yet possess active floodplains. In these valleys, the surface exists as a nearly continuous terrace. The La Mansion Surface is more discontinuous to the south where active floodplains are developing.

Developed within the mountainous backbone of the peninsula, the Cerro Azul Surface was first recognized by the presence of a prominent topographic scarp (Figures 5d and 6). Beautifully displayed near the small town of Cerro Azul, the scarp is continuous over many kilometers, with slopes frequently greater than 45°. 150-250 m high fluvial knickpoints, are ubiquitously associated with the topographic scarp. The scarp boundary is also the start of many first-order streams and in most places is coincident with major drainage divides.

Above the scarp the Cerro Azul Surface displays markedly subdued topography when considered with respect to the mountain range as a whole. The highest point is Cerro Azul at 1019 m and the relief of the surface in this region is in excess of 500 m. Nevertheless, the difference in topography can be recognized

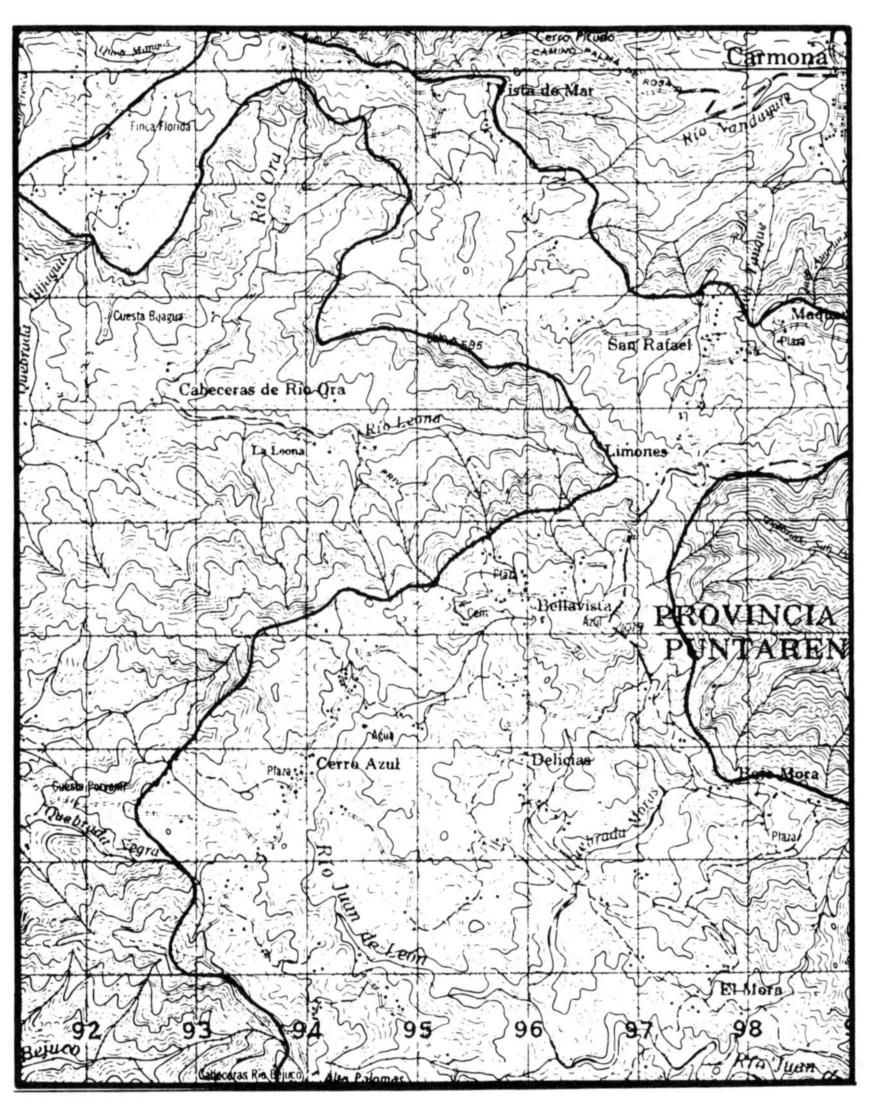


Figure 6. Portion of the Cerro Azul topographic map sheet showing the topographic scarp and fluvial knickpoints associated with the Cerro Azul Surface (block 5 in figure 4). Map gridded on a 1.0 km square with a 20 m contour interval.

by a change in contour line density (Figure 6).

The Cerro Azul Surface is further characterized by a deep weathering profile. This profile, developed on the Nicoya Complex, can be as much as 5-10 m deep. In certain cuts, only basaltic corestones remain surrounded by "structured clays" which reflect the parent rock texture. Similar soil profiles are not present on the rocky scarp face or the lower geomorphic surfaces.

The boundary of the Cerro Azul Surface (Figure 4) was delineated on 1:50,000 topographic maps based on the following criteria: (1) topographic scarp, (2) fluvial knickpoints, and (3) subdued topography. Following the suggestions of Rich (1938), elevation was not used as a criterion. Because of limited accessibility to the mountainous regions, the presence of a thick weathering profile could not be used.

The total area of the Cerro Azul Surface is approximately 265 square kilometers. The surface is not continuous but exists as numerous fragments which lie on a linear trend parallel to the axis of the peninsula. Except in the southern part of the peninsula, the fragments lie along the Pacific Ocean-Gulf of Nicoya drainage divide (Figure 4). Fragments can be grouped into "blocks" because some are separated by short distances over narrow ridges. The surface has been divided into six blocks which are separated by major trunk streams: the Finca Carrera Buena, Montana, Esperanza, San Isidro, Cerro Azul, and Once Estrella blocks (Figure 4).

There are three reasonable mechanisms for formation of the Cerro Azul Surface: (1) differential erosion of resistant horizons within the Nicoya Complex, (2) erosional stripping of an overlying, less-resistant sedimentary cover, or (3) erosion while the Nicoya landmass was lower with respect to a steady oceanic base level. The first hypothesis is discounted because extensive resistant horizons are absent within the Nicoya Complex. Because the Cerro Azul Surface is everywhere developed in the Nicoya Complex, the second hypothesis would require complete stripping of an overlying cover with concurrent erosional consumption of the basaltic core. No erosional remnants of a Tertiary cover are present within the boundaries of the Cerro Azul Surface (figure 2). Furthermore, the deep weathering profile, suggestive of an older age, is also problematic with such a hypothesis. The third hypothesis best explains the observed morphology.

Distribution of the Cerro Azul Surface remnants suggests genesis by fragmentation of an uplifted erosional surface into blocks separated by major rivers (Figure 7). Following the ideas of Glock (1931) and Schumm and Parker (1973), it is suggested that post-uplift rejuvenation travelled quickly through the fluvial systems by headward elongation. In this way the old surface was first subdivided by rapid elongation along major rivers into blocks that were subsequently divided into smaller erosional fragments during elaboration of the rejuvenated system. These stages are

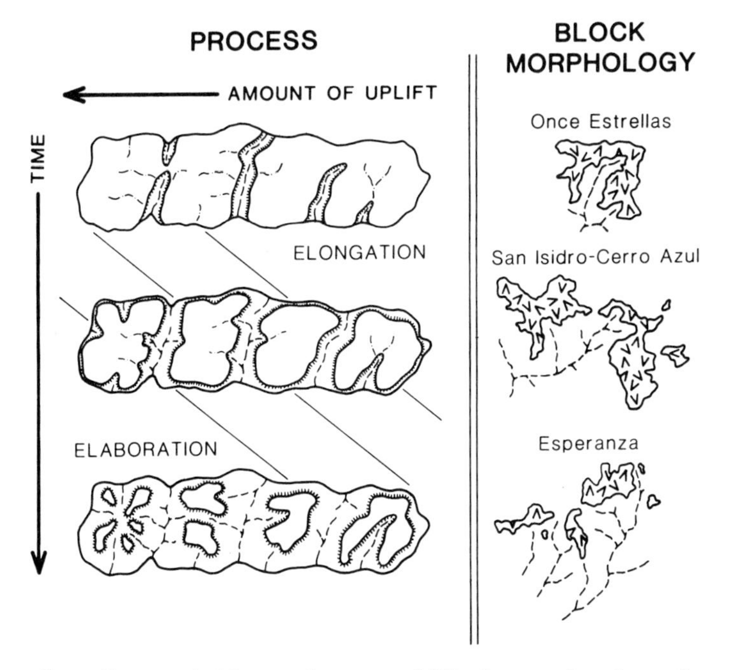


Figure 7. Fragmentation of an uplifted erosional surface as rejuvenated fluvial systems undergo headward elongation along major drainage lines followed by elaboration. Actual blocks of the Cerro Azul Surface in various consumptive stages are shown on right. Diagonal lines represent isostage lines.

reflected in the present distribution of the Cerro Azul Surface. A wave of sedimentation following the elongation phase may have led to the formation of the La Mansion Surface as part of a complex response to uplift. Conversely the La Mansion surface may represent another period of accelerated uplift. Recognition of the Cerro Azul Surface as an uplifted erosional surface implies that uplift of the peninsula has not been continuous and/or that the rate of uplift has not been constant.

## DEFORMATION OF THE CERRO AZUL SURFACE

### Assumptions and complications

Assuming that the Cerro Azul Surface represents a time line--that it was at one time everywhere "graded" to a single sea level stand--the elevation of the surface may reflect tectonic deformation. However, the presence of significant initial topography introduces several complications. For example, the elevation of the surface would gradually decrease after uplift due to erosion which continues at previous rates above the migrating knickpoints (Figure 8a). Consumption of the Cerro Azul Surface by the rejuvenated fluvial system results in preservation of higher portions of the surface. This tends to increase the "average" elevation of erosional remnants through time while decreasing the aerial extent of the surface (Figure 8b). Thus, elevation cannot be used as a reliable indicator of the absolute uplift which has occurred since formation of the Cerro Azul Surface.

Another complicating factor is the paleotopography of the peninsula prior to uplift. The shape of a landmass can lead to differences in "average" elevation due to the effect of distance to baselevel. Fortunately, the Nicoya Peninsula is more-or-less rectangular in plan at present and distance to baselevel for different drainage basins may have been nearly constant prior to uplift.

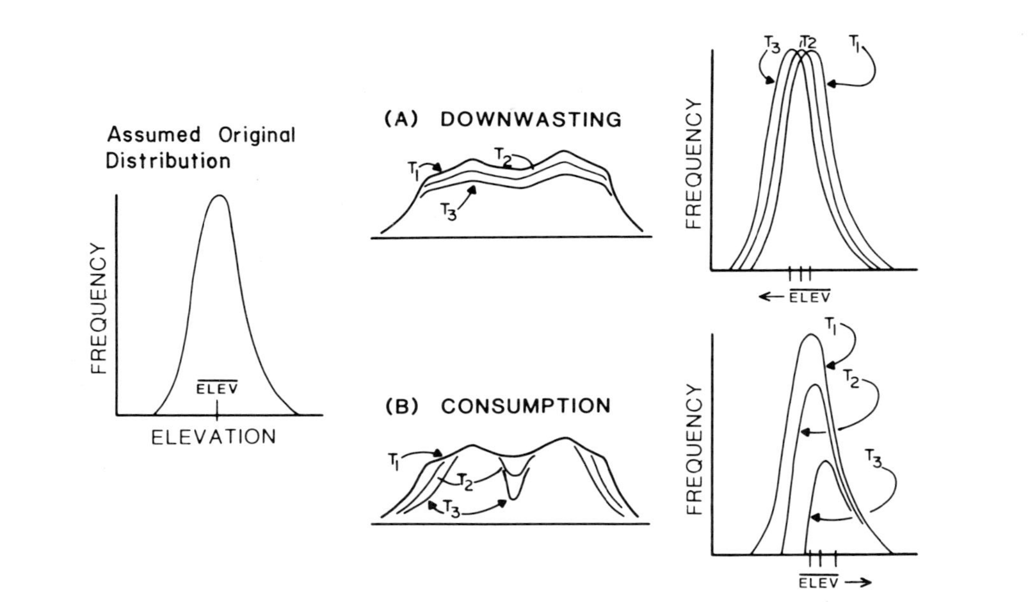


Figure 8. Influence of the processes of downwasting and consumption on the "average" elevation of a fragment of the Cerro Azul Surface through time. Assumed original frequency distribution modified from the San Isidro and Cerro Azul blocks.

## Statistical analysis

These complications are surmountable if a statistical approach is used to define elevation trends of the Cerro Azul Surface. Data were collected on a 0.5 km square grid superimposed on the surface. At each of the 1017 grid nodes, four variables were recorded: X1, X2, ELEV and BLOCK where X1 and X2 represent nodal position with respect to the north-south/east-west reference axes, ELEV the elevation of the surface, and BLOCK the block of the surface for use as a class variable.

Erosional remnants of the Cerro Azul Surface lie on a linear trend, hereafter referred to as the remnant trend, which must be determined quantitatively before elevation trends parallel and perpendicular to the remnant trend can be studied. Because no dependent positional variable can be assumed, an equation relating X1 and X2 is calculated using the reduced major axis method (Miller and Kahn, 1962):

$$X2 = 87.622 - 0.669\ (X1)$$

$$R^2 = 0.918$$

The resultant remnant trend (123.8°) is subparallel to the Middle America Trench off the Nicoya Peninsula (110°-140°). Rotation of the original reference axes about the origin creates two new transformed position variables, T1 and T2, such that T1 represents distance along the remnant trend and T2 distance across the remnant trend. The transformation involves rotation alone with no rescaling of axes or change in origin.

Summary statistics for the six blocks of the Cerro Azul Surface suggest systematic changes in elevation along the remnant trend (Figure 9). Mean, minimum, and maximuxm elevation of the blocks (Figure 4) increase sharply from the Finca Cerrera Buena block (1) to the Esperanza block (3) and then gradually decrease to the Once Estrellas block (6).

Systematic variation in elevation along the remnant trend suggests differential uplift of the Cerro Azul Surface. As a first approximation, this deformation can be modelled using linear regression analysis. By modelling the Finca Carrera Buena to Esperanza blocks and the Esperanza to Once Estrellas blocks separately, highly significant statistical relationships were determined (Figure 9) which can be used to estimate plunges

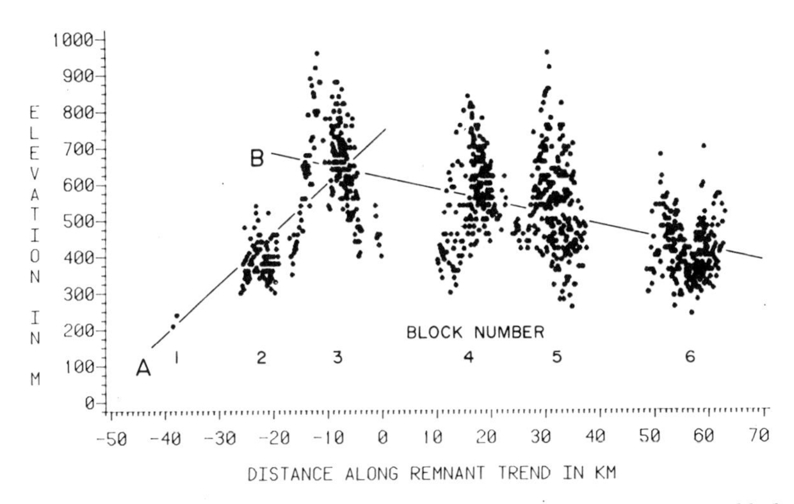


Figure 9. Scatter plot of elevation versus distance parallel to the remnant trend of the Cerro Azul Surface with regression lines shown [A: ELEV = 741.82 + 13.88 (T1), n = 302, $R^2$ = 0.422, F-stat. prob. = 0.0001; B: ELEV = 619.67 -3.33 (T1), n =921, $R^2$ = -/329, F-stat. prob. = 0.0001]. Block numbers refer to Figure 4.

associated with uplift of the surface. Relatively low coefficients of determination result from the relative relief within the Cerro Azul Surface. Estimates of 0.80° and 0.19° were calculated for northwest and southeast plunge respectively. Though small, these angular deformations result in a difference in mean elevation of 220 m between the Esperanza and Once Estrellas blocks and 410 m between the Esperanza and Finca Carrera Buena blocks. Data imply that uplift of the Cerro Azul Surface has been greater for the southern portion of the peninsula (Once Estrellas block) than the northern portion (Finca Carrera Buena block) with greatest uplift occurring in the central part of the Nicoya Peninsula (Esperanza block).

Cross-trend plots of elevation for each block (Figure 10) are used to determine deformation of the Cerro Azul Surface perpendicular to the remnant trend. Most significant variation occurs between the Montana and Esperanza blocks. Visual inspection and statistical testing suggest no cross-trend deformation of the Montana block whereas broad arching is suggested for the Esperanza block. Existence of a major structural discontinuity between these two blocks is further supported by the following:

(1) Both the Montana and Esperanza blocks have northeast linear trends (070° and 065° respectively) coincident with a highly preferred LANDSAT lineament direction (figure 11) (060°-075°).
(2) A prominent zone of lineament concentration (Montana lineament zone) is present between the Montana and Esperanza blocks.
(3) A dramatic shift in the Pacific Ocean -Gulf of Nicoya drainage divide parallels the northwestern edge of the Esperanza block (Figure 4).

A major LANDSAT lineament between the Finca Carrera Buena and Montana blocks (Limones-Canas lineament) marks a second structural break further suggesting that deformation along the length of the Nicoya Peninsula has not been purely flexural but is interrupted by major structural breaks.

Trend surface analysis (Lustig, 1969; Davis, 1973) can be used to model more complex deformations of the Cerro Azul Surface. Briefly, surfaces of different complexity are fit to three-dimensional data using least-squares methods. Thus, a first order surface models an inclined plane, a second order surface models a paraboloid, and higher orders more complex surfaces.

The Statistical Analysis System (SAS) was used to generate trend equations and diagnostic statistics for surfaces up to sixth order (with ELEV dependent on X1 and X2). The trend surfaces (Figures 12 and 13) are significant based on F-statistic probabilities and have R-squared statistics above that expected from random data (Howarth, 1967).

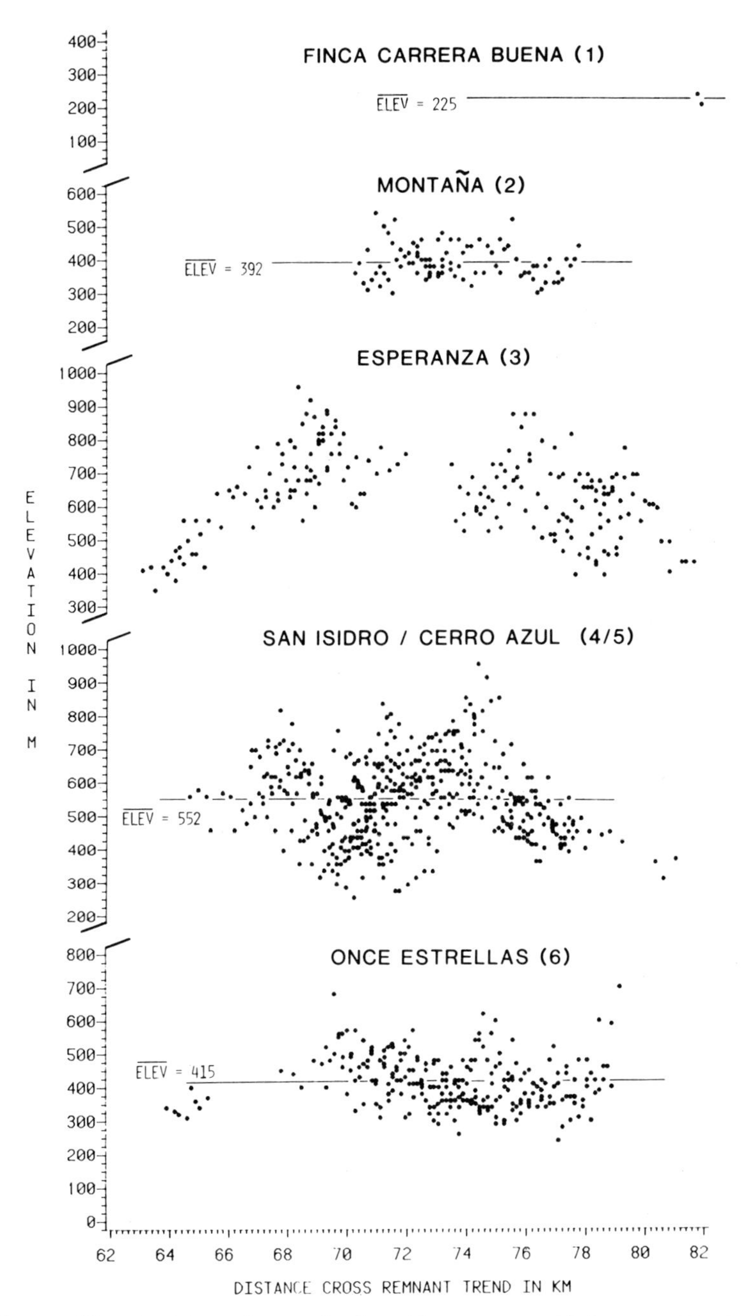


Figure 10. Scatter plots of elevation versus distance perpendicular to the remnant trend of the Cerro Azul Surface. Mean block elevations are indicated where slope of linear regression models are not significantly different from zero. Numbers to the right of each block refer to figure 4.

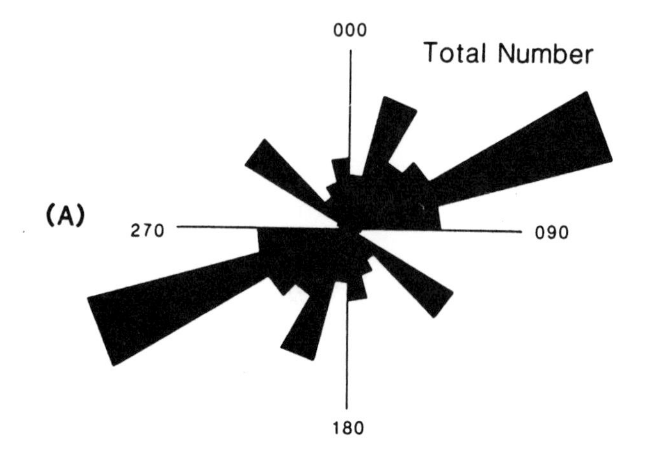


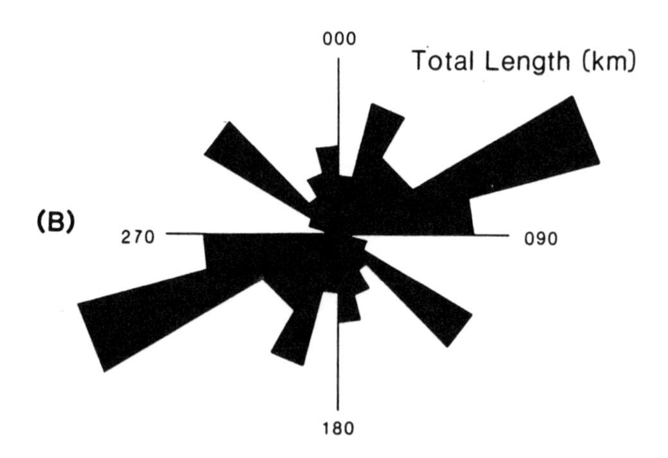


Figure 11. Rose diagrams based on total number of lineaments (A) and total length of lineaments (B) found in the LANDSAT analysis of the Nicoya Peninsula-Tempisque Valley-Gulf of Nicoya region.

Trend surfaces were fit to the entire data set ignoring possible structural discontinuities (Figure 12). The second order surface resembles an elongate dome with a maximum elevation of 600m located just northwest of the San Isidro block. The fourth order

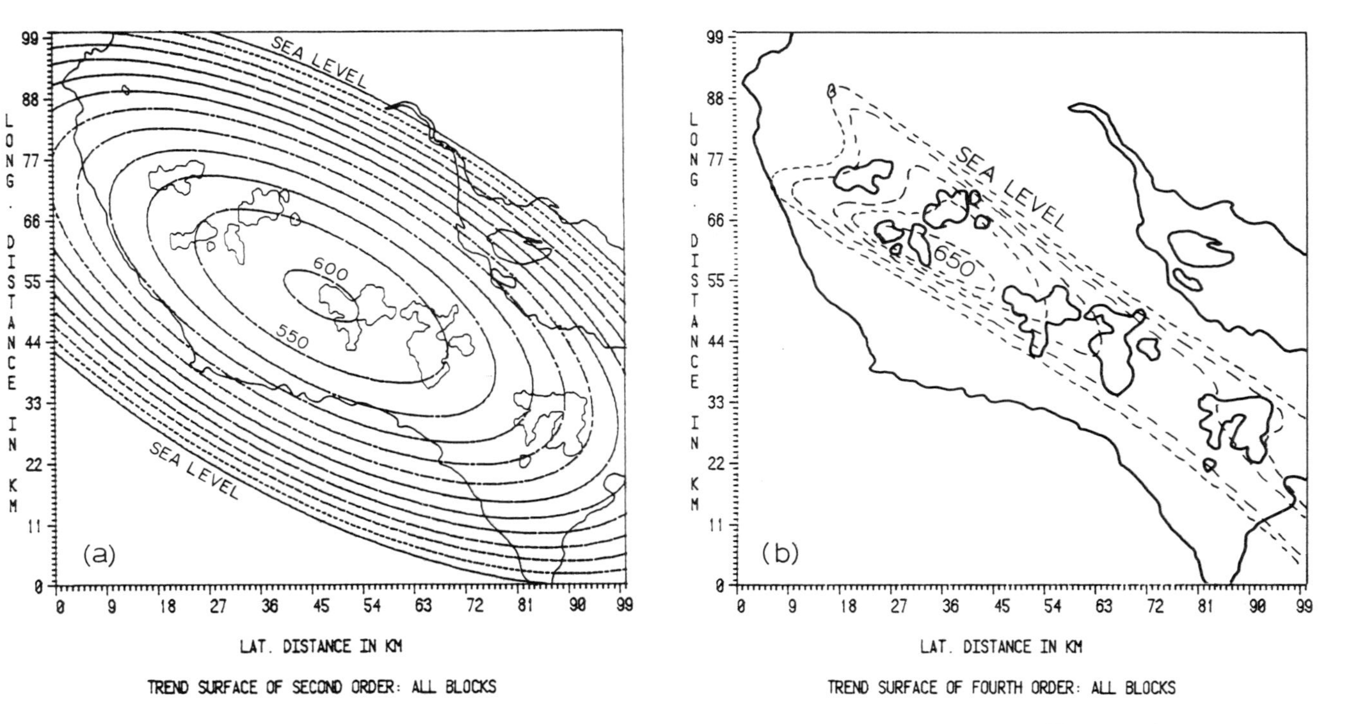


Figure 12. Trend surfaces for all blocks of the Cerro Azul Surface with (a) second order trend [n = 1017, $R^2$ = 0.284, F-stat. prob. = 0.0001] and (b) fourth order trend [n = 1017, $R^2$ = 0.419, F-stat. prob. = 0.0001]. Block outlines are located on figure 4.

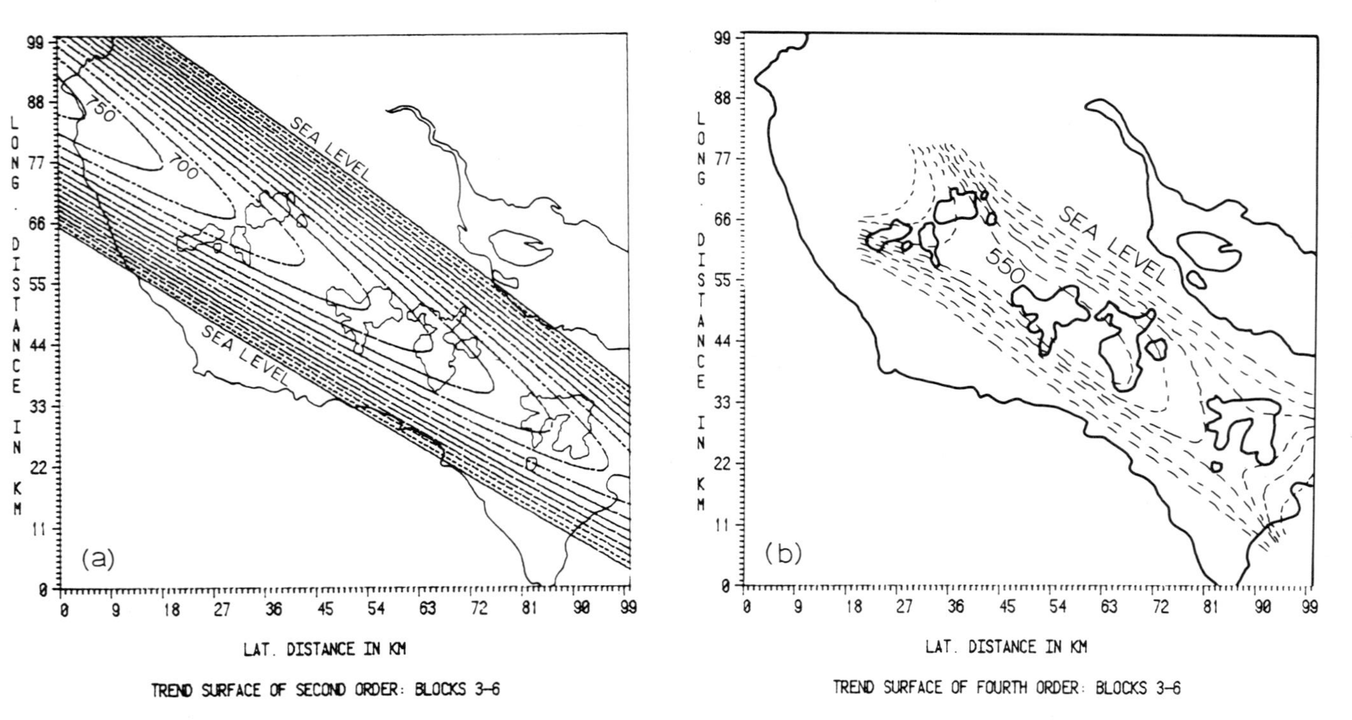


Figure 13. Trend surfaces for the Esperanza, San Isidro, Cerro Azul, and Once Estrellas blocks of the Cerro Azul Surface with (a) second order trend [$n = 921$, $R^2 = 0.399$, F-stat. prob. = 0.0001] and (b) fourth order trend [$n = 921$, $R^2 = 0.456$, f-stat. prob. = 0.0001]. Block outlines are located on figure 4.

surface also shows an elongate dome with a crest elevation of over 650 m located at the Esperanza block. The fourth order trend surface suggests that cross-trend arching is present only from the Esperanza to Cerro Azul blocks; the Once Estrellas block is not part of an actual elongate dome.

Trend surfaces were also calculated for the Esperanza to Once Estrellas blocks (Figure 13) to allow incorporation of the structural discontinuities suggested by LANDSAT and regression analyses. The second order trend surface models a southeast plunging antiform whereas the fourth order surface models a southeast plunging antiform which dies out at the Once Estrellas block. Maximum elevation of the antiform is 725 m at the Esperanza block.

## DEFORMATIONAL MODELS

Neotectonic deformation proposed by Kuijpers (1980) for the Nicoya Peninsula involves a broad Nicoya Dome, elongate in the NNW-SSE direction along the length of the peninsula. Dips associated with doming are inferred to be generally less than 5°. If this purely flexural model is assumed correct (Figure 12a), then topographic analyses of the Cerro Azul Surface can be used to refine the deformational model associated with the Nicoya Dome. Elevation data would support that doming because formation of the Cerro Azul Surface has involved steeper plunge to the northwest than southeast (0.80° and 0.19° respectively from linear regressions). Other possible estimators yield similar results such as minimum block elevations which indicate a northwest plunge of 0.27° and a southeast plunge of 0.10°. The broad, cross-trend arching of the Esperanza block would suggest that the elongate dome of Kuijpers is symmmetric.

However, incorporation of structural discontinuities indicated by previous analyses, allows refinement of Kuijpers' Nicoya Dome. Structural breaks between the Finca Carrera Buena (1), Montana (2), and Esperanza (3) blocks suggest a faulted half-dome model (Figure 14).

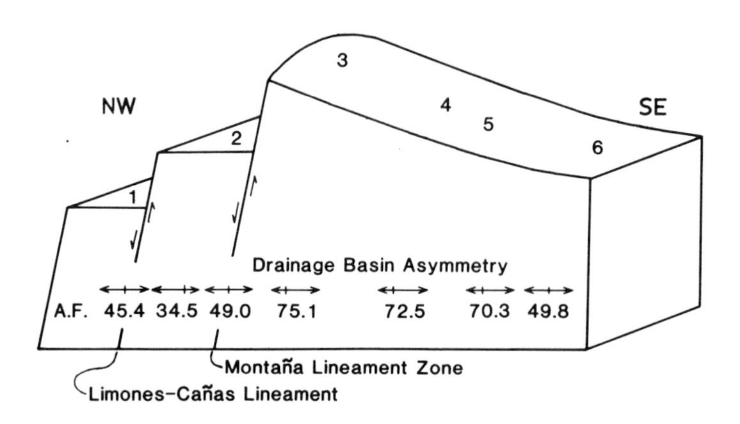


Figure 14. Faulted half-dome model developed from statistical analyses of the Cerro Azul Surface and LANDSAT lineament analysis showing two structural discontinuities northwest of the Esperanza block. Drainage asymmetry factors (A.F.) also shown in their relative spatial positions for major rivers draining westward to the Pacific Ocean. Numbers on upper surface refer to block locations in figure 4.

Vertical displacements along the two breaks can be estimated by using several parameters. Displacement along the Limones-Canas lineament, between the Finca Carrera Buena and Montana blocks with minimum elevations of (210 m and 300 m respectively) suggesting 90 m of vertical displacement whereas average block elevations (225 m and 392 m respectively indicate 167 m. Use of regression line A (Figure 9) with average block position results in 219 m of estimated vertical displacement.

Due to cross-trend arching of the Esperanza block, vertical displacement varies along the Montana lineament zone. Maximum displacement of 275 m has been estimated using the minimum elevations of the Montana block and the Esperanza block in the vicinity of the arch crest. Use of the second order trend surface for Esperanza to Once Estrellas blocks (Figure 13a) and average elevation of the Montana block suggests a maximum displacement of 285 m.

South of the structural break between the Montana and Esperanza blocks (figure 14) , deformation is modelled as a plunging antiform. Plunge is to the southeast at 0.19° as estimated by regression lines or 0.2° as suggested by the second order trend surface on Esperanza (3) to Once Estrella (6) blocks. The difference in uplift between the Esperanza and Once Estrellas blocks has been estimated at 220 m by average block elevation, 110 m by minimum block elevation, 216 m using regression equations, and 250 m by the second order trend surface for the four southern blocks. The fourth order trend surface indicates that the antiformal deformation diminishes near the Once Estrellas block.

The deformational model formed from statistical results is supported by additional stratigraphic and geomorphic evidence. The preliminary edition of the Mapa Geologico de Costa Rica (1982) (figure 2) shows that younger sedimentary (Miocene to Pliocene) formations outcrop only on the southern tip of the peninsula. Furthermore, large outcroppings of the Cretaceous Rivas Formation occur with the Nicoya Complex in the mountains of the southern peninsula. The difference in stratigraphic levels currently exposed indicates that central parts of the peninsula have experienced greater uplift in the past than southern parts of the Nicoya Peninsula.

The consumptive stage (degree of erosional fragmentation) of the Cerro Azul Surface is not constant but varies along the peninsula (Figure 7). The Once Estrellas block remains unfragmented and wraps around the Rio Ario Valley. The San Isidro and Cerro Azul blocks to the north surround the valley of Rio Ora and have only recently been separated. The Esperanza block in the central part of the peninsula has been severely fragmented in a direction normal to major drainage lines. The degree of fragmentation may be related to the amount of uplift indicating greatest uplift has occurred at the Esperanza block with progressively less uplift to the south.

Floodplain development, terrace height, and drainage basin asymmetry also vary systemmatically along the peninsula. Current floodplain development along major rivers increases from the central to southern parts of the peninsula and is reflected in the degree of continuity of the La Mansion Surface. Moreover, the height of the La Mansion terrace apparently decreases to the south.

Both tendencies could be due to less uplift in the southern region than in the central part of the peninsula.

Morphometric analysis of major fluvial systems reveals important variation in drainage basin asymmetry. Drainage basin asymmetry is calculated from an asymmetry factor (A.F.) based on area due to the inaccuracies in stream delineation (Morisawa, 1957):

$$A.F. = 100(Ar/At)$$

where Ar is the drainage area on the downstream right of the main drainage line and At is the total drainage area. Though the asymmetry factor varies in a downstream direction for any given river, southward asymmetry occurs only for rivers between the Esperanza (3) and Once Estellas (6) block such as Rios Quiriman, Ora, and Bongo (Figures 1 and 14). To the south, the asymmetry factor suggests that Rio Ario is an aerially balanced system; to the north of the Esperanza block the asymmetry factor indicates balanced systems or systems asymmetric to the north. These findings are in agreement with results of the regression and trend surface analyses of the Cerro Azul Surface.

## REGIONAL VERTICAL NEOTECTONICS

Based on tectonic deformation, drainage network morphometry, topographic characteristics, and the distribution of geomorphic surfaces, the Nicoya Peninsula-Tempisque Valley-Gulf of Nicoya region can be subdivided into four geomorphic provinces (Figure 4). Northwest of the Limones-Canas lineament, province I is characterized by the Santa Cruz Surface. Topography is much lower and less rugged than central and southern portions of the peninsula (province IV). Large floodplains have developed on most rivers and those streams flowing toward the Pacific are often separated from Rio Canas drainage by broad, low divides. Major drainage lines of the well-integrated network tend to be intermittent during the dry season. Vertical neotectonism of the northern part of the Nicoya Peninsula has been characterized by relative stability and current subsidence is possible.

Province II, located north of the Rio Morote Valley (Figure 4) is characterized by entirely intermittent and very poorly-integrated drainage with areas of internal drainage and sag ponds. Major LANDSAT lineaments (120°-135°, figure 11) occur which parallel the Tempisque Valley-Gulf of Nicoya topographic depression, a trend uncommon to lineaments mapped in other provinces. This province is coincident with exposures of the Barra Honda Formation, a massive limestone which forms numerous isolated buttes which drop in elevation toward Rio Tempisque. Neotectonic subsidence is indicated north of the Rio Morote Valley, a major structural break, with subsidence increasing to the northeast across possible normal faults.

Province III is coincident with the Cobano Surface on the southern tip of the Nicoya Peninsula. This province is characterized by accordant summits, rectangular drainage, and maximum relief of approximately 200 m. Several marine benches and associated fluvial terraces exist below the Cobano Surface which indicate intermittent uplift of this region.

Located in the central and southern portion of the peninsula, province IV is bounded by the Cobano Surface to the south, Rio Limones to the northwest, and the Rio Morote Valley to the northeast (Figure 4). Both the Cerro Azul and La Mansion surfaces characterize this geomorphic province. The topography is rugged and mountainous and major drainage divides are relatively sharp. The drainage network is well-integrated and composed mostly of perennial streams. Numerous barbed tributaries and elbows of capture are present as well as superimposed meanders along several major rivers.

The Nicoya Peninsula, especially central and southern portions, has probably been the site of uplift since Oligocene-Miocene emergence. However, uplift has been differential and discontinuous. The marine terraces below the Cobano Surface indicate that uplift has been intermittent. Moreover, the presence of the Cerro Azul and Cobano Surfaces, which probably formed during tectonically quiet times, leads the authors to suggest that uplift in this region occurs in major phases with each phase characterized by intermittent uplift. Neotectonic uplift decreases north of the Esperanza block of the Cerro Azul Surface across two structural discontinuities. The coincident trends of the Tempisque Valley-Gulf of Nicoya depression, axis of uplift of the Cerro Azul Surface, and the Middle America Trench seaward of the Nicoya

Peninsula suggest that deformation is in response to intermittent subduction of the Cocos Plate.

## ACKNOWLEDGMENTS

Financial support for field work was provided by the Standard Oil Company of California, the Chevron Family of Companies. Field transporation and logistical support was provided by the Instituto Geografico Nacional of Costa Rica. Computational funding was provided by the College of Earth and Mineral Sciences, The Pennsylvania State University.

## REFERENCES

Alt, J.N., Harpster, R.E., and Schwartz, D.P., 1980, Late Quarternary deformation and differential uplift along the Pacific Coast of Costa Rica [abs.]: Geological Society of America, Abstracts with Programs, v. 12, p. 378-379.

Ashley, G.H., 1935, Studies in Appalachian Mountain sculpture: Geological Society of America Bulletin, v. 46, p. 1395-1436.

Aubouin, J., Stephan, J.F., Renard, V., Roump, J., and Lonsdale, P., 1981, Subduction of the Cocos Plate in the Mid America Trench: Nature, v. 294, p. 146-150.

Bergoeing, J.P., 1983, Dataciones radiometricas de algunas muestras de Costa Rica: Informe Semestral, Instituto Geografico Nacional, p. 71-86.

Campbell, M.R., 1933, Chambersburg (Harrisburg) peneplain in the piedmont of Maryland and Pennsylvania: Geological Society of America Bulletin, v. 44, p. 553-573.

Case, J.E., 1974, Oceanic crust forms basement of eastern Panama: Geological Society of America Bulletin, v. 85, p. 645-652.

Davis, J.C., 1973, Statistics and Data Analysis in Geology: New York, John Wiley and Sons, 550 p.

Davis, W.M., 1889, The rivers and valleys of Pennsylvania: National Geographic Magazine, v. 2, p. 183-253.

__________, 1899, The geographic cycle: Geographical Journal, v. 14, p. 481-504.

deBoer, J., 1979, The outer arc of the Costa Rican Orogen (oceanic basement complexes of the Nicoya and Santa Elena peninsulas): Tectonophysics, v. 56, p. 221-259.

Dengo, G., 1962, Tectonic-igneous sequences in Costa Rica: Geological Society of America, Petrographic Studies, A Volume to Honor A.F. Buddington, p. 133-161.

Dixey, F., 1942, Erosion cycles in central and southern Africa: Geological Society of South Africa Transactions, v. 45, p. 151-181.

Doornkamp, J.C. and Temple, P.H., 1966, Surface drainage and tectonic instability in part of southern Uganda: Geographical Journal, v. 132, p. 238-252.

Fair, T.J.D. and King, L.C., 1954, Erosional land-surfaces in the eastern marginal areas of South Africa: Geological Society of South Africa Transactions, v. 57, p. 19-26.

Fischer, R., 1980, Recent tectonic movements of the Costa Rican Pacific coast: Tectonophysics, v. 70, p. T25-T33.

Galli-Olivier, C., 1979, Ophiolite and island-arc volcanism in Costa Rica: Geolological Society of America Bulletin, v. 90, p. 444-452.

Glock, W.S., 1931, The development of drainage systems: a synoptic view: Geographical Review, v. 21, p. 475-482.

Herron, E.M., 1972, Sea-floor spreading and the Cenozoic history of the east-central Pacific: Geolological Society of America Bulletin, v. 83, p. 1671-1692.

Holden, J.C., and Dietz, R.S., 1972, Galapagos Core, NazCoPac triple junction, and Carnegie/Cocos ridges: Nature, v. 235, p. 266-269.

Howarth, R.J., 1967, Trend-surface fitting to random data --an experimental test: American Journal of Science, v. 265, p. 619-625.

Johnson, D., 1931, Stream Sculpture on the Atlantic Slope: Columbia University Press, New York, 142 p.

Judson, S. 1975, Evolution of Appalachian topography: in Melhorn, W.N. and Flemal, R.G. (eds.), Theories of Landform Development, 6th Annual Geomorphology Symposia Series, Binghamton, N.Y., p. 29-45.

King, L.C., 1947, Landscape study in southern Africa: Geological Society of South Africa Proceedings, v. 50, p. 23-52.

__________, 1951, The geomorphology of the eastern and southern districts of southern Rhodesia: Geological Society of South Africa Transactions, v. 54, p. 33-64.

Kuijpers, E.P., 1980, The geologic history of the Nicoya ophiolite complex, Costa Rica, and its geotectonic significance: Tectonophysics, v. 68, p. 233-255.

Larson, R.L. and Chase, C.G., 1970, Relative velocities of the Pacific, North America, and Cocos plates in the Middle America region: Earth and Planetary Science Letters, v. 7, p. 425-428.

Lundberg, N., 1981, Evolution of the slope landward of the Middle America Trench, Nicoya Peninsula, Costa Rica: in Leggett, J.K. (ed.), Forearc Geology, Geological Society of London Special Publication, p. 431-447.

Lustig, L.K., 1969, Trend-surface analysis of the Basin and Range Province, and some geomorphic implications: U.S. Geological Survey Professional Paper 500-D, 69 p.

Mapa Geologico de Costa Rica (preliminary edition), 1982, Nicoya, Costa Rica (1:200,000): Direccion de Geologia, Minas, y Petroleo.

Miller, R.L. and Kahn, J.S., 1962, Statistical Analysis in the Geological Sciences, New York, John Wiley & Sons, Inc., 483 p.

Miyamura, S., 1975, Recent crustal movements in Costa Rica disclosed by relevelling surveys: Tectonophysics, v. 29, p. 191-198.

Molnar, P., and Sykes, L.R., 1969, Tectonics of the Caribbean and Middle America regions from focal mechanisms and seismicity: Geological Society of America Bulletin, v. 80, p. 1639-1684.

Morisawa, M., 1957, Accuracy of determination of stream lengths from topographic maps: American Geophysical Union Transactions, v. 38, p. 86-88.

Quennell, A.M., 1956, The structure and geomorphic evolution of the Dead Sea Rift: Quarterly Journal of the Geological Society of London, v. 114, p. 2-18.

Rich, J.L., 1938, Recognition and significance of multiple erosion surfaces: Geological Society of America Bulletin, v. 49, p. 1695-1722.

Schmidt-Effing, V.R., 1979, Alter und genese des Nicoya-Komplexes, einer ozeaneschen Palaokruste (Oberjura bis Eozan) im sudlichen Zentralamerika: Geologische Rundschau, v. 68, p. 457-494.

Schumm, S.A. and Parker, R.S., 1973, Implications of complex response of drainage systems for Quarternary alluvial stratigraphy: Nature Physical Science, v. 243, p. 99-100.

# 5
# Large-scale tectonic geomorphology of the Southern Alps, New Zealand

*John Adams*

ABSTRACT

The Southern Alps are the result of plate tectonic convergence across the Alpine Fault that has caused the edge of the Pacific plate to be upturned and rapidly upthrust. Erosion of the upthrust plate edge is fast enough to balance the uplift rate, so that the mountains are in a dynamic steady state. The steady state balance is maintained by dynamic relationships between uplift rate, relief, rainfall, and erosion rate that tend to restore the balance. The effects of lithology, storm intensity, earthquake shaking, and vegetation are of secondary importance, and mankind and glaciation appear unimportant factors. The restoring relationships require a minimum pertubation to be triggered, so that only in parts of the world with rapid tectonic uplift is the balance likely to be attained rapidly.

Although the individual mountains of the Southern Alps change from being flat-topped, time-dependent landforms to spiky time-independent ones, the range as a whole, and the type, lithology and height of mountains in each part remain the same, with the range resembling a cuesta that is dynamically renewed by uplift. The dynamic cuesta model has important implications for the tectonic geomorphology and the economic influence of the Southern Alps.

---

Contribution of the Earth Physics Branch No. 1130

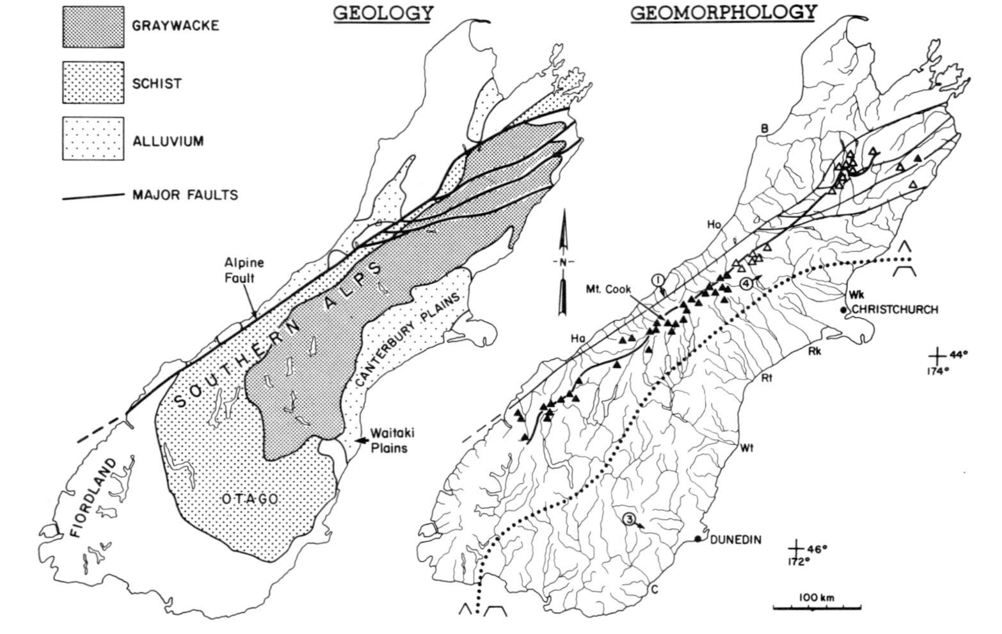


FIGURE 1. Maps of South Island showing left: simplified bedrock geology of the Southern Alps, the Alpine Fault and the chief alluvial deposits, and right: the main rivers and lakes, the drainage divide (heavy line) and the eastern boundary between the spiky and the flat-topped mountains (dotted line). Solid triangles represent mountains > 2500 m; open triangles in northern Southern Alps, mountains > 2200 m. Circled numbers with arrows indicate photograph location and direction. Letters at river mouths indicate B-Buller, C-Clutha, Ha-Haast, Ho-Hokitika, Rk-Rakaia, Rt-Rangitata, Wk-Waimakariri, and Wt-Waitaki Rivers.

## PHYSICAL AND TECTONIC ENVIRONMENT

The Southern Alps of New Zealand form the backbone of the South Island and run diagonally NE-SW across it (Fig. 1). They are a triangular prism 400 km long, 80 km wide and 3 km high with a much steeper western than eastern face. The mountains rise at their highest point, Mt Cook, to 3764 m, and in the range as a whole there are many peaks over 3000 m. The drainage divide lies near the crestline and divides the range asymetrically (Fig. 1); as a result 3-km high mountains are within 40 km of the western coast (Fig. 2), but 100 km from the eastern.

FIGURE 2. Photograph of the steep western slopes of the Southern Alps showing the Waiho River flowing from the present Franz Joseph Glacier (centre), and an old terminal moraine (loop in foreground) marking a temporary 11 000-year-old advance of the glacier. The Alpine Fault crosses the centre of the photograph (arrows) and separates the mountains from the plains. The mountains are schists upthrust on the fault and about 3000 m high (note apparent summit height accordance); the plains are formed from gravel eroded from the mountains and are 100 m above sea level and only 18 km from the highest mountains. Note how the narrow, bedrock rivers in the mountains become broad, braided rivers on the plains.

Along their western edge the Southern Alps are bounded by the Alpine Fault, which separates schists and graywacke upthrown on the eastern side from Paleozoic granites and Tertiary sediments on the western side. Conglomerates west of the fault show that the schist of the Southern Alps was exposed and eroded for the first time only 2.5 m.y. ago (Wellman, 1979), and the conglomerates are therefore thought to date the beginning of rapid uplift on the Alpine Fault and to nearly date the origin of the Southern Alps.

The Southern Alps lie across the main westerly winds that bring most precipitation to New Zealand. Runoff (precipitation minus evaporation) is more than 4 m $yr^{-1}$ everywhere on the western slopes, and exceeds 11 m $yr^{-1}$ (more than one inch per day on average) in the highest central alps (Griffiths and McSaveney, 1983). The eastern slopes are in a relative rain shadow, and the runoff decreases exponentially eastward, halving every 12.5 km away from the crestline (Coulter, 1973). Much land above 2400 m is in permanant snow and ice. The tree line is at about 1100 m and drops slightly from north to south. The forest is still largely undisturbed by mankind and introduced animals, and consists of subtropical evergreen trees (southern beech) adapted to the cold temperate climate.

The region is one of severe but widespread frontal storms accompanied by high winds. There have been large historical earthquakes at the northern and southern ends of the Southern Alps but not in the middle. Erosion rates are extremely high, as is obvious from the muddy, gravel-choked rivers flowing from the mountains.

The Southern Alps owe their origin to compressional movement on the Alpine Fault. The Alpine Fault is the surface expression of the boundary between the Indian and Pacific plates. Plate movement vectors calculated for the last few million years indicate dominant dextral (right-lateral) slip on the fault, accompanied by an increasing amount of compression in more geologically recent times. The contemporary slip vector indicates about 40 mm $yr^{-1}$ of dextral slip and 22 mm $yr^{-1}$ of compression normal to the fault (Allis, 1981; Wellman, 1983).

The origin of the Alpine Fault as a pure strike-slip fault probably explains its unusual straightness and length, while the more recent compression across the fault explains the uplift and origin of the mountain range. The compression is expressed as thrusting on an east-dipping plane that exposes a crustal section on end and produces a very narrow zone of localized uplift (Wellman, 1979). The schists exposed near the fault are metamorphosed graywacke, and are thought to underlie all the graywacke at depth.

In this paper I discuss the restoring relationships that constrain the size, shape, and position of the Southern Alps mountain range and result in a steady-state landform in a dynamic balance between uplift and erosion, and describe how the landforms have responded to the tectonic environment. This leads to some significant features of the tectonically-controlled geomorphology that are at a larger scale than those usually considered, and to a discussion of how young mountain ranges exist and evolve through time.

## LONG-TERM RELATIONSHIPS THAT ENSURE MAINTANENCE OF A STEADY STATE LANDFORM

The Southern Alps are in a steady-state balance with rapid tectonic uplift of the mountains due to compression along the Alpine Fault being balanced by rapid erosion (Adams 1980a). Over the past decade there has been a general appreciation that many young mountain ranges are in this type of steady-state; some examples are the European Alps, the Himalayas, Taiwan, Japan and New Guinea (Table 10 in Adams, 1980a). While tectonic uplift drives the cycle, the rapid erosion in each range may be attributed to a few conditions common to all. In each, relief is high and annual rainfall heavy. Each experiences high intensity storms and large earthquakes more frequently than the world average, and landslides are considered part of the normal geomorphic process. Vegetation is dense because of the high rainfall. In each region the steady state is attained through a balance among the above parameters. Exactly what sort of balance, and what nature of mountain range results, depends on the relative importance of each parameter. The discussion below shows how the balance is achieved for the Southern Alps and how external changes trigger restoring changes in the other parameters, the whole acting to maintain the mountain range in its present form.

## First Order Effects

If the relief of the Southern Alps were suddenly reduced - say by a cosmic bulldozer - but the plate compression and tectonic uplift rates remained unchanged, the erosion rate would be greatly lowered in the short-term. The erosion would be lower partly because mountain slopes would be less steep and less prone to landsliding and partly because channel gradients would be reduced until they were unable to carry away the original debris load. A further immediate effect of a suddenly lowered Southern Alps would be a decrease in orographic precipitation, leading to reduced runoff and aggravating the problem of debris transport.

The results would be these: slower slope erosion and an accumulation of debris in the valleys and along the mountain front. With the erosion rate decreased, erosion would no longer balance uplift and the excess uplift would increase the elevation and relief. Eventually the relief would increase until the erosion and the uplift were again in balance as at the present day, where the mountain range would remain in a dynamic steady state.

Uplift rates of 10-20 mm/yr in the Southern Alps (Adams, 1979; Wellman, 1979) mean that even if the Alps were planed to sea level, only 0.3 m.y. would be needed (in the absence of counter-acting erosion) to attain their present height. Of course, erosion will increase the time needed substantially, and the crude modelling of the erosion-uplift-rainfall-relief relationship by Adams (1980a) suggests that 3x-6x as long might be needed. Therefore at these uplift rates any balance could be attained in about one million years, a very short time period geologically.

If rainfall was suddenly reduced - say by a sudden change in oceanic and atmospheric circulation or by a glacial cooling - but continued at the same relative intensity, erosion would initially be reduced. The extra, unbalanced uplift would serve to increase the relief and steepen the mountain slopes and river channels. The increased debris supplied from the steeper slopes to the channels would eventually be carried by the reduced runoff at a higher concentration than before. In part, the initial reduction in rainfall and erosion would be mitigated by the trapping of a larger fraction of the passing atmospheric moisture by the newly-higher land. Overall, the reduction in rainfall would result in attainment of a new steady state with higher mountains. Similarly an increase in rainfall would result in the opposite feedback-relations and result in lower steady state mountains. For the Southern Alps, the crude relationships suggested by Adams (1980a p. 96-99) indicate a simple inverse relationship between the height of 1.7 - 2.2 km high mountains and runoff, so that a halving of runoff would result in a doubling of mountain height.

## Second Order Effects

If storm intensity was reduced there would be no long-term change in the erosion rate. However the fewer storms of a given intensity would shift more sediment and the short-term fluctuations in erosion would be larger.

If earthquakes did not trigger a large fraction of the sediment supply to the rivers through extensive though infrequent landslides, erosion would be reduced, the land would increase in elevation, the slopes would become steeper and less stable, and eventually storm-caused landslides would increase the supply rate until erosion again accounted for the uplift.

If vegetation did not cover lower parts of the alps, erosion might be more rapid on the bare rock, but the possible protection and binding of the rock must be weighed against the accentuated chemical weathering under vegetation and the loosening effects of tree roots. I infer that vegetation does not necessarily change the long-term ($10^3$ yr) erosion rate on steep slopes. It may however affect the style of erosion: unvegetated slopes erode steadily, vegetated slopes erode by infrequent but sudden slope failure beneath the vegetation.

Other Perturbations

If lithology changed, so might the height of the mountains. In a steady-state mountain range, mountain height is controlled largely by river spacing and the steepness of the straight mountain slopes. In strong rock, the slopes are steeper (and the river spacing may be greater also), and so the mountains are higher. In the Southern Alps, the graywacke and schist have similar strength (in the mountain-forming sense), with the schists being slightly weaker ([84 $\pm$ 18]% of the strength of the graywacke based on 313 measurements of maximum slopes for 500 m lengths that were taken at 1000 m grid intersections from 1:63,360 scale maps, Adams, unpub. 1976). A change in lithology from graywacke to schist occurs as the graywacke is eroded away and the underlying schist is exposed. The less resistant schist results in slightly lower mountains, but the Southern Alps as a whole changes little because the eroded mountains are renewed from the east by the tectonic shortening (see below).

Two further effects, mankind and glaciations, have not affected all mountain ranges, and the balance attained between erosion and uplift is thought to be largely independant of both. Many northern hemisphere studies have suggested that present erosion is much more rapid than the geological rate because of agricultural activities. Trimble (1977) suggests that 1.5 mm $yr^{-1}$ of soil (say 0.8 mm $yr^{-1}$ of rock equivalent) may be lost from the eastern United States through improper farm management. Erosion in the Southern Alps is ten times faster and is occurring in areas with no agriculture development. A side effect of mankind in New Zealand has been the release of introduced animals (deer, pigs, rabbits and opossums) that have multiplied without natural controls and become noxious by overgrazing forest vegetation. Much erosion in New Zealand has been unfairly blamed on these animals despite evidence that even virgin forest does not prevent erosion, and observation that deer could not prevent revegetation of the landslides caused by the 1929 Buller earthquake (Wellman, pers. comm. 1978). Locally the short-term erosion rate may increase, but this may be considered as reducing the erosion that would have happened catastrophically at some later date. With or without introduced animals the long-term erosion rate will be the same.

Glaciations may radically modify landforms in mountainous regions, but most valleys on the western slopes of the Southern Alps look remarkably unglaciated because about 100 m of postglacial erosion in the soft schist rock has destroyed most of the glacial landforms. Glacial erosion deepens and widens valleys to classic 'U' shape, but may remove little rock from the valley slopes; interglacial erosion may concentrate on the oversteepened valley walls. Erosion may be more rapid during glaciations than between them, but insufficient work has been done to confirm and quantify this in regions of very high erosion rates. Long-term erosion rates would probably adjust to any such variation by compensating glacial and interglacial rates. Thus neither mankind nor valley glaciations can change the long-term erosion rates, though their effects may cause short-term fluctuations.

The way that the parameters above were perturbed could have been reversed with the opposite effects but same end result: either the long-term rate is not changed, or if changed, changes so as to restore and then maintain the initial conditions, namely a balance between erosion and uplift. A certain minimum change is needed before the balance between uplift and erosion is upset and needs restoration, as it is a macro balance that applies for times greater than about $10^4$ yr and uplifts greater than about $10^2$ m. Once attained the balance is then insensitive to repeated short-term changes in either uplift or erosion. The overall agreement between erosion and uplift rates in young mountain ranges is not fortuitous but is a reflection of the most stable state that can exist in regions of rapid tectonic uplift.

## GEOMORPHIC RESPONSE

The rapid erosion in the Southern Alps is readily apparent in the gravel choked rivers, steep slopes, landslides and thin soils. In this section I point out how the steady-state balance is reflected by the tectonic geomorphology at varying scales from the individual mountains through to the entire mountain range taken as a whole.

## Time-dependant Landforms

Achievement of a steady state is dependent on sufficient time and uplift to allow landform development. Where uplift is rapid but the amount of uplift is small there will be no restoring relationship triggered and erosion need not reflect uplift rates. For example, uplift rates after deglaciation may reach 10 mm $yr^{-1}$ but total uplift may only be 100 m. The uplift is enough to cause channel downcutting but not enough to steepen basin slopes appreciably and so increase the erosion rate. A second example comes from the geodetic leveling of the U.S.S.R. and North America which reveals that there are large areas of supposedly stable continents that are being uplifted at 1 to 5 mm $yr^{-1}$ (e.g., Burnett and Schumm, 1983). Such uplift, if continued, would rapidly build mountain ranges within stable continents, contrary to accepted geological understanding, and instead it is likely that the levelings record transient or oscillatory vertical movements with a period of about $10^3$ - $10^4$ years (e.g., Adams, 1980c) that will not continue long enough to build mountains or affect the erosion rate.

The time needed for erosion and uplift to achieve a balance depends on the uplift rate. In areas of slow uplift there may not yet have been sufficient time for the balance to be attained, and so erosion will be less than uplift, and uplift of the land will continue until erosion balances the uplift.

In Japan (Table 1) the central zone with uplift of 2.2 mm $yr^{-1}$ has attained balance, the outer zone of southwest Japan at 1.5 mm $yr^{-1}$ has almost attained balance, and the northeast and inner southwest zones are not in balance (Yoshikawa, 1974). The uplift rates and the degree to which erosion rate approaches uplift rate suggest that uplift in all zones may have started at the same time, but that in the zones with the slower rates the landforms are still responding to uplift and no restoring relationships yet dominate. The inner zone where uplift is six times as rapid as erosion is described by Yoshikawa (1974) as a highland region dissected by narrow gorges with rapidly eroding steep slopes that separate highlands of low relief (the Chugoku Mountains) that are not being so intensely eroded. The gorges are presumably downcutting apace with uplift, but the highlands are eroding more slowly than the uplift and are presumably rising, so that the landform is growing and changing with time.

TABLE 1 Average erosion and uplift rates for sub-regions of Japan (after Yoshikawa, 1979) and New Zealand (after Adams 1980a) where uplift and erosion are roughly equal

| | | Erosion ($mm\ yr^{-1}$) | Uplift ($mm\ yr^{-1}$) |
|---|---|---|---|
| Japan: | Central | 2.2 | 2.2 |
| | Outer Southwest | 1.3 | 1.5 |
| | Northeast | 0.32 | 1.3 |
| | Inner Southwest | 0.10 | 0.6 |
| | average | 0.85 | 1.3 |
| South Island, New Zealand | | | |
| average, whole island | | 1.7±0.4 | 1.5±0.1 |
| maximum, Southern Alps | | 10-20 | 22±2 |
| areas not in steady state: | | | |
| south Canterbury | | 0.2 | 0.1-0.4 |
| eastern Otago | | 0.07 | 0.1-0.3 |
| Fiordland | | 0.2 | 0.2-0.3 |

## Time-independant Landforms

New Zealand also has mountains with nearly flat tops separated by deep valleys. The "flat topped" mountains are in eastern Otago and the flat tops are the remains of a Tertiary peneplain (Fig. 3). The peneplain is best preserved at low altitudes near the coast, but as it is traced inland it rises and the width of each summit narrows. As the mountains have been uplifted, channels have more or less kept pace with uplift, and as they have downcut, slope erosion has cut back from the channel to narrow the mountain summit. The more rapid the uplift, the higher the mountain, the faster the channel downcutting, the wider the valley, and the narrower the flat summit. Eventually traces of the peneplain surface disappear from mountains that are about 2 km above sea level and the hillslopes intersect to give a sharp ridge and a "spiky" mountain (Fig. 4). Most mountains that rise above 2 km in the Southern Alps are spiky mountains (Wellman, 1979).

FIGURE 3. Uplifted peneplain and incipient flat-topped mountains near Lawrence, Otago.

The mean erosion in eastern Otago is 0.07 mm $yr^{-1}$ (determined from river sediment load, Adams, 1980a), while the mean uplift of the region is about 0.1 - 0.3 mm $yr^{-1}$ (determined from a 4 m.y. nominal age for the peneplain, Wellman, 1979), so that the flat topped mountains in Otago are still growing. Like the Chugoku Mountains of Japan, the flat topped mountains of Otago are being eroded at a fraction of their uplift rate.

Once the peneplained summit has been eroded from the flat topped mountains and the hillslopes on either side have intersected as the sharp ridge, the shape of the mountain will change but little with time. The hillslopes are essentially

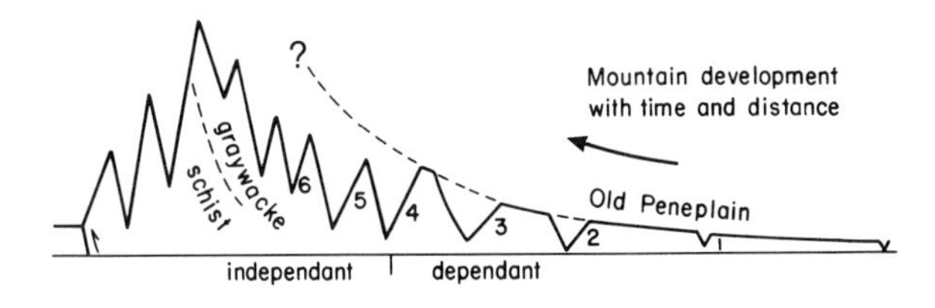


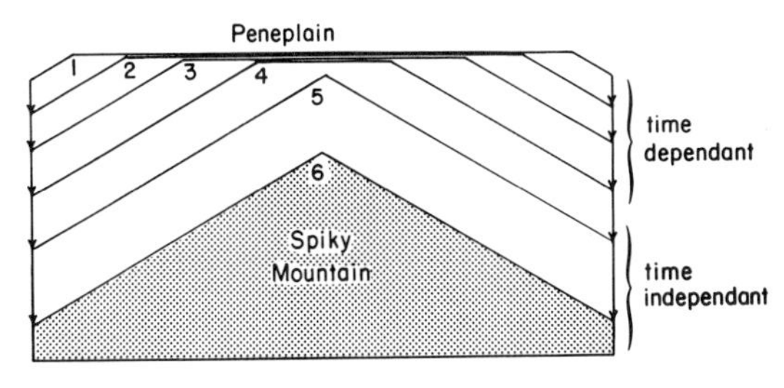


FIGURE 4. Schematic cross-section across the Southern Alps showing relationship between flat-topped and spiky mountains and how the flat-topped ones evolve in space into spiky mountains. Diagram below shows the evolution of a single mountain with time. After stage 4 the spiky mountain is a time-independent landform and nothing can be said about the original elevation of the peneplain.

straight and the height of the mountain is fixed by the spacing of the river channels and the steepness of the slopes (which may be lithology dependent). The spiky mountains will grow until the local erosion rate equals the local uplift rate, and at this point the mountain will be eroding as fast as it is being uplifted and it will maintain its height and shape through time. Spiky mountains are therefore a time-independent ("equilibrium" or "steady-state") landform and can be contrasted with the flat topped mountains which are time-dependent landforms. The division of mountains into these two types is fundamental since mountain shape indicates whether erosion has been keeping pace with uplift or not.

For extremely soft rocks, the "mountains" in steady-state may not be very high at all, especially if the uplift rate is slow. A New Zealand example may be the hill area northeast of Wanganui

(39.5°S, 175°E) where the strongly dissected soft mudstones attain an elevation of 700 m, but may be in the same sort of steady-state as the grander Southern Alps.

Like the spiky mountains, the straight hillslopes are time-independent landforms. Schumm (1963, p. 8), in an early paper often quoted out of context, criticized the belief in time-independent landforms from the basis that landscape denudation comprised two parts: channel erosion which can be accomplished by channel incision at a rate equal to the uplift, and hillslope erosion which did not seem able to balance uplift. Schumm considered that hillslope erosion was much less effective than channel erosion and so convex hillslopes that are not time-independent landforms would be formed. He recognised the theoretical need for straight slopes if time-independent landforms were to be formed, but concluded incorrectly that they probably did not exist. Schumm failed to appreciate that most hillslopes in areas of rapid erosion, like the slopes in the Southern Alps, are straight, and that they do represent a time-independent land form where erosion and uplift are in balance.

The eastern boundary between spiky and flat-topped mountains is shown on Fig. 2. When compared to Wellman's (1979) uplift map of the South Island, the boundary corresponds approximately to the 0.5 mm $yr^{-1}$ uplift contour. At lower uplift rates the landform is time-dependent and erosion is slower than uplift (Table 1). These areas amount to about half of the South Island. For the remainder, which includes all the areas of rapid uplift, erosion balances uplift, and for the South Island as a whole erosion roughly equals uplift (Table 1). The plate collision zone is thus in a steady state of erosional equilibrium that can continue almost indefinitely with the upthrusting of the Pacific plate being balanced by equal erosion of the plate edge, the Southern Alps. The Southern Alps are time-independent or "steady state" mountains because of the balance and their growth to steady state can be inferred from present rates of uplift and erosion.

## Evolution of Drainage Along the Southern Alps

The young uplift of the Southern Alps has resulted in a rather simple drainage system with rivers draining northwest and southeast away from the divide. Various authors, notably Wellman and Willett (1942) have remarked upon the regular patterns of drainage on the western slopes. The chief departure comes south of Mt. Cook where the schist is exposed southeast of the divide and the southeast slope rivers flow south, parallel to fold axes.

The drainage divide closely follows the crestline of the Southern Alps - that is, the headwaters of the rivers and the highest parts of the mountains coincide (Fig. 1). This contrasts with some other mountain ranges such as the Andes of southern Chile where the crestline lies 150 km downstream of the headwaters. In such ranges eastward headward erosion is encouraged by steep, wet western flanks and gentle eastern flanks, and in many there has been considerable time for headward erosion by the western rivers, leading to capture of east-draining rivers and to eastward migration of the divide. However the Southern Alps are extremely young and their uplift has been very rapid. Furthermore, the shortening against the Alpine Fault means west-draining rivers must erode headward at 15 km/m.y. merely to maintain a constant length. Thus the present length of the rivers reflects a dynamic balance between eastward headward erosion and westward crustal shortening. Separation of the drainage divide and crestline occurs mainly due to capture of east- or south-draining rivers as at Lewis Pass, Haast Pass, the Hollyford, and Arthur's Pass; these provide trans-mountain routes of considerable economic importance.

## The Southern Alps as a Fault-scarp and Influence of Lithology

On a large scale we can envisage the western slopes of the Southern Alps as a giant fault scarp consequent on uplift along the Alpine Fault. Although much larger than commonly cited examples of fault scarps, its morphology and the consequent drainage pattern are similar. Slope retreat, typified in the schist rocks by the consequent, west-draining streams, is slowed by the resistant overlying graywackes which can be thought of as a "caprock" (Fig. 5). Hence the more gentle eastern slope resembles the dip slope of a cuesta.

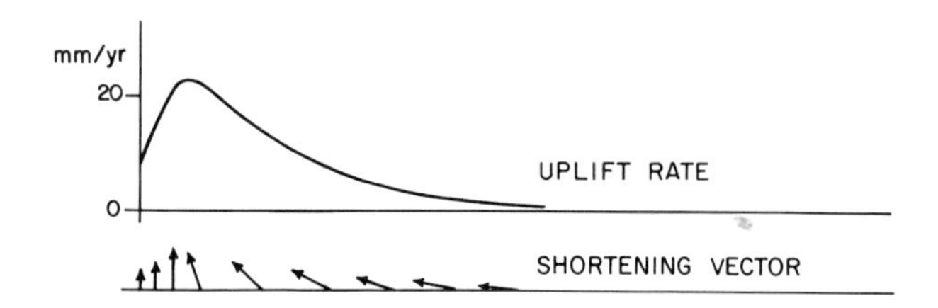


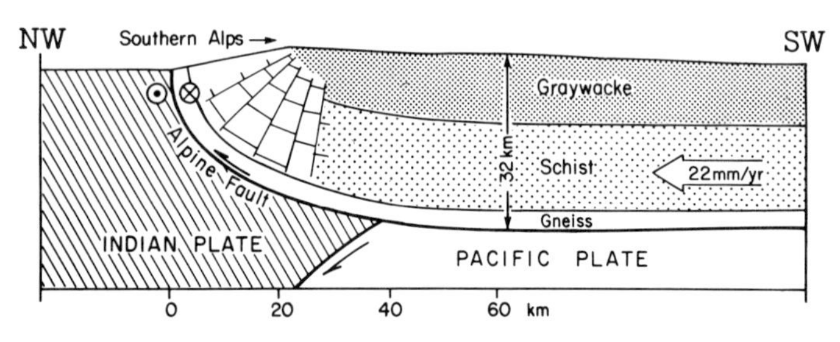


FIGURE 5. Cross section at equal horizontal and vertical scales showing the suggested dip of the Alpine Fault and its relationship to the schist and graywacke rocks (modified after Wellman, 1975, 1979), shortening vectors of 22 mm $yr^{-1}$ and their inclination near the Alpine Fault (including the decrease due to drag, Adams, 1979), and a corresponding uplift rate profile (after Adams 1980a). Some thickening within the schist is due to thrusting on west-dipping faults (Wellman, 1979) that are shown schematically.

My current understanding of the active tectonic and geomorphic processes suggests that this "cuesta" is self-perpetuating. As fast as the schist is eroded from the western slopes and the graywacke cap-rock is eroded, "undermined", and forced to retreat, the schist and graywacke rocks are being uplifted and, more importantly, shortened against the Alpine Fault; as a consequence the crest remains in the same place. Similarily, as discussed above, the length of the west-draining rivers remains constant despite their active headward retreat. The outcrop of the schist-graywacke boundary therefore remains fixed as a geometric constraint: it is at the point where the plate shortening vector dips 45° so that as downward erosion balancing uplift (vertical shortening component) causes an eastward slope retreat, the horizontal shortening component moves the boundary an equal distance westward. Hence while the individual spiky mountains are steady-state landforms, so is the entire mountain range; though the individual mountains grow, erode and are moved towards the fault by the shortening, the mountain range as a whole, and the type, lithology and height of the individual mountains in each part, remain the same.

A first modelling of the growth to steady state of the Southern Alps is summarized by Fig. 6 which reproduces the constant position of the divide. By contrast, during erosion of a cuesta in a tectonically stable environment the divide migrates towards the dip slope. The Southern Alps thus forms a special exception to Gilbert's 1877 "law of equal declivities" which suggests that drainage divides migrate until their east and west slopes are equal (Bloom, 1978, p. 276).

## CONSEQUENCES OF THE TECTONIC GEOMORPHOLOGY

The simple steady-state cuesta model for the evolution of the Southern Alps leads to some interesting long-term geological and economic implications. These suggest the important, but largely unstated, role that tectonic geomorphology has played in New Zealand's human affairs.

### Erosion and Continental Recycling

The shortening against the Alpine Fault means that points distant from the fault move closer with time, and as they do so, their uplift rate increases (shortening vectors on Fig. 5). Therefore rock eroded from the eastern slopes of the Southern Alps and deposited on the Pacific plate will eventually be carried westwards by the plate shortening and uplifted and eroded again. Thus a recycling of continental crust is occurring with the

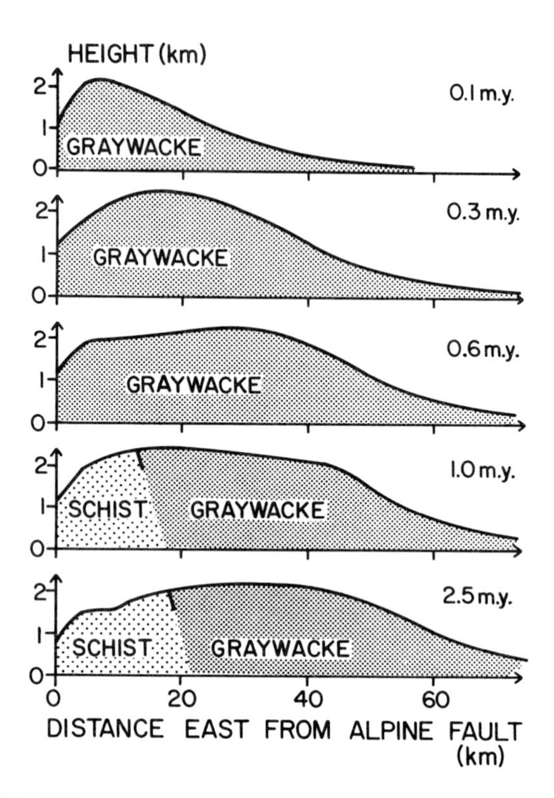


FIGURE 6. Profiles normal to the Alpine fault showing the growth to steady state of the Southern Alps. Vertical exaggeration 6.7 times. At 0 m.y. the region was assumed to be a peneplain underlain by 12 km of graywacke and then by schist (Fig. 5). Shortening and uplift was assumed to begin suddenly at the rate shown on the top of part of Fig. 5. Unroofing of the schist took 0.6 m.y. during which time the range attained its approximate height and shape. By 1.0 m.y. the schist was fully unroofed on the western slopes and the range had achieved a steady-state. By 2.5 m.y. about 50 km of rock has been eroded from the upturned plate edge (see Adams, 1980a, p. 94-100 for a fuller description of constraints).

sediment once deposited in eastern mountain-edge basins now being eroded from uplifted intermontane deposits (cf. Wellman, 1979). However most of the rock is eroded from the western slopes and represents a direct transfer of mass from the Pacific plate to the Indian.

The pronounced lithological contrast between east and west slopes of the Southern Alps controls the nature of the rivers and the depositional landforms. The schists of the western slopes are fissile, generally soft rocks that tend to break down rapidly into sand and mud during river transport (Adams, 1980a, p. 31). By contrast the graywacke rocks of the eastern slopes are more durable and produce more boulders and pebbles. East-draining rivers like the upper Waimakariri, Rakaia, Rangitata, and Waitaki have wide mobile beds consisting almost entirely of graywacke gravel (Fig. 7); the rivers are braided and carry substantial gravel directly to the sea. By contrast west-draining rivers like the Cook and Waiho tend to have deeply incised narrow channels in the mountains with bedload partly of large resistant schist boulders and partly graywacke gravel. Indeed, although graywacke may make up only 10% of the bedrock in the west-draining catchments, graywacke gravel may dominate in the bedload. Away from the Southern Alps the bedload of the rivers declines rapidly as shown by the reduced slope of the rivers (Adams, 1980a, Fig. 13)

FIGURE 7. The braided Harper River on the eastern slopes of the Southern Alps occupies a gravel-filled valley widened during the last glaciation. Note the steep mountain slopes and the scree slides.

and is due to the rapid abrasion of the schist away from the Alpine Fault to leave graywacke predominant.

The type of bedload supplied to each coast also affects their morphology. To the west, the fine sediment abraded from the schist is washed offshore and gravels have not accumulated to any great degree. To the east, the voluminous supply of graywacke gravel has formed the Waitaki and Canterbury plains over the last 2.5 m.y. and has resulted in a large area of arable land.

## Landslide - Dammed Lakes

Through their rapid uplift and numerous earthquakes the Southern Alps provide both the rugged topography needed for landslides and rock avalanches, and a triggering mechanism (Adams, 1980a, p. 64-68). Although earthquake-triggered landslides have formed many small lakes in the northern and southern parts of the South Island (Adams, 1981), and there are many prehistoric rock avalanches in the graywacke rocks of the central Southern Alps (Whitehouse, 1983), there are few landslide-dammed lakes in the Southern Alps. The rapid erosion there would rapidly fill small lakes with sediment and breach their landslide dams. It may have been 500 yrs since the last major earthquake on the Alpine Fault (Adams, 1980b), and so the absence of such small lakes, if earthquake-formed, could be explained.

## Hydroelectric and Geothermal Energy

The tectonic forces that formed the Southern Alps lead to the high relief and large amounts of runoff needed for hydroelectric development. Large west-draining rivers like the Haast and Hokitika have annual runoffs of 7 m $yr^{-1}$ and a fairly even monthly distribution of flows, and are typical of the numerous steep rivers flowing northwest out of the Southern Alps. There are a few small hydro plants for local use, but the development of large rivers like the Haast and Buller is hindered by their large sediment loads and their remoteness from electricity demand.

On the western slopes of the Southern Alps there was heavy precipitation, a steep gradient, and a narrow coastal plain and so the Pleistocene valley glaciers extended beyond the present coastline. Their arcuate terminal moraines are now submerged beneath the sea, and the lakes formerly impounded are completely filled with sediment from the rapidly eroding mountains. On the eastern slopes, the lower gradients and lower precipitation meant that none of the glaciers extended far beyond the foothills, and their terminal moraines now impound large glacial lakes. These provide important water storage and critical sediment storage for the large downstream developments on the Waitaki and Clutha rivers, benefits not available on the western slopes.

The rapid uplift on the Alpine Fault is moving hot rock to the surface faster than it can cool, so that the heat flow is very high. Hot springs suggest temperatures of perhaps 200°C within 1 or 2 km of the surface (Allis et al., 1979) over a region 400 km long and 7 km wide, making it the largest potential geothermal resource in New Zealand. Because of the economic problems of extracting energy from the largely unfractured hot rock and its remoteness from demand, development in the near future is unlikely.

## Gold

The cuesta model for the central Southern Alps implies that only graywacke will be eroded from the eastern slopes of the Alps. Williams (1965) has shown that metamorphism of these rocks to Chlorite II grade (as has occurred in the schists) and above is necessary before substantial segregation of the disseminated gold occurs, so that the rocks eroded from the eastern slopes should be barren of gold. Indeed only minor bedrock gold was found (at Wilberforce in Canterbury) on the eastern slopes and placer gold was never important, even though substantial gold was found only 35 km to the west in the schist-draining western coast rivers.

Canterbury is a former province of New Zealand that extended westward from the coast to the divide of the Southern Alps. Its western boundary, at the graywacke/schist boundary, will probably remain the same as the Southern Alps evolve. Hence Canterbury lacks present bedrock and placer gold and will do so for the forseeable geological future. (Note: with a western boundary a constant distance from the provincial capital, Christchurch, crustal shortening would eventually (in 1-2 m.y.) mean that Canterbury would extend to include the gold-bearing rock and alluvial deposits of Westland).

In north Westland the terminal moraines are onland and the rivers have concentrated the gold into rich placers that were mined in the 1860's. By contrast, in south Westland the terminal moraines are offshore and substantial gold-bearing gravels are now submerged and have not been worked. Hence while most of the geological conditions for the formation of giant placer fields (Henley and Adams, 1979) are present in south Westland, the gold remains inaccessible until the next 100-m drop in sea level allows further reworking of the deposits.

## YOUNG STEADY STATE MOUNTAIN RANGES AND THEIR TECTONIC GEOMORPHOLOGY

In this paper I have chosen to look in some detail at one of the very large geomorphic features of the earth's surface. The Southern Alps, being athwart the roaring forties and in a special tectonic environment, are an unusual mountain range, and simpler than most. Extremely rapid uplift has produced landforms that reflect a balance between uplift and erosion. Many of the young mountain ranges around the world have a similar balance between uplift and erosion, and hence the ideas described here for the Southern Alps may apply to them.

Recently there has been an increasing appreciation in the literature that natural erosion rates can be very rapid indeed. From the sediment transport side, the sediment loads of short, mountain rivers have rarely been properly measured (as is thoroughly discussed in Adams, 1980a, p. 43-51). There has also been little appreciation that areas of rapid uplift may be very restricted, so that basin-wide average erosion rates need not reflect the balance between uplift and erosion in each place. Now that such data and appreciations are available it is becoming evident that the tectonically active and rapidly uplifted areas of the world supply almost all of the river sediment load (Milliman and Meade, 1983).

The study of the tectonic geomorphology of the Southern Alps suggests how other mountain ranges can and should be studied. Spiky mountains are a good indication of rapid uplift and rapid erosion and a distinctive set of landforms - e.g., spiky mountains, straight slopes, smoothly graded rivers - signify a steady state mountain range. Estimates of uplift and erosion need to be made and compared for these mountain ranges in order to understand how the landforms and the mountain range as a whole were initiated and changed with time. Such an understanding must use geological, geomorphic, and process information to produce an integrated description of the evolution of these large-scale landforms.

## ACKNOWLEDGEMENTS

Ideas presented here were developed during and subsequent to my Ph.D. research at Victoria University of Wellington, and many of them arose during discussions with my thesis supervisor H.W. Wellman. Wellman's perceptive comments on the tectonic geomorphology of the Southern Alps were first made 40 years ago (Wellman and Willett, 1942) and have stood the test of time well. I thank the organizers and participants of the 1983 Penrose Conference on Tectonic Geomorphology for an early discussion of this work, and G. Rogers, R. Parrish and H.W. Wellman for critical reviews.

## REFERENCES

Adams, J., 1979, Vertical drag on the Alpine Fault, New Zealand *in* Walcott, R.I. and Cresswell, M.M., eds., Origin of the Southern Alps: Royal Society of New Zealand Bulletin 18, p. 47-54.

________, 1980a, Contemporary uplift and erosion of the Southern Alps, New Zealand: Geological Society of America Bulletin, Part II, v. 91, p. 1-114.

________, 1980b, Paleoseismicity of the Alpine fault seismic gap, New Zealand: Geology, v. 8, p. 72-76.

________, 1980c, Active tilting of the United States midcontinent: Geodetic and geomorphic evidence: Geology, v. 8, p. 442-446.

________, 1981, Earthquake-dammed lakes in New Zealand: Geology, v. 9, p. 215-219.

Allis, R.G., 1981, Reply to comments on "Continental underthrusting beneath the Southern Alps of New Zealand": Geology, v. 9, p. 490-491.

Allis, R.G., Henley, R.W., and Carman, A.F., 1979. The thermal regime beneath the Southern Alps in Walcott R.I. and Cresswell, M.M., eds., Origin of the Southern Alps: Royal Society of New Zealand Bulletin 18, p. 79-85.

Bloom, A.L., 1978, Geomorphology: Englewood Cliffs, N.J., Prentice-Hall, 510 p.

Burnett, A.W., and Schumm, S.A., 1983, "Alluvial-river response to neotectonic deformation in Louisiana and Mississipi: Science, v. 222, p. 49-50.

Coulter, J.D., 1973, A water balance assessment of the New Zealand rainfall: Journal of Hydrology (New Zealand), v. 12, p. 83-91.

Griffiths, G.A., and McSaveney, M.J., 1983, Hydrology of a basin with extreme rainfalls - Cropp River, New Zealand: New Zealand Journal of Science, v. 26, p. 293-306.

Henley, R.W., and Adams, J., 1979, On the evolution of giant gold placers: Institute of Mining and Metallurgy Transactions, v. 88, p. B41-B50.

Milliman, J.D., and Meade, R.M., 1983, World-wide delivery of river sediment to the oceans: Journal of Geology, v. 91, p. 1-21.

Schumm, S.A., 1963, The disparity between present rates of denudation and orogeny: United States Geological Survey Professional Paper 454-H, 13 p.

Trimble, S.W., 1977, The fallacy of stream equilibrium in contemporary denudation studies: American Journal of Science, v. 277, p. 876-887.

Wellman, H.W., 1975, The obduction-subduction part of the Australian-Pacific plate boundary in New Zealand: International Union of Geodesy and Geophysics, Commission on Crustal Movements Circular, v. 10, p. 50.

____________, 1979, An uplift map for the South Island of New Zealand and a model for the uplift of the Southern Alps, in Walcott, R.I., and Cresswell, M.M., eds., Origin of the Southern Alps: Royal Society of New Zealand Bulletin 18, p. 13-20.

____________, 1983, New Zealand horizontal kinematics: in Hilde, T.W.C. and Uyeda, S., eds., Geodynamics of the Western Pacific-Indonesian region, American Geophysical Union Geodynamics Series, v. 11, p. 423-457.

____________ and Willett, R.W., 1942, The geology of the West Coast from Abut Head to Milford Sound - Part 1: Royal Society of New Zealand Transactions, v. 71, p. 282-306.

Whitehouse, I.E., 1983, Distribution of large rock avalanche deposits in the central Southern Alps, New Zealand: New Zealand Journal of Geology and Geophysics, v. 26, p. 271-279.

Williams, G.J., 1965, Detrital gold, in Williams, G.J., ed., Economic geology of New Zealand, Australian Institute of Mining and Metallurgy, 8th Commonwealth Mining and Metallurgy Congress, v. 4, p. 69-86.

Yoshikawa, T., 1974, Denudation and tectonic movement in contemporary Japan: Tokyo University, Department of Geology Bulletin, v. 6, p. 1-14.

# 6
# Correlation of flights of global marine terraces

*William B. Bull*

## ABSTRACT

Two decades of detailed studies of marine terraces on New Guinea, Hawaii, Barbados, Bermuda, Japan, California, and many other seacoasts provide valuable details about glacio-eustatic high stands of sea level. These studies reveal diversity in the formation factors of (1) altitudes of formation and (2) intervals between times of formation for shore platforms and coral reefs. Diverse formation factors result in unique altitudinal spacings of terraces for each uniform uplift rate. Such spacings provide a basis for terrace correlations in areas of assumed uniform uplift.

Correlations involve making field surveys and/or topographic map analyses of an undated flight of terraces to define the altitudinal spacings of former major high stands of sea level. Then, inferred uplift-rate graphs identify (1) the ages of each global marine terrace present in the local flight, and (2) the late Quaternary uplift rate. Altitudes-ratio graphs are a useful crosscheck to ascertain if terraces at several localities are part of the same population, and whether the local terraces are part of the population of global marine terraces.

## INTRODUCTION

### Time Lines in Tectonic Geomorphology

Many studies in tectonic geomorphology involve landforms that are sensitive to vertical and horizontal tectonic displacements. Examples include horizontal offsets of stream channels and terrace risers, and the effects of vertical movements on the segmented topographic profiles of fault scarps and alluvial fans. Furthermore, certain landforms, although formed during time spans of $10^2$-$10^4$ yr, have abrupt terminations of formative processes and thus may be regarded as time lines within landscape assemblages. Fill terraces result when climate-controlled valley aggradation abruptly changes to degradation as the threshold of critical power (Bull, 1979) is crossed. Straths beveled into bedrock by streams record periods of equilibrium adjustments between independent and dependent variables of fluvial systems. Tectonic or climatic perturbations may cause accelerated downcutting by a stream that leaves remnants of the strath as a terrace -- a time line within the landscape.

Shore platforms also are geomorphic subsystems in equilibrium where a balance exists between available energy, sediment amount and size being transported across the platform, and hydraulic factors such as the width and roughness of the platform surface. Maximum widths of shore platforms and coral reefs are attained during those times when sea level remains constant relative to a landmass altitude. Times of rapid relative sea-level rise and fall are not conducive for sufficiently stable conditions to allow shore platforms to develop. Instead, shore platforms typically form during high stands of sea level characterized by only minor fluctuations in sea level, such as during the last 6 ka (ka is kilo anno; 1 ka = 1,000 yr). Shore platforms that are more than 500 m wide may form in soft sandstone and mudstone in less than 1 ka (Kirk, 1977; Berryman and Hull, 1984). Subsequent decline of sea level terminates the constructional phase of the shore-platform time line and incorporates these coastal landforms into preexisting fluvial systems. Shore platform or coral reef time lines may occur intermittently along $10^2$ km of rocky coasts.

Marine terrace time lines are directly tied to the chronology of major fluctuations in sea level. Although mid and late Holocene sea levels may have fluctuated mildly (Fairbridge, 1961; Berryman and Hull, 1984) or remained constant (Clark et al, 1978), and the viscoelastic properties of the earth in response to melting ice sheets may have caused spatial variation of Holocene sea levels (Walcott, 1972; Chappell, 1974b; Mörner, 1976; Clark et al, 1978; Kidson, 1982), it is the worldwide synchroneity of late Quaternary major high stands in sea level that are of paramount importance when using marine terraces in tectonic geomorphology studies. For example the major high stands that occurred 120 ka and 6 ka ago may be considered times of worldwide formation of shore platforms and coral reefs. Similar arguments may be presented for each of nine sea-level maxima that occurred between 120 ka and the Holocene. Dating of terraces along coastlines in many different parts of the world clearly indicates a general synchroneity of major high stands of sea level (Veeh, 1966; Ku et al, 1974; Machida, 1975; Fairbanks and Matthews, 1978; Lajoie et al, 1979; Bloom, 1980; Pillans, 1980; Chappell, 1983). For example more than 30 localities in the warm oceans of the world have dated coral indicating a high stand of sea level at +2 to +12 m altitude at roughly 120 ka. The prior studies clearly suggest that entire flights of marine terraces may be part of an overall population of global marine terraces. This article tests this hypothesis.

### Purpose and Scope

Materials for dating are difficult to obtain from buried shore platforms. Entire flights of marine terraces along many coastlines lack either coral or volcanic materials that permit radiometric age determinations older than the upper limit of radiocarbon dating. However, when considered as sequences instead of single terraces, local flights may be correlated with the overall population of dated global marine terraces.

The purpose of this article is to discuss the basis and procedures for correlation of undated flights of marine terraces with the flights of dated terraces in New Guinea, which are regarded as being representative of the overall population of global marine terraces. It will be shown that unique differences in the number and altitudinal spacings of marine terraces occur in a diagnostic manner for each uniform uplift rate. The procedure involves preparation of altitudinal frequency-distribution plots and inferred uplift-rate graphs. Several types of pitfalls may be expected, but they can be identified, and crosschecks other than radiometric dating can be applied to test the results.

## IS IT REALLY A BAFFLING ARRAY?

The student of emergent late Quaternary marine terraces along a rising coastline is likely to find what appears to be a lack of spatial continuity even between terraces of the same apparent age. He or she may be tempted to work only with marine deposits and/or beach ridges, but should remember that such deposits merely record formation at some distance below or above mean sea level. In contrast, the equilibrium landform of shore platforms approximates mean sea level (Kirk, 1977). Estimates of the altitude of the inner edge of the shore platform at the base of the former sea cliff are the most desirable datum. Unfortunately shore platforms may be cut on headlands while concurrent deposition of beaches results in progradation in the intervening bays. Cliff retreat associated with shore-platform beveling commonly removes one or more preexisting higher shore platforms, so a complete sequence of terraces at a single transect is rare. Exposures of shore platforms generally are not common; they typically seem to be masked by rain forests and cover-bed stratigraphy of beach gravels, stream deposits, colluvium, loess, and landslide deposits. Present and past cliff retreat also induces landslides that displace shore platforms.

Even when the above problems are resolved by persistent fieldwork, it may appear that correlation of shore platforms is improbable because of the general lack of consistency of heights above present sea level for several study sections along 100 km of coastline. So in the absence of dates, or faunal studies such as those done by Lajoie et al (1979), the cautious observer is tempted to refrain from correlating terraces along a given coast, and even more from considering correlations of a global nature.

Actually, it is the initially baffling differences in spacings and numbers of marine terraces that are the vital key for correlation, either locally or globally. Another favorable control on long distance correlation of marine terraces is that interplate tectonic stresses commonly result in uniform overall rates of uplift at a given site when one considers time spans of more than 30 ka. Even for the few sites considered in this article, it will be fairly apparent that uniform uplift most likely has prevailed for time spans of 50 to 300 ka. Two periods of different but uniform uplift can be demonstrated at some sites (Bull and Cooper, in preparation).

## FORMATION FACTORS AND UNIQUE TERRACE SPACINGS

Careful dating of 20 marine terraces on New Guinea is summarized in the publications of Bloom et al (1974) and Chappell (1983). The chronology is based largely on $^{234}U/^{230}Th$ dating of carefully selected coral as old as 336 ka (table 1). Global synchroneity of times of coral terrace formation is suggested by even a single comparison between sites in the Atlantic and Pacific Oceans. Ages of raised coral reefs on Barbados (Butzer, 1983, table

2) of 61, 82, 105, 124, 212, and 242 ka have correlative terrace ages on New Guinea (see table 1 below).

Chappell's (1983) summary of the ages and times of formation has been adopted with only minor modifications. A common New Zealand terrace has been added at 320 ka rather than having the 336 to 320 ka high stand represented by only one terrace; even at 0.5 mm/yr, 8 m of uplift would occur in 16 ka. The dates suggested by Chappell for terrace 3C and 3D of 58 and 62 ka consistently result in the 62 ka inferred uplift point occurring above the uniform rate line when assigned to New Zealand flights in the 1300 km between sites E and F of figure 3. A solution is suggested by four New Guinea dates (Bloom et al, 1974). Three dated corals are in close agreement; 57, 58, and 61 ka. A fourth sample dated at 66 ka but was included in the younger group because of the ±4 ka estimated error overlap. Subsequently, Chappell (1983) demonstrated the presence of two distinct terraces in the field. Like Bloom et al (1974, p. 200) I have used the termination of postglacial eustatic rise (regarded as 6 ka) and the assigned age of 4 ka for reef I to arrive at a "premature emergence correction" of about 2 ka. Thus in table 1 the age of terrace 3C is listed as 57 ka and terrace 3D as 64 ka.

Rather than use local (New Guinea) terrace numbers, the global marine terraces are assigned numbers of the marine oxygen isotope stages of Shackleton and Opdyke (1973, p. 49). Their interpolated sedimentation-rate ages were based on an age of 700 ka for the Brunhes-Matuyama magnetic epoch boundary. The table 1 interpolated ages are based on a 730 ka age, per a suggestion by Opdyke (oral commun., 5 April 1984). Capital letters designate separate high stands of sea level associated with global marine terraces on New Guinea and elsewhere.

Spacings of terraces within a flight depend partly on (1) altitudes of formation of terraces and (2) intervals of time between the consecutive high stands of sea level. Neither of these formation factors has remained constant with time (table 1). The most simple hypothetical terrace spacing would occur on a uniformly rising coastline periodically notched by shore platforms at equal time intervals and at the same absolute sea level. The resultant constant altitudinal spacings of terraces would be merely a function of the uniform uplift rate, and there would be no hope of identifying the age of a particular terrace within the global sequence.

TABLE 1 Ages and altitudes of formation New Guinea marine terraces (Bloom et al, 1974; Chappell, 1983).

| Number | Age, ka | Interval between times of terrace formation, ka | Departure from mean interval of 17 ka between times of terrace formation, ka | Altitude formed, m | Departure from mean altitude of formation of -20 m m |
|---|---|---|---|---|---|
| 1A | 6 | | | 0 | 20 |
| | | 24 | 7 | | |
| 2A | 30 | | | -42 | 22 |
| | | 10 | 7 | | |
| 3A | 40 | | | -37 | 17 |
| | | 6 | 11 | | |
| 3B | 46 | | | -37 | 17 |
| | | 11 | 6 | | |
| 3C | 57 | | | -29 | 9 |
| | | 7 | 10 | | |
| 3D | 64 | | | -26 | 6 |
| | | 12 | 5 | | |
| 4A | 76 | | | -46 | 26 |
| | | 7 | 10 | | |
| 5A | 83 | | | -13 | 7 |
| | | 11 | 6 | | |
| 5B | 94 | | | -20 | 0 |
| | | 9 | 8 | | |
| 5C | 103 | | | -10 | 10 |
| | | 17 | 0 | | |
| 5D | 120 | | | + 6 | 26 |
| | | 13 | 4 | | |
| 5E | 133 | | | + 5 | 25 |
| | | 43 | 26 | | |
| 6A | 176 | | | -21 | 1 |
| | | 26 | 9 | | |
| 6B | 202 | | | -17 | 4 |
| | | 12 | 5 | | |
| 7A | 214 | | | - 3 | 17 |
| | | 28 | 11 | | |
| 7B | 242 | | | -28 | 8 |
| | | 44 | 27 | | |
| 8A | 286 | | | -46 | 26 |
| | | 19 | 2 | | |
| 8B | 305 | | | -27 | 7 |
| | | 15 | 2 | | |
| 9A | 320 | | | + 4 | 24 |
| | | 16 | 1 | | |
| 9B | 336 | | | + 4 | 24 |
| mean values | | 17.4 | 8.3 | -20.2 | 14.8 |

Fortunately the great diversity of formation factors summarized in table 1 shows that each terrace starts at a different sea level and that the time intervals between sea-level highs vary markedly. Such temporal changes in the formation factors assure that there will be a different altitudinal spacing of shore platforms for each uniform uplift rate. Therefore it is not a baffling array -- one should expect different altitudinal sequences of terraces if spatial variations in uniform uplift rate are present along a given coastline.

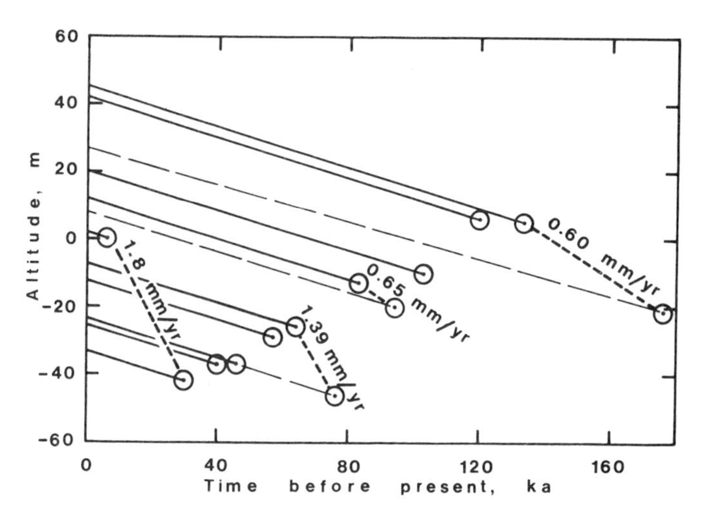


FIGURE 1 Altitudes and times of glacio-eustatic high stands of sea level associated with the formation of coral reef marine terraces on Huon Peninsula, New Guinea. The altitudes and times are mainly from Chappell's (1983) revision of the original study by Bloom and others (1974). The parallel lines depict altitude changes for each terrace for a hypothetical uniform uplift rate of 0.3 mm/yr; unbroken lines where not submerged, and long dashes where submerged by the next sea-level high stand. Short dashed lines are rates needed to escape submergence. 1 ka = 1,000 yr.

The relative importance of formation factors and rate of uplift in determining terrace altitudes along a given coast partly determines the ease of correlation of local flights of terraces with dated sequences of global marine terraces. The formation factors may be combined in the denominator of a ratio, and when uplift rates are expressed in $m/10^3$ yr, a dimensionless number results.

$$\frac{\text{uplift rate in } m/10^3 \text{ yr}}{\text{formation factors in } m/10^3 \text{ yr}} = \frac{\text{uniform uplift rate in } m/10^3 \text{ yr}}{\begin{array}{l}\text{(mean departure from average} \\ \text{altitude of formation in m)} \div \\ \text{(mean departure from average} \\ \text{interval between times of terrace} \\ \text{formation in } 10^3 \text{ years)}\end{array}}$$

The uplift rate/formation factors ratio ranges from less than 0.2 along slowly rising coasts to more than 5.0 for rapidly rising coastal mountains.

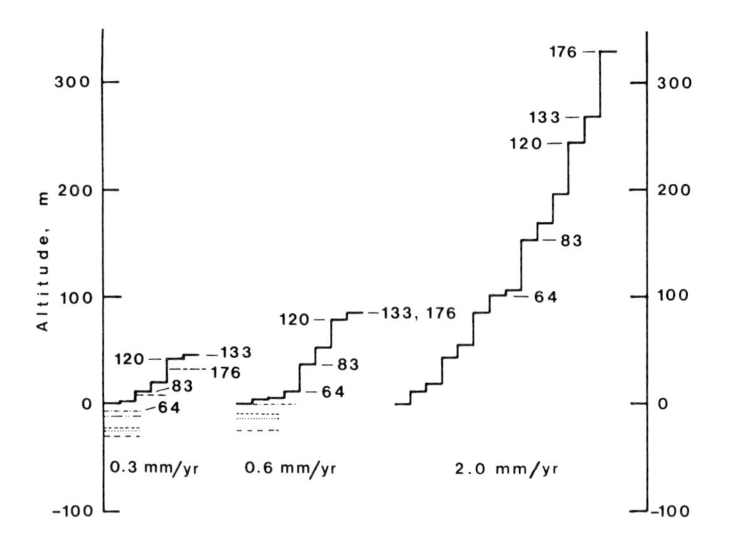


FIGURE 2 Altitudinal spacings of global marine terraces for three different hypothetical uniform uplift rates. See table 2 for ages, amount of uplift and present altitude for each terrace.

Terrace correlations are easier where formation factors are more important relative to uplift rate in determining spacings.

The altitudes and times of formation of the New Guinea terraces during the last 176 ka are shown in figure 1. The altitudes of formation since 120 ka record overall decline of sea-level highs until the Holocene. This progression of successively lower sea-level maxima is an important reason for preservation of most of the terraces. Although the terraces may be destroyed by fluvial erosion, cutting of younger sea cliffs and landslides, they were not beneath Pleistocene seas since their formation. Two exceptions may be the terraces formed at 94 and 76 ka. In areas of moderate to low uplift rates these terraces have been submerged by younger sea-level rises and then have emerged during still younger sea-level declines.

Post-formation inundation by the sea does not occur when uplift rates exceed a minimum value. For example terraces formed at 176 ka would have to be uplifted at rates greater than 0.6 mm/yr in order not to be submerged by the 133 ka sea-level rise. Terraces formed at 94, 76, and 30 ka would have to be uplifted faster than 0.65, 1.4, and 1.8 mm/yr, respec-

TABLE 2 Altitudes of global marine terraces along a hypothetical coast with three different rates of uniform uplift during the last 176 ka. - signs for present altitudes indicate terrace levels below present sea level. ( ) indicate those terraces that are not likely to be prserved because of submergence and re-emergence subsequent to their formation.

| | | | 0.3 mm/yr | | 0.6 mm/yr | | 2.0 mm/yr | |
|---|---|---|---|---|---|---|---|---|
| No. | Age, ka | Altitude formed, m | (1) | (2) | (1) | (2) | (1) | (2) |
| 1A | 6 | 0 | 2 | + 2 | 4 | + 4 | 12 | + 12 |
| 2A | 30 | -42 | 9 | (-33) | 18 | (-24) | 60 | + 18 |
| 3A | 40 | -37 | 12 | (-25) | 24 | (-13) | 80 | + 43 |
| 3B | 46 | -37 | 14 | (-23) | 28 | (- 9) | 92 | + 55 |
| 3C | 57 | -29 | 17 | (-12) | 34 | + 5 | 114 | + 85 |
| 3D | 64 | -26 | 19 | (- 7) | 38 | +12 | 128 | +102 |
| 4A | 76 | -46 | 23 | (-23) | 46 | ( 0) | 152 | +106 |
| 5A | 83 | -13 | 25 | +12 | 50 | +37 | 166 | +153 |
| 5B | 94 | -20 | 28 | (+ 8) | 56 | (+36) | 188 | +168 |
| 5C | 103 | -10 | 31 | +21 | 62 | +52 | 206 | +196 |
| 5D | 120 | + 6 | 36 | +42 | 72 | +78 | 240 | +246 |
| 5E | 133 | + 5 | 40 | +45 | 80 | +85 | 266 | +271 |
| 6A | 176 | -21 | 53 | (+32) | 106 | +85 | 352 | +331 |

(1) Uplift, m; (2) Present altitude, m

tively, in order to escape inundation. The intersections of the 0.3 mm/yr uplift-rate lines with the ordinate portrays the present altitudinal spacing of the flight of terraces. Many of the terraces are submerged below present sea level and those above it may be closely or widely spaced.

The altitudinal spacings at the left side of figure 1 are compared with spacings associated with larger uplift rates in figure 2 and table 2. For a uniform uplift rate of 0.3 mm/yr all of the Pleistocene terraces formed since 83 ka would be still below present sea level. Closely spaced terraces at +12 and +21 m altitude should be discernible under favorable conditions, but the +8 m terrace most likely would have been destroyed by subsequent submergence and emergence, or it may not be possible to separate it from the +12 m terrace in the field. The terraces at +42 and +45 m might be considered the same terrace in many field settings, and the +32 m terrace is unlikely to have been preserved even locally for the same reasons noted for the +8 m terrace. Thus the obvious features of coastlines that have been raised at 0.3 mm/yr are a couple of low terraces and a substantially higher dominant terrace formed during the interglacial at 120 ka. Figure 1 shows that even if the 76 ka terrace were to survive multiple submergences, at an uplift rate of 0.3 mm/yr it would coincide with the high stand of sea level during the formation of the 46 ka shore platform, and therefore would be indistinguishable from the younger shore platform. Such complicating coincidences are not rare. Combinations of large numbers of different times and altitudes of formation commonly result in altitudinal overlap that results in the omission of a terrace from a sequence. Such complications should be expected, but can be evaluated by preparation of altitude-age plots, such as the prediction diagram of figure 1.

The altitudinal spacings and numbers of shore platforms associated with a uniform uplift rate of 0.6 mm/yr are markedly different. The three late Pleistocene terraces younger than 57 ka would be still below sea level and the 76 ka terrace (if it has survived subsequent submergence and reemergence) would not be recognizable because it coincides with the level of shore platforms being cut by modern sea level. Likewise the 176 ka shore platform would coincide with the 133 ka shore platform, and a similar situation prevails for terraces formed at 94 and 83 ka. Thus the potential terrace sequence consists of two closely spaced terraces formed at high stands of sea level at 64 and 57 ka, prominent terraces formed at 83 and 120 ka and less conspicuous terraces formed at 103 and 133 ka.

The number of prominent terraces increases with increases in uniform uplift rate as is illustrated by the 2 mm/yr plot. All of the terraces are above present sea level and none of them have been beneath the sea since their formation. Therefore, their numerical and age sequence is not scrambled. The 2 mm/yr uplift rate separates general groups of terraces that on one hand tend to obliterate and overlap one another from more complete flights of terraces that may not have as much contrast in heights between adjacent terraces because of larger values of the uplift rate/formation factors ratio. The increase in the number of global terraces that are present with increases in uplift rates is accompanied by increase in the uplift rate/formation factors ratio which for the examples of figure 2 and table 2 increases from 0.17 to 0.34 to 1.14. The decrease in the relative importance of the formation factors on terrace spacing is readily apparent in the next section which deals with the methodology of terrace correlations.

## CORRELATIONS OF GLOBAL MARINE TERRACES

The correlation procedures are based on data from four New Zealand sites (figure 3) that encompass an order of magnitude variation in uplift rates. The procedures and crosschecks described here can be used for other flights of global marine terraces. Undated New Zealand terraces will be correlated with the dated terraces of Huon Peninsula, New Guinea (table 1).

### Important Assumptions

A large number of important assumptions must be met in order for correlations of global marine terraces to succeed. That they succeed at all is in large part due to the exceptionally careful work of those earth scientists who have studied late Quaternary terraces and sea-level fluctuations. The main assumptions are:

(1) Uniform uplift at both the New Guinea and New Zealand study transects.

(2) Accurate measurement of terrace heights both on New Guinea and New Zealand.

(3) Accurate dating of the New Guinea marine terraces.

(4) Accurate estimations of the altitudes and times of formation for each New Guinea terrace.

(5) Although heights of the marine geoid vary spatially (Marsh et al, 1982; McAdoo and Martin, 1984), there have been minimal changes in relative geoid heights between marine-terrace localities at sea-level high stands during the last 400 ka. A corollary is that the altitudes of formation of marine terraces are the same in New Zealand and New Guinea.

All of the assumptions dealing with the New Guinea work seem quite valid and have been previously discussed by Bloom et al (1974) and Chappell (1974a, 1983). Temporal distortions in the shape of the marine geoid as postulated by Mörner (1976) do not appear to have occurred during the short geologic time spans considered in this article.

The correlation procedure involves determination of the present altitudes of the terraces, making inferred uplift-rate plots to locate the position of the local terraces in the sequence of global marine terraces, and checking the results with altitudes-ratio graphs. Most inferred uplift-rate graphs

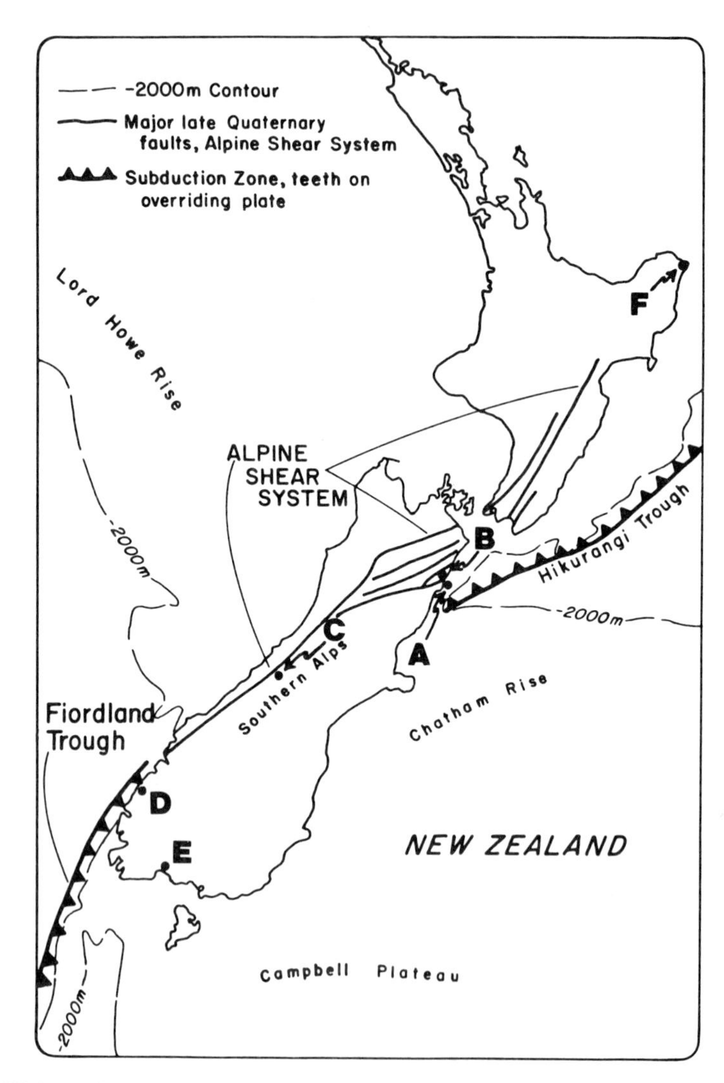


FIGURE 3 Tectonic setting of New Zealand and locations of study sites at Kaikoura Peninsula (A), Seaward Kaikoura Range (B), Franz Josef (C), and Caswell Sound (D).

clearly show that uniform uplift is at least one possible solution when correlating flights of undated with dated terraces. Furthermore examination of overall landscape assemblages makes it readily apparent whether or not different parts of terrace sequences are associated with different uniform uplift rates. For example, Bull and Cooper (in preparation) note that the major escarpment along the northwestern side of the Southern Alps is associated with a doubling of uplift rates as compared to less rugged terrain formed before 140 ka. Along the east coast of the North Island deposits on shore platforms can be dated by tephra (Yoshikawa and others, 1980). Inferred uplift-rate graphs indicate a six-fold increase in uplift rates since 30 ka ago. However the overall aspect of the fluvial landscape -- rolling hills and broad valleys -- is still suggestive of moderate to low rates of uplift rather than of rapid rates. Still another complication involves spatial variations in uplift rates. Although uplift rates may be uniform at any given point on a fold or near a fault, nonuniform uplift rates will be indicated by terrace analyses if the transect studied crosses geologic structures with varying degrees of tectonic activity.

**Shore-Platform Surveys**

Figure 2 and table 2 depict hypothetical situations of known terrace ages and uniform uplift rates that demonstrate unique spacings and numbers of uplifted marine terraces for each uplift rate. Figure 1 may be regarded as a prediction diagram where the present altitudes of marine terraces are predicted if one assumes a certain uplift rate for a terrace in the flight and assumes the same uniform uplift rate for other terraces.

We now turn our attention to the field setting at Kaikoura Peninsula (fig. 3) where we know the altitudes of eight raised shore platforms, but do not have absolute dating control. Predictions in the form of inferred uplift-rate graphs were used to determine whether or not uniform uplift is one of several possible solutions. Shore-platform altitudes were surveyed from sea level; traverses to the higher terraces ended at a triangulation station on the highest terrace. The most prominent terrace is associated with the 39 m ± 2 m shore platform; it is wide and rises abruptly above the dated mid-Holocene terrace (Kirk, 1977). The 39 m terrace appears to be one of the global marine terraces, but lack of dating prevents an absolute age determination. The attempted correlations of table 3 and figure 4 merely indicate some possible age assignments, and provide a basis for evaluation

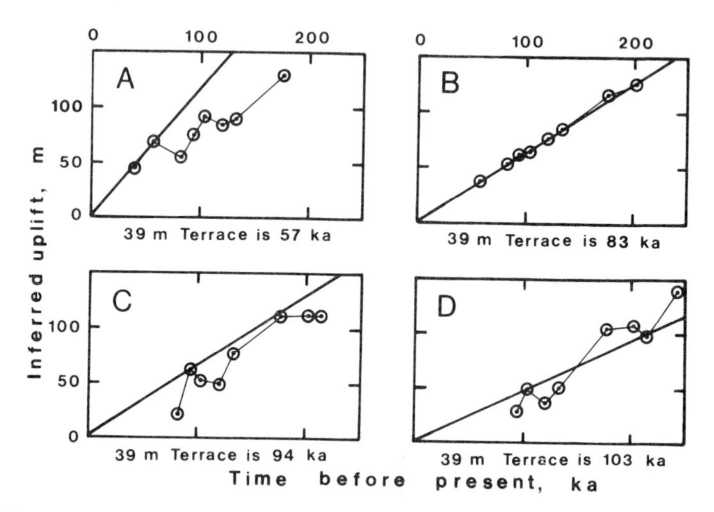


FIGURE 4 Attempted correlations of the Kaikoura Peninsula marine terraces with the global marine terraces. The heavy lines depict four predictions of inferred uniform uplift of other terraces in the flight when four different ages are assigned to the 39 m terrace. The light lines depict variations in uplift rate between assigned times of terrace formation.

of the most probable ages of the 39 m and other terraces. In figure 4A the 39 m terrace is assigned an age of 57 ka, and a set of inferred-uplift values for the other terraces is generated when the altitudes of formation are subtracted algebraically from the present altitudes (table 3). The heavy line drawn through the origin and the point for the 39 m terrace provides a graphical prediction of how accurately this age assignment predicts the inferred uplift of the other global marine terraces in the flight. The prediction for the 8 m terrace seems good, but the points for all the other terraces depart substantially from the uniform uplift rate indicated by the heavy line. Similar attempts to predict inferred uplifts of other terraces in the flight can be made by assigning ages of 83, 94, 103, and 120 ka to the 39 m terrace. All attempts failed except the 83 ka plot shown in figure 4B. When the 39 m terrace is assigned an age of 83 ka, the heavy line drawn through the origin and the point of the 39 m terrace accurately predicts the inferred uplifts of all the other terraces in the flight. Thus uniform uplift appears to be one possible solution, which if accepted results

TABLE 3 Attempted correlations of the Kaikoura Peninsula, New Zealand, marine terraces with the dated sequence of global marine terraces at New Guinea.

| New Guinea Terraces | | New Zealand Terraces | | | | | | | |
|---|---|---|---|---|---|---|---|---|---|
| Age, ka | Altitude formed, m | (1) | (2) | (1) | (2) | (1) | (2) | (1) | (2) |
| | | See 4A | | | | | | | |
| 40 | -37 | 8 | 45 | See 4B | | | | | |
| 57 | -29 | 39 | 68 | 8 | 37 | See 4C | | | |
| 83 | -13 | 42 | 55 | 39 | 52 | 8 | 21 | See 4D | |
| 94 | -20 | 55 | 75 | 42 | 62 | 39 | 59 | 8 | 28 |
| 103 | -10 | 82 | 92 | 55 | 65 | 42 | 52 | 39 | 49 |
| 120 | + 6 | 90 | 84 | 82 | 76 | 55 | 49 | 42 | 36 |
| 133 | + 5 | 95 | 90 | 90 | 85 | 82 | 77 | 55 | 50 |
| 176 | -21 | 109 | 130 | 95 | 116 | 90 | 111 | 82 | 103 |
| 202 | -17 | | | 109 | 126 | 95 | 112 | 90 | 107 |
| 214 | - 3 | | | | | 109 | 112 | 95 | 98 |
| 242 | -28 | | | | | | | 109 | 137 |

(1) Present altitude, m; (2) Inferred uplift, m

in age assignments for every terrace in the flight and an inferred uplift rate of 0.64 mm/yr. However, other factors need to be considered.

Plot 4B does not preclude the possibility of the other plots being correct interpretations, so it is important to assess (1) variability of uplift rates, and (2) whether or not all terraces have been included in the analysis. Variability is shown by the thin lines connecting individual points. Segments that slope upward from the origin represent periods of uplift and segments that slope downward reflect subsidence. For plot 4B there is little variation in apparent uplift rate between the eight times of terrace formation. The other plots all show substantial variation in uplift rates and each of them has two periods of subsidence. For example, in order for plot 4A to be the correct interpretation, one would have to agree that there was a 26 ka interval of 0.5 mm/yr tectonic subsidence between 83 and 57 ka. It seems unreasonable that the local tectonic stress field would reverse so dramatically on such a small time scale. Furthermore, evidence for submergence such as marine deposits is totally lacking at Kaikoura Peninsula.

Other important stratigraphic information includes the numbers of loess units and their associated paleosols in the cover-bed stratigraphy. The same arguments apply to plots 4C and 4D and to the plot for the 120 ka age assignment (not shown in figure 4) for the 39 m terrace.

It is also important to assess whether important global marine terraces have been omitted. For example, in plot 4D most of the points fall within 20 m of the inferred uniform uplift rate line. However if one were to accept this as being the most probable correlation, one would have to accept the corollary that all of the global marine terraces younger than 94 ka were never formed and preserved at Kaikoura Peninsula despite an inferred uplift rate of 0.47 mm/yr. At this uplift rate one would surely expect to find the 83 ka terrace which has been dated as a prominent terrace at many localities in the world.

The contrast between the correlation attempts of figure 4 is striking in large part because the uplift rate/formation factors ratio is low, being only 0.36. When correlation attempts are made for flights of terraces along high uplift rate coasts, the best solution is less obvious because intervals of apparent subsidence do not occur.

Thus, on the basis of all topographic and stratigraphic data, correlations A, C, and D are rejected and correlation B is accepted.

## Topographic-Map Analysis

Inferred uplift-rate graphs may also be made from data collected from topographic maps where field checking shows that the altitudes of notched spur ridges and flat summits are representative of shore-platform remnants. New Zealand topographic maps have scales of 1 inch = 1 mile and 1 cm = 500 m, and contour intervals of 100 feet and 20 m. Numerous spot altitudes are printed on the maps and field checks show that the maps have great detail. Most contours are closely spaced at 15 to 30 per inch. In such rugged terrain the abrupt flattening of ridgecrests that is associated with remnants of former shore platforms is readily identifiable. Field surveys by Bull and Cooper (in preparation) show that the shore-platform remnants are associated with rounded quartz beach pebbles to altitudes as high as 1700 m and that the remnants are downslope from steep slopes suggestive of degraded former sea cliffs.

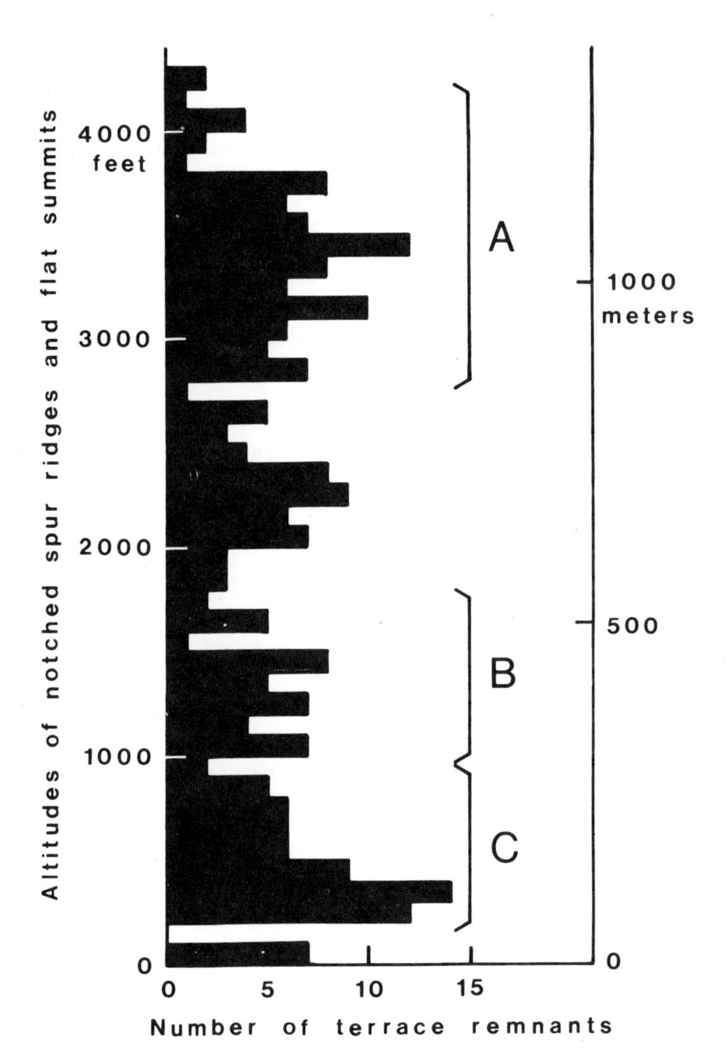


FIGURE 5 Altitudinal distribution of notched spur ridges and flat summits indicative of remnants of uplifted marine terraces near Caswell Sound.

Map analysis involves recording several hundred altitudes of apparent shore-platform remnants on adjacent ridgecrests. These data are used to make a frequency distribution plot. Landform elements indicative of remnants of uplifted marine terraces occur at virtually all altitudes near Caswell Sound but are much more common at some altitudes than others. The inferred uniform uplift rate is 3.7 mm/yr. In part A of figure 5 there is good separation of peaks because the uplift rate is rapid and the intervals between times of high sea level are long. Because the contour interval of 100 feet places a lower limit of 100 feet on classes of altitude distributions, a minimum separation of peaks of frequency occurs at every other hundred feet, such as in part B of figure 5. Topographic map analyses are insufficient to determine how many terraces are represented in broad compound peaks such as part C of figure 5. Field surveys, such as those made at Kaikoura Peninsula, are needed to desribe more accurately the altitude ranges of closely spaced terraces. Surveys at Caswell Sound determined the altitudes of five distinct terraces represented by the part C complex peak. Where uplift rates are more rapid the five terraces may be represented by separate peaks.

The next step is to make inferred uplift rate plots in the same manner as for Kaikoura Peninsula (table 3 and figure 4). A large number of points may not be available from topographic map analyses when evaluating uplift rates of blocks of terrane between specific faults. Only 133 altitudes were used in the figure 6 analysis. Young marine terrace remnants are not present because the lowest part of this structural block is higher than 400 m. The 6 mm/yr inferred uplift for this part of the Seaward Kaikoura Range appears to be uniform back to at least 340 ka.

## PITFALLS AND CROSSCHECKS

Potential problems with the above correlation procedure center about (1) whether or not all the global marine terraces are preserved in a local terrace flight, and (2) whether the age assignments truly represent a correlation with the global marine terrace sequence. Terraces may be missing as a result of coastal erosion, and landslides may displace parts of terraces creating the appearance of extra terraces.

The figure 1 prediction diagram provides a variety of useful crosschecks. For different assigned uniform uplift rates one can readily see which terraces should be missing from a given flight because their time of emer-

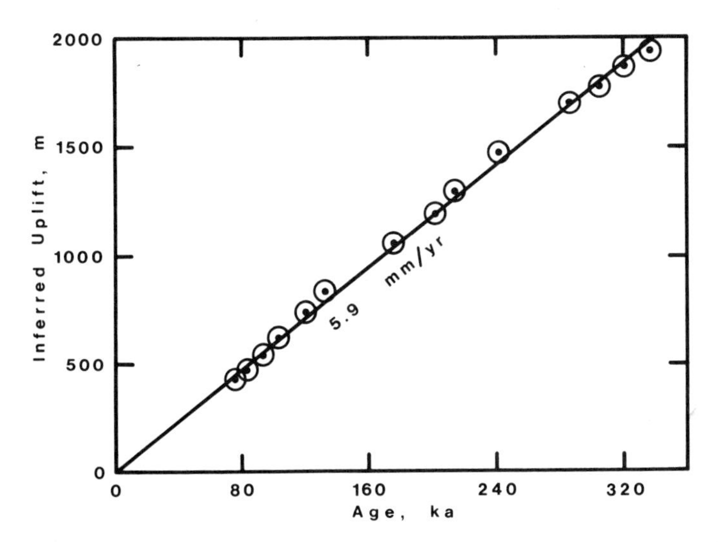


FIGURE 6 Inferred uplift-rate graph for the Seaward Kaikoura Range northwest of the Kekerengu fault.

gence from the sea coincides with the time of formation of a younger terrace. Missing shore platforms generally were formed at low altitudes, because shore platforms formed at high altitudes generally have not been submerged by subsequent sea-level rises. Prediction diagrams portray a parallel family of lines except where a misidentification of a terrace has been made. In cases of missing or additional terraces a line drawn between an assigned terrace formation altitude and its known present altitude will have a slope that departs from that of the family of parallel lines.

Inferred uplift-rate graphs based on topographic map analyses will have offset, but parallel, lines where extra or missing terraces complicate the interpretation (fig. 7). The uniform uplift rate line is offset upward where a terrace is missing, and is offset downward to a new parallel line if there is an additional terrace in the local flight.

Altitudes-ratio plots are a most useful tool for verifying correct correlation with the global marine terraces in areas where absolute ages are not available. Consider the South Island of New Zealand where uplift rates are extremely variable. For different widely spaced sites (such as A, B, C (two

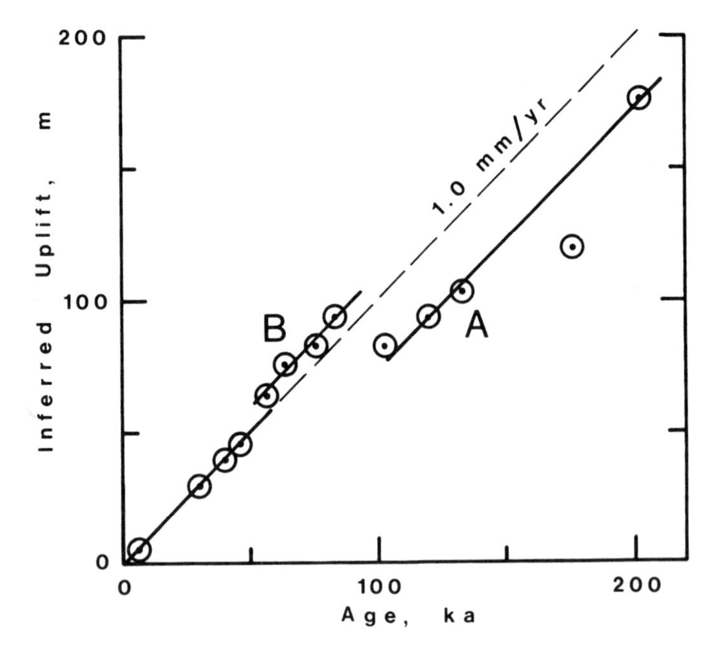


FIGURE 7 Hypothetical distortion of uniform inferred uplift-rate graph of global marine terraces for situations of (a) additional local terraces, and (b) missing local terraces.

analyses), and D of figure 3) a given global marine terrace will start at the same altitude but its present altitude will be determined by the local uplift rate. The altitudes of two ages of terraces are plotted in order to test for a consistent relation. Minimal scatter about two such regressions (figure 8) graphically illustrates an important internal consistency for 5 flights of terraces that range in inferred uplift rate from 0.6 to 8 mm/yr. The two altitudes-ratio regressions show that the widely spaced terrace flights are all part of one general population.

An additional important crosscheck can be made to determine whether the New Zealand marine terrace population is the same as the global population. This second crosscheck is done by comparing altitudes-ratio regressions from different parts of the world. This correlation method was developed by Ota et al (1968) and Miyoshi (1983) has made regressions for dated flights of terraces in New Guinea, Barbados, New Hebrides, and

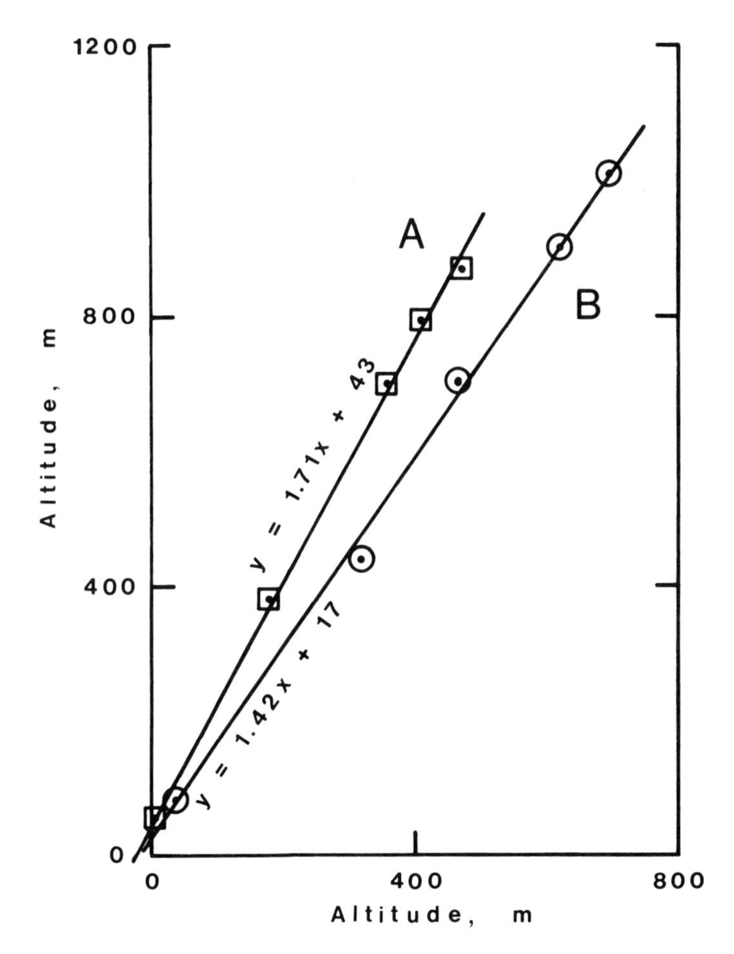


FIGURE 8 Ratios of altitudes of global marine terrace remnants whose assigned ages are based on inferred uplift-rate graphs. A. Ratio of altitudes of 57 ka and 103 ka terraces. B. Ratio of altitudes of 83 ka and 120 ka terraces. Both regression lines extend to the lower left and end at the respective altitudes of formation of the 4 global marine terraces.

Japan. When the altitudes of the 57, 83, and 103 ka terraces are compared against the altitudes of the 120 ka terrace, the New Zealand regression yields slopes of 0.51, 0.70, and 0.88, respectively; the dated flights yield ranges of regression slopes of 0.44-0.47, 0.60-0.69, and 0.77-0.83. Thus, figure 8 confirms consistent age assignments for marine terraces at the New Zealand sites, and the regression constants indicate that the New Zealand terraces may be considered a part of the global marine terrace population defined in table 1. Figure 8 also provides additional support for the assumption that uniform uplift rates have prevailed at the New Zealand sites, at least since 120 ka.

## CONCLUSIONS

This New Zealand study shows that flights of undated marine terraces can be correlated with the global marine terraces. In view of the assumptions that must be met in order for such correlations to succeed, it is apparent that through the efforts of many workers a high quality data base has been developed during the last two decades. The rigorous and thoughtful studies by other workers are an outstanding example of interdisciplinary scientific progress. The scientific base now is sufficient to permit correlations of marine terraces between coastlines that are undergoing uplift rates of 0.2 to 10 mm/yr. Variable deformation of the marine geoid by changing loads of water and ice during the Holocene glacio-eustatic sea-level rise appear to be responsible for spatial variations in Holocene sea level of small amplitude. Such minor effects do not influence the correlations discussed in this paper because (1) correlation of global marine terraces deals with major high stands of sea level and (2) hydroeustatic effects may have been roughly similar at a given site at each time of major sea-level rise.

Concerns expressed by Stearns (1976), Cronin (1982), and Butzer (1983) about the lack of progress in defining late Quaternary eustatic sea level curves, and of apparent overlap in radiometric dates of coral now seem unduly conservative. The ages and altitudes of high stands of sea level must be approximately correct in order for numerous correlations to be made between marine-terrace flights situated on distant coastlines.

Tectonic uplift, instead of being regarded as a confounding overprint, should be regarded as useful. Uplift is responsible for marine terraces being above present sea level, and different rates of uniform uplift provide a means of crosschecking correlations of terrace flights that are either adjacent to one another or separated by thousands of kilometers. All local flights of marine terraces should be regarded as possibly correlative with the global population of marine terraces.

It appears that uniform uplift is a widespread phenomenon for time spans of 50 to 300 ka. Therefore marine-terrace correlations should be a standard tool for tectonic geomorphologists who are curious about uplift rates associated with specific coastal landscapes and geologic structures, and who would like to know more about uniform or changing styles of vertical deformation along plate boundaries.

**Acknowledgements** -- The results of my work as presented in this article have benefited substantially from discussions and suggestions provided by Art Bloom, John Chappell, Alan Cooper, Bob Kirk, Peter Knuepfer, Ken Lajoie, and Yoko Ota. This research was supported by grants from the National Science Foundation (EAR-815836 and EAR-8305892) and U.S. Geological Survey contract 14-08-0001-21882.

## REFERENCES CITED

Berryman, Kelvin, and Hull, Alan, 1984, International symposium on recent crustal movements of the Pacific region, Guidebook to the North Island Scientific Excursions, Royal Society of New Zealand, 115 p.

Bloom, A.L., 1980, Late Quaternary sea level change on South Pacific coasts; a study in tectonic diversity: in Earth rheology, isostasy and eustasy (N.A. Mœrner, editor), Wiley Chichester, p. 505-516.

Bloom, A.L., Broecker, W.S., Chappell, J.M.A., Matthews, R.K., and Mesolella, K.J., 1974, Quaternary sea level fluctuations on a tectonic coast: New $^{230}Th/^{234}U$ dates from the Huon Peninsula, New Guinea: Quaternary Research, v. 4, p. 185-205.

Bull, W.B., 1979, The threshold of critical power in streams: Geological Society of America Bulletin, v. 90, p. 453-464.

Bull, W.B., and Cooper, A.L., in preparation, Uplifted marine terraces along the Alpine fault, New Zealand.

Butzer, K.W., 1983, Global sea level stratigraphy: An appraisal: Quaternary Science Reviews, v. 2, p. 1-15.

Chappell, John, 1974a, Geology of coral terraces on Huon Peninsula, New Guinea: Geological Society of America Bulletin, v. 85, p. 553-570.

Chappell, John, 1974b, Late Quaternary glacio- and hydro-isostasy on a layered earth: Quaternary Research, v. 4, p. 429-440.

Chappell, John, 1983, A revised sea-level record for the last 300,000 years on Papua New Guinea: Search, v. 14, p. 99-101.

Clark, J.A., Farrell, W.E., and Peltier, W.R., 1978, Global changes in post-glacial sea level: A numerical calculation: Quaternary Research, v. 9, p. 265-287.

Cronin, T.M., 1983, Rapid sea level and climate change: Evidence from continental and island margins: Quaternary Science Reviews, v. 1, p. 177-214.

Fairbanks, R.G., and Matthews, R.K., 1978, The marine oxygen isotope record in Pleistocene coral, Barbados, West Indies: Quaternary Research, v. 10, p. 181-196.

Fairbridge, R.W., 1961, Eustatic changes in sea level: Physics and Chemistry of the Earth, v. 5, p. 99-185.

Kidson, C., 1982, Sea level changes in the Holocene: Quaternary Science Reviews, v. 1, p. 121-151.

Kirk, R.M., 1977, Rates and forms of erosion on intertidal platforms at Kaikoura Peninsula, South Island, New Zealand: New Zealand Journal of Geology and Geophysics, v. 20, p. 571-613.

Ku, T.-L., Kimmel, M.A., Easton, W.H., and O'Neil, T.J., 1974, Eustatic sea level 120,000 years ago on Oahu, Hawaii: Science, v. 183, p. 959-961.

Lajoie, K.R., Kern, J.P., Wehmiller, J.F., Kennedy, G.L., Mathieson, S.A., Sarna-Wojcicki, A.M., Yerkes, R.F., and McCrory, P.F., 1979, Quaternary marine shorelines and crustal deformation, San Diego to Santa Barbara, California: in Geological excursions in the southern California area (P.L. Abbott, editor), San Diego State University, San Diego, California, p. 3-15.

Machida, Hiroshi, 1975, Pleistocene sea level of South Kanto, Japan, analyzed by tephrachronology: in Quaternary Studies (R.P. Suggate and M.M. Cresswell, editors), Royal Society of New Zealand Bulletin 13, p. 215-222.

Marsh, J.G., Matin, T.V., and McCarthy, J.J., 1982, Global mean sea surface computation using GEOS 3 altimeter data: Journal of Geophysical Research, v. 87, p. 10,955-10,964.

McAdoo, D.C., and Martin, C.F., 1984, Seasat observations of lithospheric flexure seaward of trenches: Journal of Geophysical Research, v. 89, p. 3201-3210.

Miyoshi, M., 1983, Estimated ages of late Pleistocene marine terraces in Japan, deduced from uplift rate: Geographic Review of Japan, v. 56, 819-834.

Mörner, N.A., 1976, Eustasy and geoid changes: Journal of Geology, v. 84, p. 123-151.

Ota, Yoko, Kaizuka, Sohei, Kiluchi, Takao, and Naito, Hiroo, 1968, Correlation between heights of younger and older shorelines for estimating rates and regional differences of crustal movements: The Quaternary Research, v. 7, p. 171-181.

Pillans, B., 1983, Upper Quaternary marine terrace chronology and deformation, South Taranaki, New Zealand: Geology, v. 11, p. 292-297.

Shackleton, N.J., and Opdyke, N.D., 1973, Oxygen isotope and palaeomagnetic stratigraphy of equatorial Pacific core V28-238: Oxygen isotope temperatures and ice volumes on a $10^5$ year and $10^6$ year scale: Quaternary Research, v. 3, p. 39-55.

Stearns, C.E., 1976, Estimates of the position of sea level between 140,000 and 75,000 years ago: Quaternary Research, v. 6, p. 445-449.

Veeh, H.H., 1966, $^{230}Th/^{238}U$ and $^{234}U/^{238}U$ ages of Pleistocene high sea level stand: Journal of Geophysical Research, v. 71, p. 3379-3386.

Walcott, R.I., 1972, Past sea level eustasy, and deformation of the earth: Quaternary Research, v. 2, p. 1-14.

Yoshikawa, Torao, Ota, Yoko, Yonekura, Nobuyuki, Okada, Atsumasa, and Iso, Noyomi, 1980, Marine terraces and their tectonic deformation on the northeast coast of the North Island, New Zealand: Geographical Review of Japan, v. 53, p. 238-262.

# PART II

# 7
# Origin of drainage transverse to structures in orogens

*Theodore M. Oberlander*

## ABSTRACT

Streams transverse to the structures of orogens are commonly regarded as either antecedent to deformation or arbitrarily superimposed upon deformational structures from erosion surfaces covered by post-orogenic sediments. Such explanations are contradicted by successive changes in stream direction that appear to be controlled by transverse structures in the orogen. Major transverse streams frequently follow transverse structural highs ("transanticlines") and evolve by headward extension across the orogen during early periods of relief inversion. The key factor is the exposure of erodible masses of flysch-like sediments or schists, which are present in all orogens. Other mechanisms of transverse stream formation during the orogenic phase are effective on local scales. Areas discussed include the Zagros Mountains of Iran, the Himalayas, the Alps, and the Appalachian fold belt.

## INTRODUCTION

Large streams transverse to deformational structures are conspicuous geomorphic elements in orogens of all ages. Each such stream and each breached structure presents a geomorphic problem. However, the apparent absence of empirical evidence for the origin of such drainage generally limits comment upon it. Transverse streams in areas of Cenozoic deformation are routinely attributed to stream antecedence to structure; where older structures are involved the choice includes antecedence, stream superposition from an unidentified covermass, or headward stream extension in some unspecified manner. Whatever the choice, we are rarely provided with conclusive supporting arguments.

Two classes of problematical transverse streams are apparent: those rising near the topographic ridgepole of orogenic belts, and those rising beyond, or on the cratonic side of the belt of strong deformation, entering the latter in opposition to the gross structure of the entire orogen. Drainage from the ridgepole is inevitable, and becomes a geomorphic problem only as it breaches individual positive structures that could be avoided by an alternative path, whereas drainage transverse to the orogen as a whole is manifestly anomalous on the most general scale as well as in terms of local structures. The Himalayan region epitomizes the contrast, with eleven large streams collecting on the Tibetan Plateau or northern slopes of the Great Himalaya Range and subsequently crossing the full width of the range in deep gorges (Figure 1). Between these fully discordant streams are others, parallel to them, that rise on the southern slopes of the Great Himalaya Range and cleave the lesser buttressing ranges in gorges perpendicular to the deformational trend. The stream pattern in the Himalayas contrasts with that of the Andes, which are cut at low angles by smaller numbers of large streams, all of which rise within the cordillera and flow parallel to the strike throughout

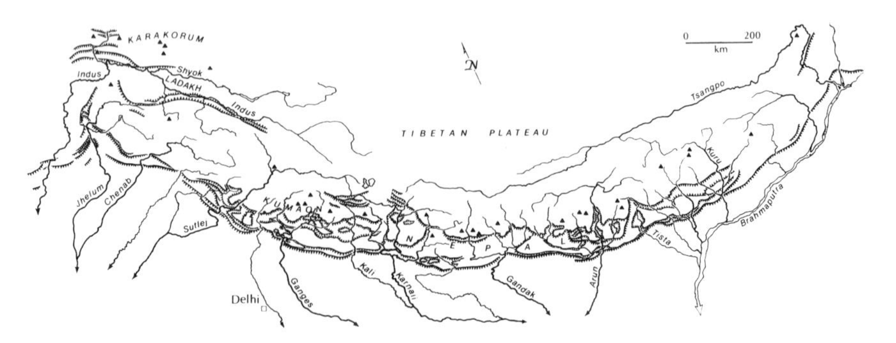


FIGURE 1. Major streams of the Himalayas. The axis of the range is indicated by the locations of the highest mountains (above 7,000 m), represented by black triangles. Margins of major overthrusts are indicated as escarpments, with hachures pointing toward overridden block. (Modified from Gansser, 1964.)

the larger part of their courses. Large areas within the Andean cordillera are blanketed with Tertiary volcanic rocks that could be expected to be a source of stream superposition on underlying structures. However, the influence of longitudinal deformational structures in guiding stream development appears to be much greater in the Andes than in the non-volcanic Himalayas, a fact that has evoked little comment from geomorphologists. Indeed, no comparative studies of stream formation in different orogens are known to this author.

Contrasting with still-active orogenic belts are those of greater age and less imposing relief, whose worn-down structures are also cut discordantly by large streams. In these instances the transecting streams often rise beyond the erosionally-reduced belt of strong deformation and cross it on their way from the associated craton to the sea. The best known example in North America is the portion of the Appalachian fold belt extending from Pennsylvania to Virginia, which stimulated many classic contributions concerning transverse streams (e.g., Davis, 1889; Johnson, 1931). A striking parallel is seen in the Cape Fold Belt of South Africa, where the Groot River is the analogue of the Susquehanna and Potomac rivers of the folded Appalachians. It is in areas of open folds, such as the Cape region and the Appalachian Ridge and Valley Province, that the discordance between streams and structure is most easily comprehended. In such areas individual anticlinal barriers and the patterns of structural culminations (anticlinoria) and depressions (synclinoria) are seen with much greater clarity than are the structures in areas of more powerful deformation, where overthrusting may intercalate marine sediments, melange complexes, and igneous and metamorphic rocks of both synorogenic and cratonic origin. In the Alps, Himalayas, or Northern Rockies we are aware of streams breaching ranges rather than structural units. Geologic maps of such regions reveal streams transverse to complex outcrop patterns, but in areas of very high relief interpretation of the details of deformational structure from mapped outcrops is not the straightforward task it is elsewhere. Accordingly, it is in the simple folds often associated with decollements and miogeosynclinal lithologies that drainage anomalies are most clearly expressed.

## THE HIMALAYAN EXAMPLE

Returning to the very ridgepole of the earth--the Himalaya Range--we find striking discordances between stream paths and structural barriers. Many observers have concurred that the transverse streams of the Himalayas are antecedent, originating on the rising Tibetan Platform prior to the main Himalayan orogeny (e.g., Wager, 1937; Gansser, 1964; Hagen, 1968; Wadia, 1968; Birot, 1970). However, Gansser (1964, p. 230), echoing Ludlow (1940), notes that the Indus and the Tsangpo (Brahmaputra) each cross the Himalaya exactly at the structural culminations--the syntaxes at the western and eastern ends of the range where the structures crowd together and pivot more than 90 degrees.

Bordet (1955) was the first to emphasize the presence of "transanticlines" crossing the structural trend in the eastern Himalayas, their axes coinciding with the Arun and Karnali rivers. The Arun transverse anticline, east of Mt. Everest, has an amplitude of some 10,000 m, into which the Arun River has incised some 15,000 meters. Birot (1970, p. 252) stated that most Himalayan streams follow transverse warps in the structure of the Himalayas, such as that of the Arun, but followed this by asserting erroneously that the other rivers of Nepal lie in the saddles between thrust arcs. Valdiya (1981) stressed the importance of recently formed transverse folds in the Kumaon Himalaya, with the antiforms best exposed in the valleys of transverse streams. All investigators indicate that such transanticlines postdate the formation of the overthrust structures of the range. (e.g., Mithal, 1968; Krishnaswamy, 1981). According to Krishnaswamy, "most streams in the Central Himalayas parallel the transverse folds", which are the latest products of deformation in the range. Gansser (1964, p. 249) while accepting the antecedence theory, simultaneously stated that the eastern Himalaya are crossed by transverse folds that "are also responsible for most of the major north-south-directed river gorges of eastern Nepal and Sikkim".

Unfortunately, neither Gansser nor Krishnaswamy specified _how_ these transverse warps influenced the river gorges, or whether the famous Arun example was the rule or the exception! One might assume that the Himalayan streams were located _between_ transverse antiforms, but, in fact, the earlier statements concerning the Arun and Karnali rivers and the Kumaon Himalaya, and observation of the general geological settings of the Tista, Kali Gandaki, and Sutlej, indicate that, where their relation to structure has been specified or is clear on geologic maps, the streams coincide with transverse uplifts. Gansser, Valdiya, and many others relate these cross-grained warps to the structural grain of the peninsular shield thrust under the Himalayan range, which suggests that transverse warping may be a general phenomenon in the Himalayas, possibly exerting a pervasive influence on the drainage of the orogen.

Thus, unless we accept the presence of gigantic "river anticlines" in the Himalaya, produced by off-loading as canyons are cut, we have a problem of reconciling so-called "antecedent" streams, that should conspicuously disregard structure, with an apparent stream preference for a particular structural environment. The factors convincing Gansser of stream antecedence in the Himalaya appear to have been the strong (12,000 m) Tertiary uplift of the range, continuing to the present at rates of as much as 5 mm per year, and the absence of convex profile breaks in the streams of the deep valleys, suggesting that they have been able to maintain uninterrupted profiles despite ongoing tectonic uplift (Gansser, 1964, p. 39; 1981, p. 103). Earlier, Wager (1937) used the convex upward generalized profiles of the Arun and Tista rivers as an argument for stream antecedence and against headward stream extension across the range. Birot felt that stream antecedence was demonstrated by tilted lacustrine deposits on the Tibetan side of the Himalayas, presumed to be the result of drainage ponded by Himalayan uplift, and by the fact that the Kali Gandaki between Annapurna and Dauligiri splits the center of a major thrust arc--a structural high.

## ANALYSIS OF TRANSVERSE DRAINAGE

Is there any evidence of the way (or ways) streams become transverse to fold structures, and to entire orogenic belts? In the following it is suggested that certain stratigraphic influences on stream formation in orogens have been ignored; that transverse stream formation has occurred so recently in certain orogens that the specific formative processes remain evident; that elsewhere visible indications of the origins of different types of transverse streams persist; and finally, that the characteristics of some transverse streams provide useful and generally neglected information concerning local tectonic history.

## Stream Behavior Entering Transverse Reach

Certain transecting streams attack structural barriers straight-on (e.g., the Danube at the Iron Gate, the Columbia entering the Cascades, the Colorado at the Gore Range); others follow a longitudinal path, then turn abruptly to cleave an adjacent uplift (the Indus and Tsangpo-Brahmaputra, the Rhine and Rhone in the Central Alps, the Alt in the Transylvanian Alps of Rumania, Utah's Sevier River at the Canyon Range); and still others move toward a barrier, are deflected parallel to it, and only then turn to split the barrier (the Green River at Lodore Canyon, the Rhine at the Taunus). It seems axiomatic that where a stream is fully antecedent to a structural barrier, it should cleave the barrier without departing significantly from the line of its approach to the barrier--that is, without obtrusive deflection of the stream course. Any significant deflection at a structural barrier obviously signifies structural influence, negating the possibility of stream antecedence to structure. The same caveat would hold for regional stream superposition. In fact, drainage systems commonly regarded as antecedent to structure, or superimposed on it, such as those of the Himalayas or central Appalachians, do repeatedly alter direction at structural boundaries.

## Relevant Stratigraphic Components of Orogens

The most obtrusive expression of discord between geologic structure and stream location is in regions of folded and overthrust miogeosynclinal and eugeosynclinal sedimentary rocks in continental-margin orogens, present and past. In analyzing drainage evolution in continental-margin orogens it is helpful to consider the contrasting deformational and erosional styles of broad groups of sedimentary units utilizing the concepts of competent, mobile, and passive stratigraphic groups originally proposed by O'Brien (1950) and applied in detail by Dunnington

(1968).

Most orogenic belts display thick carbonate sequences formed in miogeosynclinal settings on continental shelves. In the global continental-margin geosynclines of both Paleozoic and Mesozoic time, carbonates accumulated to thicknesses sometimes exceeding 10,000 m. In areas of crustal plate convergence these carbonate sequences folded harmonically and were competent to transmit compressive stress across the width of the orogen. Marine carbonates constitute the principal competent group in most geosynclinal prisms. Carbonate rocks of the competent group form most positive relief features throughout the Eurasian mountain system from the Pyrenees to eastern Asia, creating the majority of the mountain ranges that are breached by transverse streams.

Within the mass of the competent group of carbonate rocks there are likely to be clastic interbeds of deep water turbidites or flysch sequences of thin-bedded sandstones, shales, and marls, as well as shallow-water lagoonal sediments of similar character, possibly also including evaporites (salt and gypsum). Where these clastics are sandwiched between much thicker masses of miogeosynclinal carbonates, they do not affect fold geometries significantly. However, these marine clastics are often incompetent to transmit compressive stress, and where they are present in thicknesses of thousands of meters, as is frequently the case, and especially where they contain important evaporite sequences, their internal failure under compression produces highly intricate structures that are disharmonic to those of the competent group. Thus thick accumulations of eugeosynclinal turbidites, flysch, and lagoonal sequences containing evaporites, can be regarded as a "mobile group".

Where the mobile group is well developed and overlies the competent group, as is normally the case, the pattern of folding at depth may not be carried through to the surface. The mobile group tends to become attenuated over the crests of rising anticlines in the competent group by tectonic creep into the adjacent synclines, producing an independent deformational pattern in the process.

Strong compressive deformation in an emerging orogen commonly causes a thick mass of lagoonal facies of the mobile group to be capped by orogenic sandstones and conglomerates, such as the alpine molasse and the Siwalik beds of the Sub-Himalayas. These coarse clastics flood into depressions in the deforming surface of the mobile group, and are themselves deformed as orogeny continues. This clastic mass can have enormous primary thickness variations reflecting the amplitude of the deformation of subjacent rocks. Since the coarse clastics ride passively on the deforming mobile group and are usually separated from the deeper competent group (except where overlying the latter along erosional unconformities), the coarse clastics have been regarded by Dunnington and O'Brien as a "passive group". The passive group commonly forms a zone of young anticlinal and synclinal folds on the margins of greatly upheaved orogens as in the cases of the alpine molasse and the Siwalik Hills.

The point to be stressed is that in the typical continental-margin orogen the initial fluvial systems form on the surfaces of the mobile and passive groups. However, in all but the youngest of orogens, the structures most conspicuously breached by anomalous transverse streams are those formed at greater depth by deformation of the competent group or by upheaval and overthrusting of plutonic and metamorphic rocks. In most existing orogens, the passive group and the underlying lagoonal facies of the mobile group have been removed by erosion during broad uplift in the "morphogenetic phase" of orogenic development, being preserved only in the present foothill zone and occasionally in interior synclinoria or individual synclinal mountains. As such, the influence of these stratigraphic units on drainage formation has not received the attention it merits.

## The Zagros Example

The roles of the passive, mobile, and competent stratigraphic groups in the evolution of structurally-discordant drainage is nowhere clearer than in the Zagros Mountains of western Iran (Figure 2), where the principal deformation began in Pliocene time, and continues to the present (Berberian, 1976, 1981). Although the Zagros Range is one of the earth's youngest orogenic develop-

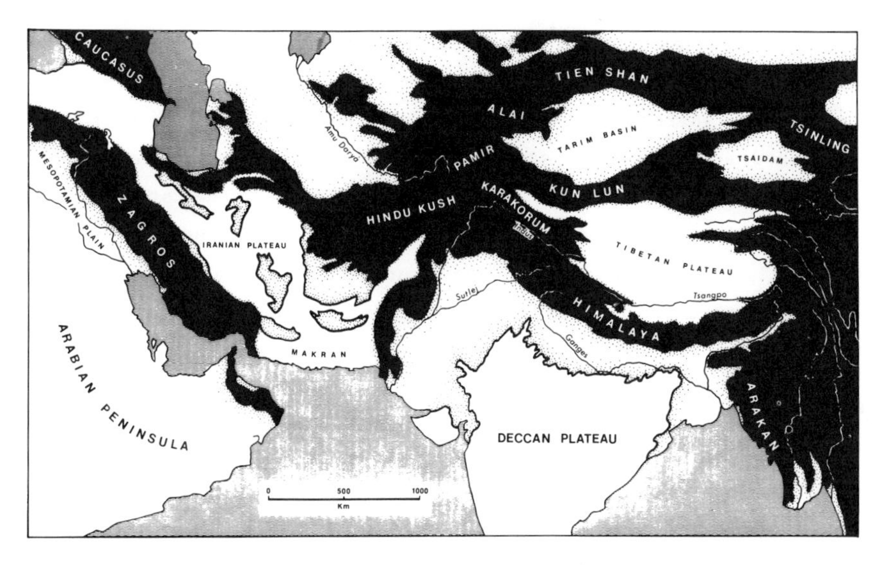


FIGURE 2. Mountains of central and western Asia, showing the setting of the Zagros Range in relation to surrounding structural elements.

ments, it exhibits as great a discord between drainage pattern and geological structure as can be seen in any orogen regardless of age or character (Oberlander, 1965). The Zagros orogen is dominated by folded and faulted Mesozoic and Tertiary miogeosynclinal carbonate rocks and flysch-type clastics, capped by a great depth of Mio-Pliocene neritic beds including evaporites, with a Pliocene blanket of molasse-like coarse clastics reflecting the onset of strong deformation in the orogen. The outer ranges of the Zagros orogen are simple anticlines that gain amplitude and elevation moving eastward to a zone of broken, imbricated folds, succeeded still farther to the east by large-displacement thrust faults, arguably including overthrusts. Various views of Zagros tectonics are found in Falcon (1969, 1974), Stocklin (1977), Alavi (1980), and Jackson et al. (1981). The stratigraphy of the Iranian Zagros is described in detail by James and Wynd (1965).

The morphology of the Outer Zagros consists of textbook examples of the progressive erosion of folds, including completely intact anticlines of Miocene Asmari limestone, partially unroofed anticlines, large anticlinal basins in Cretaceous-Eocene flysch-like clastics, anticlinal mountains resurrected on resistant Mesozoic carbonates encircled by Asmari limestone hogbacks, and seas of ridges and peaks composed entirely of the resistant Mesozoic limestone series. All portions of the Zagros fold belt are breached by transverse streams, which vary greatly in catchment area, discharge, degree of penetration across the strike, and course geometry. Some cut obliquely across the structural grain, some zig-zag their way to the Mountain Front, and some cleave straight through bundles of closely-packed anticlinal mountains. As recently as 1970, Birot dismissed these streams as "sans doute" antecedent to the folding (Birot, 1970, p. 240).

The Zagros Range presents an unusual opportunity to analyze the origin of transverse streams because the geomorphic landscapes in different portions of this extremely young fold belt have evolved to varying stages related to the relative intensities of uplift from place to place within the range. In the less powerfully uplifted portions of the range the earliest phases of transverse stream formation are still in progress. This makes possible an ergodic (space-for-time) mode of analysis, which exposes the importance of stratigraphy in the formation of transverse drainage in emerging orogens.

## OBSERVED MODES OF TRANSVERSE DRAINAGE FORMATION

Four different modes of transverse drainage formation can be identified clearly in the Zagros fold belt (Figure 3). There is no evidence of drainage antecedence to the entire fold belt, nor of stream superposition from a covermass unconformable on folds truncated by an erosion surface--the two explanations most commonly invoked to explain anomalous transverse streams.

### Role of the Mobile and Passive Stratigraphic Groups

The most conspicuous process causing streams to transect anticlinal structures in the Zagros fold belt is drainage superposition from the surface of the depositionally conformable but disharmonically folded Mio-Pliocene mobile and passive groups. These form a blanket as much as 15,000 m thick over the competent Miocene limestones that create the outer ranges of the Zagros fold belt (Figure 4, left side of sections). The surface relief of

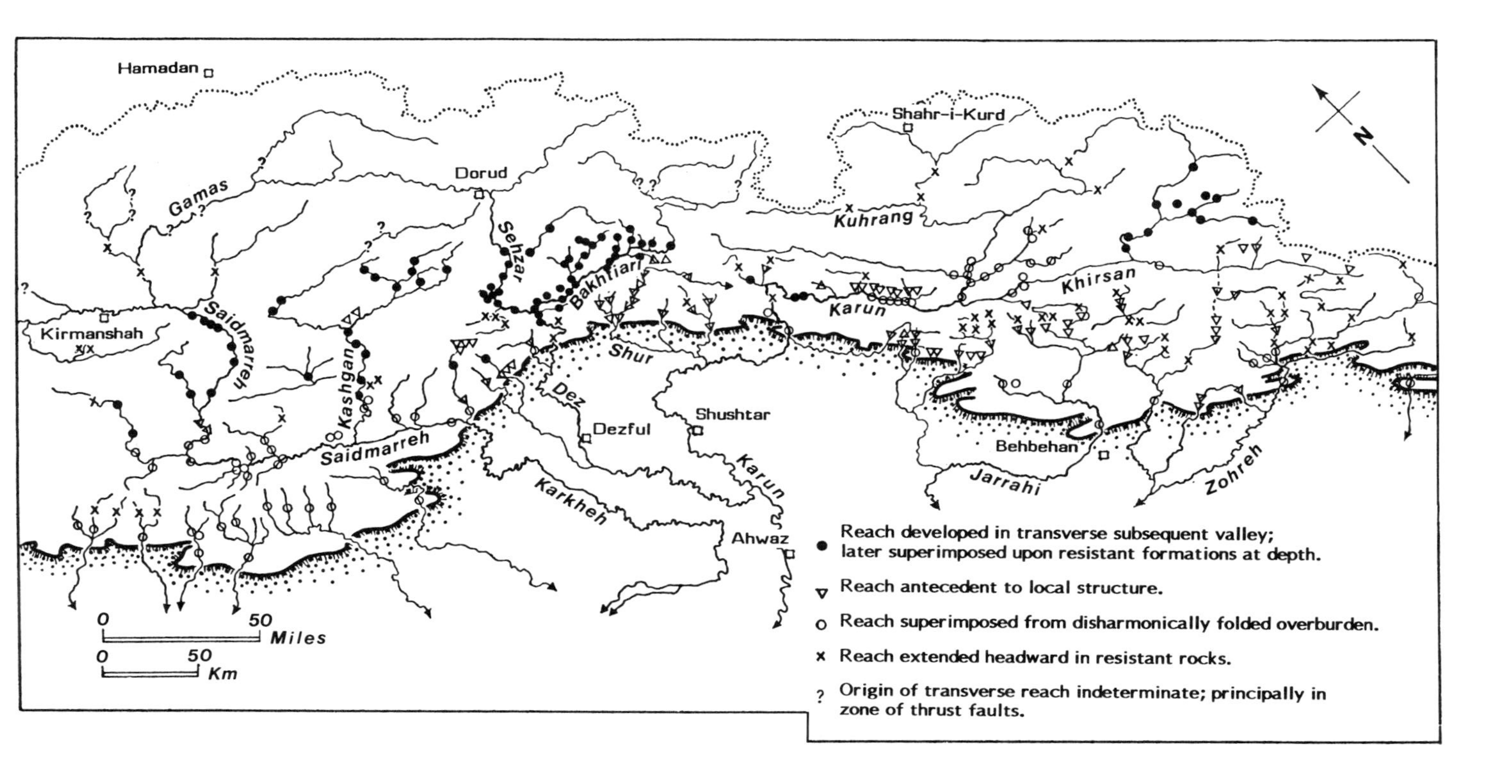


FIGURE 3. Most probable origins of stream reaches transverse to anticlinal structures of the central Zagros fold belt.

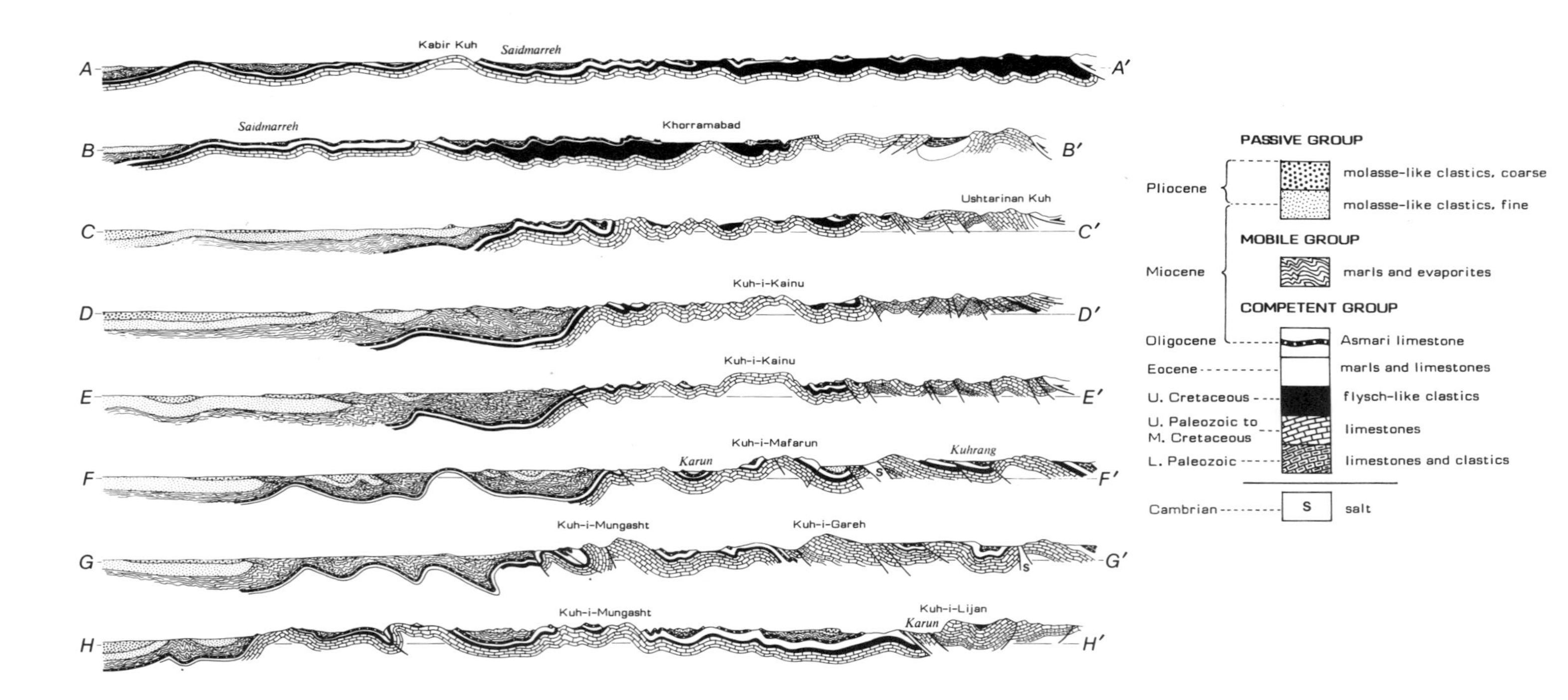


**FIGURE 4**. Cross sections of the fold and imbricate zones in the central Zagros Mountains. Vertical and horizontal scales are equal; length of each section is about 150 km. (Adapted from British Petroleum Company, 1956.)

these superjacent beds consists of badland terrain and fluvial straths in the mobile group, overlooked by mesas of passive group conglomerates. In no way does this relief express the giant anticlinal and synclinal structures present beneath it. Hence, the producing structures of the Iranian oilfields have no surface expression. Only in the youngest of orogens will this initial cover be preserved and its active role in the creation of drainage anomalies be evident.

In the outer portion of the fold belt the mobile mass effectively separated stream evolution and the configuration of the land surface from the fold structures forming in the competent group below. Streams cutting through the badland topography of the mobile group are even now "finding" the limestone crests of competent group anticlines, and are either incising the anticlinal crests (Figure 5) or migrating laterally down the anticlinal plunge by paring back the superjacent series. The latter process

FIGURE 5. Initial exposure of an anticline in the competent group below the mobile group, with superposition of drainage onto the Asmari limestone fold carapace. The anticline at the right has been unroofed as a consequence of earlier stream superposition from the mobile group, exposing the erodible Cretaceous-Eocene flysch-like clastics beneath the Asmari caprock. (Aerofilms and Aero Pictorial, Ltd.)

seems to be limited by the height of the wall of the mobile group created in the process of down-plunge migration. Lateral shifting of a superposed stream appears to give way to vertical incision when a wall of some 200 m of the incompetent beds has been created. Presumably this occurs as a consequence of repeated obstruction of the stream by mass movements triggered by the undercutting of the incompetent mass.

Transverse streams that have been superposed from both the mobile and passive stratigraphic groups originate on the higher exposed competent group anticlines and cleave straight through adjacent lower anticlines. The transections may be closely spaced, thoroughly segmenting the adjacent arch. The transecting streams have a pattern that appears to be consequent upon the higher portions of the large primary folds with their thin cover of clastic sediments, and arbitrary with respect to all formerly buried structures of lesser magnitude, whether secondary folds or the lower portions of the primary folds. Particularly clear signs of superposition occur in the form of synclinal streams that occasionally loop into the flanks of the larger anticlines in profound gorges and then return to the open syncline.

With continued uplift, all of the erodible disharmonically-folded mobile and passive groups will be removed. In their absence, surviving transverse streams superimposed from them might well be attributed to drainage antecedence to folding or stream superposition from an unconformable cover. The result is misinterpretation of the tectonic and erosional history of the orogen. It must be stressed that the mobile group, though deformed disharmonically, is conformable on the competent group. Although the larger anticlines of the competent group may have begun to form during deposition of the mobile group, the former were deeply buried and no period of erosion has interposed between the deposition of the two masses.

### Stream Antecedence to Structure

The question of antecedent streams in the Zagros fold belt is entangled with the phenomenon of stream superposition from the mobile group. The more advanced state of erosion of the larger ranges suggests that they are indeed older than their lower, less-eroded neighbors, and that the drainage is consequent on an early deformational pattern and antecedent to later structures. Alternatively, these relative states of preservation could reflect the varying spans of time different anticlines of the same age have been exposed to denudation after the stripping of the mobile group, which attained greater depth over the lesser anticlines. On the basis of morphology alone, a clear distinction between stream superposition and local stream antecedence may not be possible. However, stratigraphic studies indicate that primary thinning of the mobile group occurred at the crests of the largest anticlines of the competent group (Allison and Slinger, 1948, Berberian, 1976). This indicates that the largest structures began to form well before the main Pliocene folding. In addition, the mobile group thins in the inner portion of the fold belt, so that the molasse-like coarse orogenic clastics of the passive group rest unconformably on the uppermost limestones of the competent group. In these areas one continues to find transections of lesser satellite folds by undeflected low order streams rising on the giant anticlines. The latter are not cut by transverse streams.

All stream reaches that could be interpreted as antecedent in this extremely young mountain complex are quite short, and no extensive portions of any of the present major streams of the fold belt could be antecedent, as all repeatedly reflect structural controls. This is true despite the ideal nature of the Zagros orogen for the development of antecedent streams. Here, as in the Himalayas, the range protaxis (earliest deformational structures) and the zone of greatest tectonic dislocation are in the interior, close to or coinciding with the sources of the major transverse streams. In both the Zagros and the Himalayas the zone of deformation has expanded over time (and is doing so still) toward a foreland in the same direction as that followed by the present drainage. By contrast, in the Pennsylvania Appalachians, the protaxial zone and belt of most violent disturbance is in the present downstream area, with the main drainage divide in the gently warped foreland--a fact that as early as 1889 convinced Davis that stream antecedence was not an appropriate explanation for drainage anomalies in the northern portion of the Appalachian fold belt (Davis, 1889).

## Discordant Superposition of Transverse Subsequent Streams

The most striking discord between geological structure and the stream pattern in the Zagros Range is seen in the strongly uplifted anticlinorial portion of the fold belt dominated by tightly compressed en echelon anticlines of Mesozoic limestone (Figure 6). The gorges of this anticlinorium differ from those created by local stream antecedence or superposition from the clastic blanket and mobile group in that each successive anticlinal transection is approximately normal to the fold axis, but offset from the transection of the adjacent anticline. The successive offsets in their gorges through individual anticlines cause the courses of the transverse streams (the Sehzar and Bakhtiari rivers) to parallel closely the transverse structure of the fold belt created by clusters of en echelon anticlines. The principal streams are influenced in a consistent way by the patterns of the folds through which they cut, being drawn toward the successive points at which adjacent anticlines press closest to

FIGURE 6. Anticlinal mountains of Mesozoic limestones in the Bakhtiari anticlinorium. Synclinal mountain at right capped by Lower Miocene Asmari limestone. Dark beds are the upper portion of the Cretaceous-Eocene flysch-like clastics. (Aerofilms and Aero Pictorial, Ltd.)

one another (Figure 7). In the synclines these streams repeatedly flow up the plunge to remain amid the oldest exposed rocks of the competent group, although more erodible flysch-like materials are not far distant down the plunge.

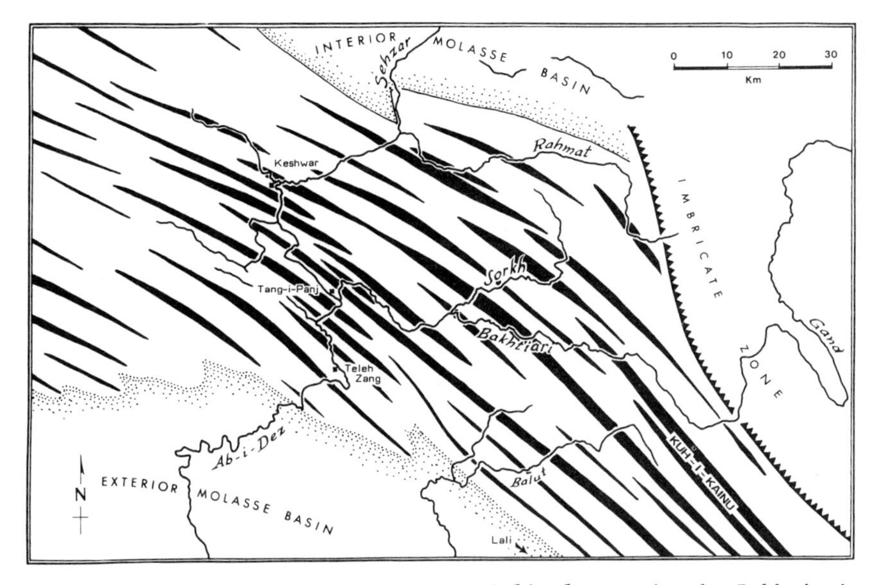


FIGURE 7. Relation of streams to anticlinal axes in the Bakhtiari anticlinorium. The Sehzar-Bakhtiari confluence is situated at the point of maximum upheaval (oldest rocks exposed) where thick flysch-like interbeds formerly produced a large erosional basin. To the southeast, where the flysch thins, there is no association between structural highs and transverse streams.

While this relationship makes no sense in the present landscape, in which these streams appear to seek the most difficult path, the relationship is easily explained if we imagine the prior landscape in this region. The Mesozoic limestone fold "cores" were exposed by erosion through the uppermost (Asmari) limestone of the competent group and a thick blanket of flysch-like Upper Cretaceous-Eocene shales and marls above the Mesozoic limestones. If we go back in time, and undo a thousand meters of denudation, the exact areas now occupied by the resistant Mesozoic limestones would have exposed only the highly erodible Cretaceous-Eocene flysch-like clastics. In the present landscape these clastics form lowlands between facing scarps of the capping Asmari limestone or between Asmari limestone homoclines and the

Mesozoic limestones. Today's anticlinal mountains were preceded in time by anticlinal valleys in the flysch-like clastics, which were preceded in turn by anticlinal mountains with Asmari limestone carapaces. Where the flysch-like clastics were sufficiently thick, the depressions they created merged into transverse "subsequent" lowlands with cuspate margins (Figure 8). Continuous outcrops of erodible beds extending across the fold belt to the High Zagros (the zone of thrust faults) could hardly escape being occupied by transverse subsequent streams. As the anticlinorium rose, these streams, reaching across the entire fold belt, were automatically superimposed upon the deeper Mesozoic rocks that presently resurrect anticlinal mountains where basins formerly existed (Figure 8).

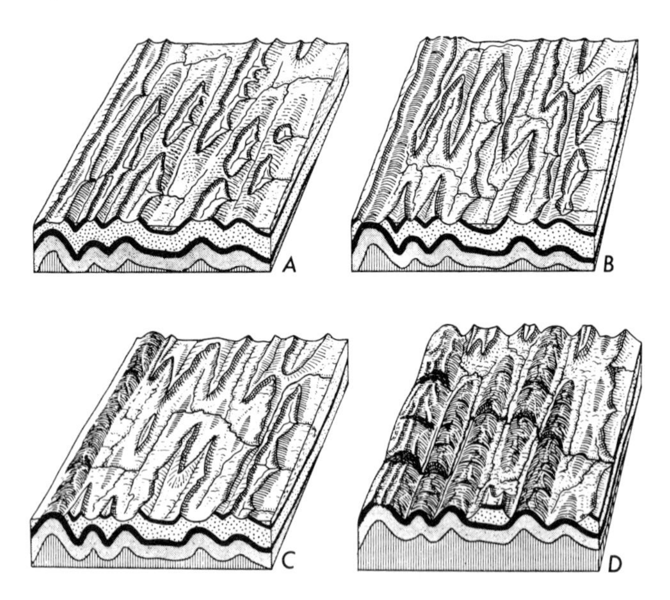


FIGURE 8. Model of drainage evolution in open folds in which the depth of erodible interbeds between competent formations exceeds fold amplitudes.

a) Initial erosion of anticlines exposes erodible interbeds, so that axial basins begin to expand.

b) Expanding axial basins begin to merge, creating outcrops of erodible materials extending across the strike.

c) Through-flowing transverse subsequent streams occupy continuous outcrops of erodible interbeds between resistant formations, and are superposed upon lower competent formations emerging in anticlinal cores.

d) Anticlinal mountains resurrected on lower resistant formation are breached by superposed transverse subsequent streams.

It is instructive to note that through-flowing transverse drainage developed only where the flysch-like clastics were thick enough to permit adjacent anticlinal basins to merge before the underlying Mesozoic limestones were exposed in their centers. Earlier emergence of the resistant Mesozoic carbonates precludes the formation of transverse subsequent streams. For transverse subsequent streams to extend across the entirety of a fold belt, the thickness of the erodible interbeds between massive resistant groups must exceed the amplitude of the folding, as in Figure 4, A-A', B-B', where a continuous planar surface could develop in the erodible mass between the Asmari and Mesozoic limestones, and where future transverse subsequent stream development can hardly fail to occur.

The principle to be derived from this aspect of the Zagros drainage is the importance, after the initial stage of uplift and erosion, of _past_ outcrop patterns in controlling stream formation, and the relevance of the present geological pattern only insofar as it suggests prior patterns that could influence drainage evolution. What is conspicuous in the Zagros orogen is the uncanny homing instinct of the larger streams, which "find" all the anticlines they can, while open structural spillways often gape on either side. Could it be mere coincidence that this also appears true of the streams in different regions of the Himalayas noted by Bordet, Gansser, Birot, and Valdiya? Or is there a general tendency for streams in fold belts (including folded overthrusts) to be attracted to cross-grained structural highs? In the Zagros fold belt the latter seems to be true wherever the clastic blanket is thick enough to permit open anticlinal basins to merge on a large scale at some stage in the geomorphic evolution of the range.

## Headward Extension of Drainage on Fold Limbs

In the Zagros region scattered anticlinal transections by headward stream extension are also evident. The initial incision of limestone fold envelopes in the Zagros is facilitated by the intersecting shear fractures that are conspicuous on intact anticlines even at the scale of Landsat images. These oblique fractures, resulting from compressive stress, control the formation of flatirons as the folds are denuded. They also assist in the enlargement of anticlinal basins where the erodible flysch-like clastic interbeds are present. By contrast, antecedent and superposed streams, identified by their undeflected approach to their gorges, ignore these fractures, which occasionally create rather prominent gashes oblique to fold axes. Similar fractures are probably present in the upper (less confined) competent layers of most folded sequences, and are a factor abetting transverse stream development in the earliest stages of landscape evolution in fold belts.

In the Zagros and elsewhere, drainage that exploits these shear fractures eventually expose weaker substrates in which the original fracture control is no longer detectable. Where the flysch-like interbeds are present in depth, fracture-controlled channels rapidly extend across fold axes in the erodible beds below fold envelopes. But even where resistant Mesozoic limestones are exposed not far below the anticlinal carapace, the persistence of shear fractures in the folded limestones assists erosional defiles to work headward to the synclines on the far side. Captured synclinal streams turn sharply into gorges resulting from headward stream extension, distinguishing such transections from those of undeflected antecedent or superposed drainage. Headward extension of defiles in resistant Mesozoic limestones is a source of scattered anticlinal transections by low order streams, and does not appear to be a factor in the location of large streams or continuous gorge systems.

### Permanence of Superposed Transverse Subsequent Streams

If the principal streams of the Zagros were determined by erodibility variations within the fold belt at a particular point in time, it might well be objected that the drainage pattern should continue to adjust to the migrating outcrop patterns that characterize any orogen. Had the uplift been too gradual to generate strong erosional relief, continual drainage adjustment could have occurred. But once certain channels become paramount *and deeply incised*, attracting all of the runoff from large areas within a strongly rising orogen, it is difficult, if not impossible, for lesser streams with smaller discharges and less competence to undercut and divert these channels.

In the Zagros, as in the Himalayas, the present rate of uplift is rapid and perhaps is increasing, as seen by unusual seismicity (Berberian, 1981), slot-like stream defiles with V-shaped upper stories, hanging tributary valleys, and the presence of flights of strath terraces on both sides of major streams, with older valley floors much wider than those of the present (Oberlander, 1965). In the Himalayas, the terraces themselves are warped to slopes of 5 to 10 degrees and include boulders much larger than those delivered by Plio-Pleistocene streams (Gansser, 1964, p. 175). Rates of anticlinal uplift measured on coastal anticlines in the southern Zagros are as much as 7.4 mm per year (Vita Finzi, 1981); on the western margin of the fold belt Lees (1955) and Lees and Falcon (1952) have reported anticlinal uplifts of up to 12 mm/yr. Thus the accelerating rate of uplift in active orogens may fix and perpetuate the drainage pattern originally developed in response to circumstances that may no longer be ascertainable.

## DISCUSSION

In the preceding we have considered the example of a young and actively expanding orogen crossed by streams that rise along the protaxis of the orogen and drain toward the foreland, a perfect relationship for the development of antecedent streams. Yet drainage antecedence (the Himalayan hypothesis) is precluded by the clear structural influence on the drainage pattern. The Zagros fold belt is far too young and too active to have experienced drainage superposition from a covermass over erosionally truncated folds (the Appalachian hypothesis); nevertheless, two types of stream superposition can be seen occurring in the Zagros at present, and explain both local anticlinal transections by lesser streams and the locations and specific patterns of the large through-flowing rivers. One superposing medium is a mass of stratigraphically conformable but mobile and disharmonically-folded sediments covering the uppermost competent marine formation in the fold belt, and the second is the mass of conformable interbeds between the uppermost competent formation and the deeper mass of Mesozoic limestones. The first results in arbitrary stream superposition and both direct and deflected anticlinal transections. The second effects automatic superposition of through-flowing streams onto resistant rocks at structural highs, creating defiles that slant or zigzag across the fold belt in strict accordance with transverse structural culminations created by groups of en echelon anticlines.

These two superposing media are not unusual, stratigraphically. Thick masses of erodible turbidites or flysch-like deposits are a feature of all continental-margin orogens, being the inevitable product of weak tectonism in or adjacent to geosynclines while marine conditions continue. Wherever it develops the flysch is likely to be capped by marine limestones or orogenic sandstones or conglomerates: competent, resistant masses that even under passive deformation will form anticlinal folds. Later unroofing of these will lead to relief inversion, and encouragement of subsequent valleys in the flysch, possibly following transverse highs, allowing drainage to find its way across the deformational zone in an early stage. These folds may subsequently be broken and overthrust, but by then the regional drainage could well be in place, embedded, and difficult to rearrange thereafter.

## Age and Origin of the Susquehanna River

Where open folds exist, the crucial antecedent outcrop patterns are clearly suggested by the present geologic map. A case in point is the classic example of the two branches of the Susquehanna in the Appalachian fold belt. Despite transecting ten anticlinal axes between Williamsport and Harrisburg, flowing against synclinal plunges, and oscillating in and out of single anticlinal structures, the two branches and main stem of the Susquehanna, in fact, have a close and consistent relation to geological structure. In the upstream areas they follow the structural slope or locate on erodible formations, and where the West Branch and the trunk channel are transverse to the structural grain they lie between--and conspicuously parallel to--the transverse axes of the major anticlinorium and synclinorium in the Pennsylvania fold belt (Figure 9).

FIGURE 9. Geologic setting of the Susquehanna River in the Appalachian fold belt. Anticlinorium exposing early Silurian and older rocks at left (crosshatched); synclinorium of Pennsylvanian-Permian rocks at right (dotted). Solid black indicates ridge-forming sandstones: St, Tuscarora (Silurian); Do, Oriskany (Devonian); Mp, Pocono (Mississippian); Pp, Pottsville (Pennsylvanian). Other formations: Su, Upper Silurian; Dho, Hamilton Group (Devonian); Dck, Catskill fm. (Devonian); Mmc, Mauch Chunk (Mississippian). (Adapted from Pennsylvania Topographic and Geologic Survey, 1960)

In the light of the Zagros example, the course of the Susquehanna suggests that this river may have evolved in a conformably folded erodible mass within the Paleozoic stratigraphic column. This mass was not thick enough to produce an uninterrupted strip across the nearby anticlinorium (unlike the Zagros streams) but produced a broad band between resistant outcrops on the east-plunging flank of the anticlinorium. If this interpretation is correct, the most appropriate mass for transverse stream formation would have been the Upper Mississippian Mauch Chunk formation, between the resistant Pottsville (L. Pennsylvanian) and Pocono (L. Mississippian) sandstones and conglomerates. At the present time these three formations outcrop mainly to the east of the Susquehanna, which is caught by superposition onto the Pocono and underlying formations (Figure 9). The question is whether the superposition was from the Mauch Chunk onto the intact Pocono crest or from an erosion surface truncating the Pocono and blanketed by an unconformable covermass as proposed by Johnson (1931).

The Mauch Chunk formation in central Pennsylvania is a 700- to 2300-meter-thick mass of shales, siltstones, and weakly cemented sandstones (Wood, 1969; Fail and Wells, 1974; Hoskins, 1976) producing the lowest elevations over large areas and rarely exhibiting more than 30 m of local relief. Resistant rocks stand 300 to 500 meters higher. The thickness of the Mauch Chunk formation exceeds local fold amplitudes in many areas, although not in the anticlinorium west of the Susquehanna River. The Susquehanna transects four homoclinal ridges of Pocono sandstone and one of Tuscarora sandstone north of Harrisburg. Stream superposition from the Mauch Chunk formation onto eastward plunging intact Pocono anticlines would require some 4,500 m of subsequent landscape lowering by erosion, making the Susquehanna River immensely older than hypothesized by Johnson, who proposed superposition from an unconformable covermass only some 400 m above present river level.

## Age and Origin of the Himalayan Streams

In the area of the Susquehanna, an ideal superposing medium exists in the appropriate position in the stratigraphic column to explain all present drainage anomalies (the Upper Devonian Catskill formation may assist at lower levels) and the transecting streams are not truly arbitrary with respect to structure, but bear a consistent relationship to it. However, the structurally-controlled superposition proposed above would make the streams of the area much older than previously believed. Conversely, either the transverse Himalayan streams are extremely old, predating the folding, thrusting, and broad uplift of the range, as in the traditional view; or, if controlled by the transverse anticlines noted by several researchers, they are very young, forming after the latest transversal phase of deformation of in situ overthrusts. The question seems to be one of considerable significance.

The relationship between structure and stream location in the Himalayas is appropriate for superposed transverse subsequent streams, given an adequately thick and erodible medium for the initial formation of such streams. Possibly the medium is the well known Indus Flysch of Upper Cretaceous age (e.g., Gansser, 1964, p. 75ff; Krishnaswamy, 1981). The Indus Flysch outcrops in a narrow longitudinal band north of the Himalayan uplift from Ladakh into Tibet, and localizes the upper portions of both the Indus and Tsangpo (Brahmaputra). However, where the flysch has not been subducted in the India-Tibet suture or tectonically overridden (e.g., Mitchell, 1984), it has been stripped from the great uplift to the south, leaving its original extent and thickness in the critical area unknown (Figure 10). Nevertheless, in the early history of the Himalayan region, before the underthrusting of the Indian plate and the obliteration of geosynclinal rocks south of Tibet, the flysch (still preserved in Ladakh) must have played its geomorphic role--possibly one quite similar to that in the much younger Zagros. There are other thick erodible masses in the Himalayan stratigraphy (e.g., Shah, 1978; Mitchell, 1984), their incompetence helping to generate the overthrusts of the range, causing them to be attenuated and overridden like the Indus Flysch. However, information concerning the structures of the region remains inadequate to relate these stratigraphic units to the problem of transverse stream formation. According to Fuchs (1982), the large windows through the overthrust complex and major

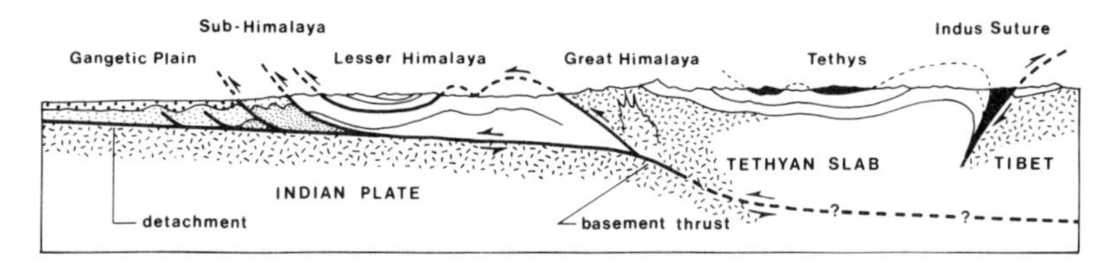


FIGURE 10. Cross section and tectonic interpretation of the Kumaon Himalayas based on surface geology and seismic evidence. Highest peak is Nanda Devi (7,820 m). Indus Flysch indicated in solid black. The Indus Suture is a former high angle subduction zone believed to have consumed most Tethyan eugeosynclinal sediments. Stippled formations at left are Siwalik molasse. Pre-Cambrian basement rocks indicated in granite pattern, intruded by Mesozoic plutons. Unpatterned layered formations are Paleozoic and Mesozoic miogeosynclinal sediments. (After Seeber et al., 1981; and Gansser, 1964.)

reentrants in thrust margins (Figure 1) are indeed determined by positive warps that are unroofed. Thus the large transverse streams that split these windows are structurally-controlled, and are not arbitrarily located.

### The Alps

The overthrust complex of the Alps is comparable to that of the Himalayas. The tectonic map of Switzerland (Spicher, 1976), which indicates the longitudinal plunges of deformational structures, reveals several streams that split the centers of major uplifts, among them the Rhone below Martigny, the Reuss above Lake Lucerne, and the Linth and Wallensee in the vicinity of Glarus. Birot's hypothesis of drainage antecedence for such relationships is denied by the many structural deflections of these streams. A more appropriate hypothesis, obviously requiring verification, is that they are structurally-controlled, and originated as transverse subsequent streams after the phases of compression and broad uplift, having been fixed in position by formerly extensive outcrops of flysch sediments or erodible schist masses. The same may be true in the Himalayas and many other mountain systems.

## CONCLUSION

The Zagros example and reconnaissance investigations of other regions of well developed transverse streams suggest that analyses of local stratigraphy and the relation of streams to the structures of orogens may well make it possible to explain transverse reaches without recourse to regional drainage antecedence or superposition from hypothetical unconformable covermasses. Such solutions are often invalidated by the nature of the stream pattern itself, which involves conspicuous structural influences. Alternative modes of transverse stream formation exist, including two forms of stream formation and later superposition from depositionally conformable units in the stratigraphic columns of orogens. Wherever analysis has been attempted, structurally-controlled lithological patterns, present or past, appear to have had a dominant influence on the location of the major transverse streams.

ACKNOWLEDGEMENTS

The author wishes to thank John Hack for his valuable suggestions regarding the orientation of this contribution, and Marie Morisawa for her expert editorial work on the manuscript.

REFERENCES

Alavi, M., 1980, Tectonostratigraphic evolution of the Zagrosides of Iran: Geology, v. 8, p. 144-149.

Berberian, M., 1976, An explanatory note on the first seismotectonic map of Iran; a seismo-tectonic review of the country: Geological Survey of Iran, Report no. 39, p. 7-142.

____________, 1981, Active faulting and tectonics of Iran: in H.K. Gupta and F.M. Delany, editors, Zagros-Hindu Kush-Himalaya Geotectonic Evolution: American Geophysical Union and Geological Society of America, p. 33-69.

Berberian, F., and Berberian, M., 1981, Tectono-plutonic episodes in Iran: in H.K. Gupta and F.M. Delany, editors, Zagros-Hindu Kush-Himalaya Geotectonic Evolution: American Geophysical Union and Geological Society of America, p. 5-32.

Birot, P., 1970, Les Regions Naturelles du Globe: Paris, Masson et Cie, 380 p.

Bordet, P., 1955, Les elements structuraux de l'Himalaya de l'Arun et la region de l'Everest: C. R. Acad. Sci. Paris, t. 240, p. 102-104.

British Petroleum Company, Ltd., 1956, Geological Maps and Sections of South-West Persia: London, E. Stanford.

Davis, W.M., 1889, The rivers and valleys of Pennsylvania: National Geographic Magazine, v. 1, p. 183-253.

Dunnington, H.V., 1968, Salt-tectonic features of northern Iraq: Geological Society of America, Special Paper 88, p. 183-227.

Faill, R.T., and Wells, R.B., 1974, Geology and mineral resources of the Millerstown quadrangle, Perry, Juniata, and Snyder counties, Pennsylvania: Pennsylvania Geological Survey, 4th series, Atlas 136, 276 p.

Falcon, N.L., 1969, Problems of the relationship between surface structure and deep displacements illustrated by the Zagros Range: in Time and Place in Orogeny, Geological Society of London Special Publication no. 3, p. 9-22.

____________, 1974, Southern Iran: Zagros Mountains: in A.M. Spencer, editor, Mesozoic-Cenozoic Orogenic Belts: Geological Society of London, p. 199-211.

Gansser, A., 1964, Geology of the Himalaya: New York: Wiley-Interscience, 289 p.

____________, 1981, The geodynamic history of the Himalaya: in H.K. Gupta and F.M. Delany, editors, Zagros-Hindu Kush-Himalaya Geotectonic Evolution: American Geophysical Union and Geological Society of America, p. 111-121.

Fuchs, G., 1982, Geologic-Tectonic Map of the Himalaya, 1:2,000,000: Vienna: Geological Society of Austria, 50 p.

Hagen, T., 1968, Report of the Geological Survey of Nepal, II, Geology of Thakkola, including adjacent areas: Mem. Soc. Helv. Sci. Nat., v. 86, 160 p.

Hoskins, D. M., 1976, Geology and mineral resources of the Millersburg 15-minute quadrangle, Dauphin, Juniata, Northumberland, Perry, and Snyder counties, Pennsylvania: Pennsylvania Geological Survey, 4th series, Atlas 146, 38 pp.

Jackson, J. A., Fitch, T. J., and McKenzie, D. P., 1981, Active tectonics and the evolution of the Zagros fold belt: Geological Society of London, Special Publication 9.

James, C. A., and Wynd, J. G., 1965, Stratigraphic nomenclature of the Iranian oil consortium agreement area: Bulletin of the American Association of Petroleum Geologists, v. 49, p. 2182-2245.

Johnson, D. W., 1931, Stream sculpture on the Atlantic slope: New York: Columbia University Press, 142 pp.

Krishnaswamy, V. S., 1981, Status report of the work carried out by the Geological Survey of India in the framework of the International Geodynamics Project: in Gupta, H. K., and Delany, F. M., editors, Zagros-Hindu Kush-Himalaya Geodynamic Evolution: American Geophysical Union and Geological Society of America, p. 169-188.

Ludlow, F., 1938, Takpo and Kongbo, S.E. Tibet: Himalayan Journal, v. 12, p. 1-16.

Mitchell, A. H. G., 1984, Post-Permian events in the Zangbo 'suture' zone, Tibet: Journal of the Geological Society of London, v. 181, p. 129-136.

Mithal, R. S., 1968, The physiographic and structural evolution of the Himalaya: in Law, B. C. editor, Mountains and Rivers of India: 21st International Geographical Congress, India, p. 41-81.

O'Brien, C. A. E., 1950, Tectonic problems of the oilfield belt of southwest Iran: Report of the 18th International Geological Congress, London, v. 6, p. 45-58.

Oberlander, T. M., 1965, The Zagros Streams: Syracuse University Geographical Series, no. 1, 168 pp.

Pennsylvania Topographic and Geologic Survey, 1960, Geologic Map of Pennsylvania, 1:250,000: Commonwealth of Pennsylvania, Department of Internal Affairs.

Seeber, L., Armbruster, J. G., and Quittmeyer, R. C., 1981, Seismicity and continental subduction in the Himalayan arc: in Gupta, H. K., and Delany, F. M., Zagros-Hindu Kush-Himalaya Geodynamic Evolution: American Geophysical Union and Geological Society of America, p. 215-242.

Shah, S. K., 1978, Facies pattern of Kashmir within the tectonic framework of the Himalaya: in Saklani, P. S., editor, Tectonic Geology of the Himalaya: New Delhi, p. 63-78.

Spicher, A., 1976, Carte tectonique de la Suisse, 1:500,000: Commission Geologique Suisse.

Stocklin, J., 1977, Structural correlation of the alpine ranges between Iran and Central Asia: Memoires, h. series, Societe de Geologie Francais, v. 8, p. 333-353.

Vita-Finzi, C., 1981, Late Quaternary deformation on the Makran coast of Iran: Zeitschrift fur Geomorphologie, Supplement Band 40, p. 213-226.

Wadia, D. N., 1968, The Himalayan Mountain; its origin and geographical relations: in Law, B.C. editor, Mountains and Rivers of India: 21st International Geographical Congress, India, p. 35-40.

Wager, L. R., 1937, The Arun River and the rise of the Himalaya: Geographical Journal, v. 89, p. 239-250.

Wood, G. H., Jr., Trexler, J. P., and Kehn, T. M., 1969, Geology of the west-central part of the southern anthracite field and adjoining areas, Pennsylvania: U. S. Geological Survey Professional Paper 602, 150 pp.

# 8
# Tectonic geomorphology of alluvial fans and mountain fronts near Ventura, California

*Thomas K. Rockwell, Edward A. Keller, and Donald L. Johnson*

ABSTRACT

Several alluvial fans in the Ventura area are being deformed by active folding and faulting. Basinward tilting beyond a fan slope threshold of about 6°-8° causes fanhead entrenchment and shifts the locus of deposition downfan. On the other hand, high rates of uplift (4-8 mm/yr) in the fan source area favors fanhead deposition. High rate of uplift of a mountain front, as for example along the San Cayetano thrust fault, also results in large quantities of debris being shed onto fans producing unusually large fans from small drainage basins. Fan tilting and rapid uplift of the mountain front results in complex cycles of fanhead deposition and entrenchment that can produce gross differences in fan morphology for adjacent fans along the same mountain front.

The relation between drainage basin area (Ad) and fan area ($A_f$) suggests two groups of fans in the Ventura area defined by the equations $A_f = 3.84\ Ad^{0.55}$ and $A_f = 0.59Ad^{0.8}$. The first equation characterizes fans associated with those mountain fronts having high rates of tectonic activity.

Mountain front sinuosity ($S_{mf}$) in the study area varies from 1.01 to 2.72. Low sinuosity, characteristic of tectonically active fronts, can hypothetically be maintained at a threshold rate of uplift of about 0.4 mm/yr in the Ventura area.

The ratio of valley width to height ($V_f$) in the Ventura area varies from 0.43 to 1.91. As with $S_{mf}$, $V_f$ is lowest for mountain fronts with high rates of uplift.

## INTRODUCTION

The Ventura basin is within the western Transverse Ranges geologic province, an east-west trending topographic and structural feature overprinted on the dominant, northwest-trending structural grain of California. The basin has a complex, geologic history characterized by Miocene extension (Yeats, 1968, 1971 and Crowell, 1976) and clockwise rotation of about 90° (Luyendyk and others, 1980) and Pliocene to present contraction. Recent study (Keller and others, 1982; Dembroff, 1983; Clark, 1982; and Rockwell, 1983) has yielded valuable information regarding rates and recency of faulting and folding on most of the major geologic structural elements within the basin. This paper will analyze the role and effects of this active tectonism on the geomorphology of alluvial fans and mountain fronts in the central Ventura basin.

An alluvial fan is a depositional landform that forms a segment of a cone radiating downslope from the point where a stream leaves the confines of a source area (Bull, 1977a). A bajada is a depositional piedmont formed by coalescing alluvial fans (Blackwelder, 1931). In the study area, many alluvial fans do not have the classical fan shape due to deformation of the piedmont and/or confinement of the deposits within an intermontane piedmont, but these fans have a convex cross-profile and most have segmented radial-profiles characteristic of many alluvial fans.

Late Pleistocene and Holocene chronology used in this paper was developed from soils work, $^{14}C$ dating and rates of faulting (see Table 1). The reader is directed to Keller and others (1982); Rockwell (1983); Rockwell and others (in preparation); and Johnson and others (in preparation) for further information.

## MOUNTAIN FRONTS, ALLUVIAL FANS, AND TECTONIC SETTING

Eight mountain fronts were delineated in the study area based on overall trend and continuity, spatial relationship to major drainages, and/or association with mapped faults or folds (Fig. 1). Alluvial fans are present along all of the fronts although the fans along Fronts 7 and 8 were not studied due to their relatively small size and lack of good topographic control.

Santa Paula Creek was used as a front boundary between Fronts 2 and 3 because of major lithologic source terraine changes

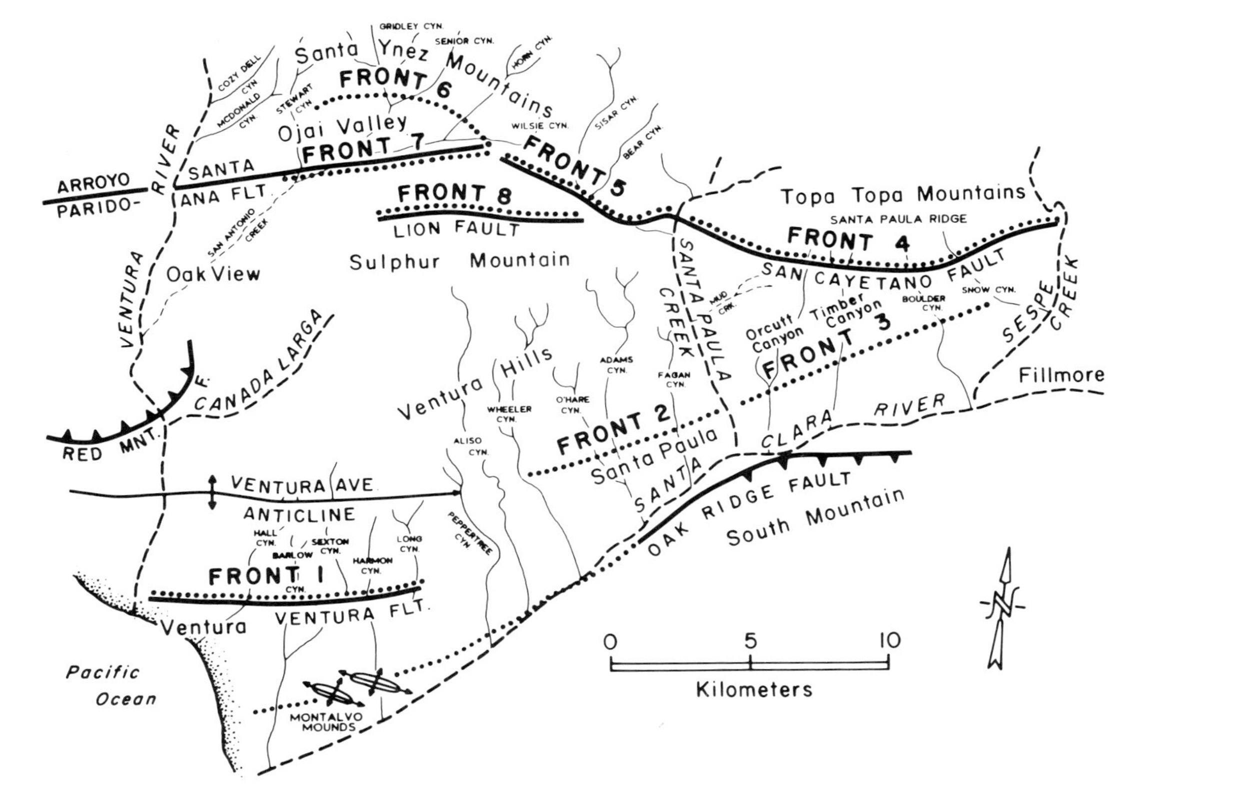


FIGURE 1. Map showing mountain fronts and major structures used for the tectonic geomorphic analysis.

TABLE 1. Measures and indices of relative age of thirty soil profiles.

| Geomorphic Surface | | Classification | Brightest moist mixed color in B horizon | | | Clay[2] XB/XA | Clayfilm Index[3] | Est. age in years before present (BP) |
|---|---|---|---|---|---|---|---|---|
| | | | Hue | Chroma | Color Index[1] | | | |
| $Q_1$ | Car body | Xerofluvent | AC profile, no B horizon | | | No B | 0 | 10-20 |
| $Q_2$ | Sespe | Fluventic Haploxeroll | AC profile, no B horizon | | | No B | 0 | 85-200 |
| $Q_3$ | Orcutt 0 | Pachic Xerumbrept | 10 YR | 3 | 4 | 0.6 | 0 | 500-5,000[5] |
| $Q_4$ | Orcutt 1 | Typic Argixeroll | 10 YR | 4 | 5 | 1.3 | 3.0 | 8,000-12,000[6] |
| $Q_{5a}$ | Honor Farm | Typic Argixeroll | 10 YR | 4 | 5 | ND | 4.0 | 15,000-20,000[7] |
| $Q_{5a}$ | Shell 2 | Typic Argixeroll | 10 YR | 3.5 | 4.5 | ND | 4.5 | 15,000-20,000[7] |
| $Q_{5b}$ | Orcutt 2 | Mollic Haploxeralf | 10 YR | 4 | 5 | 1.1 | 6.0 | 25,000-30,000[6] |
| $Q_{5b}$ | Bankamericard | Typic Argixeroll | 10 YR | 4 | 5 | ND | 7.0 | 30,000[7] |
| $Q_{6a}$ | Oak View[10] | Mollic Palexeralf | 7.5 YR | 5 | 7 | 1.4 | 7.25 | 38,000[8] |
| $Q_{6b}$ | Apricot | Mollic Palexeralf | 7.5 YR | 6 | 8 | 1.5 | 5.5 | 54,000 ± 10,000[9] |
| $Q_{6c}$ | La Vista[10] | Typic Palexeralf | 7.5 YR | 7 | 9 | 1.6 | 7.0 | 92,000 ± 13,000[9] |
| $Q_{6c}$ | Orcutt 3 | Typic Palexeralf | 5 YR | 4 | 6 | 1.6 | 7.5 | 80,000-90,000[6] |
| $Q_7$ | Timber Canyon 4 | Typic Palexeralf | 5 YR | 6 | 9 | ND | 8.0 | 160,000-200,000[6,11] |

Table 1 continued...

[1]Color index is computed by adding chroma number to hue (of moist mixed sample), where 10 YR = 1, 7.5 YR = 2, 5 YR = 3. Indices from different profiles on same geomorphic surface are averaged. To determine color, a large air-dired bulk sample was passed through a 2 mm sieve, then fractionated in a mechanical splitter, moistened, hand homogenized to a putty consistency and rolled to a sphere; the latter was then pulled into halves, and color noted from one freshly broken surface.
[2]Ratio of the mean percent of clay in B horizon to that in the A horizon (computed from particle size graphs (Keller and others, 1980).
[3]This index is based on clay film information contained in the profile descriptions and is computed by adding the percent frequency of clay film occurrence to their thickness, as follows: Percent frequency, very few = 1, few = 2, common = 3, many = 4, continuous = 5; Thickness, thin = 1, moderately thick = 2, thick = 3. For example, in the $B_{22t}$ horizon of La Vista 2 there are "... many to continuous (4.5) moderately thick and thick (2.5) clay films..." The index would be 7.0.
[4]This age is based on the inclusion of an abraded brick fragment in the C horizon of the $Q_2$ soil at Sespe Creek. A photograph taken in 1898 shows the terrace already present.
[5]This age estimate is collectively based upon tree rings of a number of mature oaks growing upon the Orcutt 0 surface, the degree of soil profile development, and a $^{14}C$ date (see Timber Canyon 1 profile description) on charcoal collected from a presumed buried soil in the lower part of the Timber Canyon profile.
[6]Age estimate based in part upon relative amount of displacement on flexural-slip faults between older surfaces in Orcutt and Timber Canyons, and one $^{14}C$ date. Also based upon soil correlation to well dated soils along the Ventura River.
[7]Age based on $^{14}C$ dates from correlative terraces along the lower Ventura River.
[8]Based on two $^{14}C$ dates on charcoal collected at the base of the Oak View Terrace below Oliva 1.
[9]Age based upon relative amount of displacement on the Arroyo Parida fault.
[10]These measures were taken from the buried soil portion of the profile; only the buried soil portion of the profiles of Oliva 1 and La Vista 3 are correlated to the $Q_6$ geomorphic surfaces.
[11]Older and more developed soils grouped with $Q_7$ have been sampled and described. Thus, a 160,000 age estimate is a minimum for $Q_7$ soils, but appears correct for Timber Canyon 4 as discussed above.

and resultant alluvial fan morphologic changes. In the area between Santa Paula Creek and Sespe Creek, two fronts were delineated; Front 4 at the San Cayetano fault and Front 3 at the approximate boundary of the highly folded older piedmont (Fig. 1).

Front 1: Ventura River to Harmon Canyon

Front 1 is bounded on the north by the actively folding Ventura Avenue anticline (Fig. 1); Pliocene through late Pleistocene sediments dated as young as 205,000 yr. b.p. (Wehmiller and others, 1978) dip 45°; younger marine terrace and alluvial fan sediments are folded lesser amounts.

The Ventura fault (Sarna-Wojcicki and others, 1976) also coincides with Front 1 and provides some uplift. Holocene fans formed at the mouths of canyons along the front are all entrenched with the present locus of deposition downfan or, in some cases, on the beach. Entrenchment may be due to southward tilting of the piedmont or seacliff retreat. Tilting is occurring as demonstrated by active folding of the Ventura Avenue anticline. Sea cliff retreat, however, appears to have caused entrenchment of the fans near the Ventura River; they are truncated by the modern seacliff, and are not segmented as are other fans along this front suggesting that when entrenchment occurred, the locus of deposition became the beach. The fans from Hall Canyon to Barlow Canyon are all segmented and entrenched but the overall surface grades to the present river or beach level, suggesting another reason for entrenchment such as tilting. Alternatively, entrenchment may simply be the result of complex response within the fan system. Only those fans and drainages not severely affected by coastal retreat were used in the tectonic geomorphic analysis; principally Hall Canyon to Harmon Canyon (see Fig. 1).

The "bedrock" in all the drainages along Front 1 is composed of upper Pliocene to middle Pleistocene Fernando formation and late Pleistocene Saugus formation. The large majority of the drainage area is composed of the easily eroded finer units of upper Pliocene to middle Pleistocene age exposed in the core and south flank of the anticline.

Front 2: Aliso Canyon to Santa Paula Creek

Front 2 consists of those drainages and fans between and

including Aliso Canyon and Fagan Canyon in the Ventura hills just east of Front 1 (Fig. 1). Along this front, the long, narrow rectangular shaped drainage basins head in Sulphur Mountain and cross the fairly uniformly southward dipping Pliocene and Pleistocene section. The eastward plunging Ventura Avenue anticline dies out just to the west of this front, affecting only Aliso Canyon which wraps around the nose of the anticline suggesting structural control of the drainage basin morphology.

The alluvial fans are all entrenched and some are segmented, grading to the present Santa Clara River floodplain. In many cases, moderately dipping older alluvial fan deposits are preserved at the mouths of the canyons indicating southward tilting. Bedrock in these drainages consists of Miocene Monterey shale, upper Miocene and Pliocene Sisquoc shale, Plio-Pleistocene Fernando shale and sand, and Saugus gravel. The gravel comprises only a small percentage of the total drainage area.

### Front 3: Santa Paula Creek to Sespe Creek

Front 3 is a continuation of Front 2 with the exception of major lithologic changes in the drainage area with resulting changes in fan composition. This front comprises all fans and drainages between Santa Paula Creek and Snow Canyon near Sespe Creek. Part of the drainage area of this front is deformed piedmont associated with Front 4 along the San Cayetano fault (Fig. 1) and is composed of older fan gravels, and steeply dipping Pliocene and Pleistocene bedrock (Fernando and Saugus formations). A very important constituent of the drainage area is the Eocene sandstone on the hanging wall of the San Cayetano fault. Although this comprises a small area with respect to the total, most of the fan deposits by volume consist of Eocene clasts indicating the activity and relative importance of Front 4. These deposits and their interpretation will be discussed further with Front 4.

### Front 4: Santa Paula Creek to Sespe Creek - San Cayetano Fault

Front 4 corresponds with the San Cayetano fault between Santa Paula and Sespe Creek (Fig. 1). The hanging wall of the fault is composed almost entirely of Eocene Matilija formation, a distinctive mottled arkosic sandstone. The piedmont contains highly folded late Pliocene and Pleistocene turbidite mudstone of the

Fernando formation, and, near the mouths of the canyons, conglomerate, all overlain by deformed alluvial fan deposits of varying ages with degree of deformation commensurate with age. Figs. 2 and 3 and Table 2 show the geology and deformation of Orcutt Canyon and Timber Canyon fans. Orcutt Canyon is deeply entrenched (Fig. 4) while adjacent Timber Canyon is filled with relatively young (Holocene) alluvium (Fig. 5). The difference between the two canyons is hypothetically related to the existence of a threshold fan slope, that if exceeded will initiate entrenchment. Given this assumption, the Orcutt Canyon fan was entrenched following late Pleistocene to Holocene deposition and tilting of Qf4-Qf5 (see Fig. 2) that increased the fan slope beyond the threshold. The Timber Canyon fan on the other hand has not entrenched since at least mid Holocene time and is now steepening due to fan head deposition and tilting associated with folding of the Santa Clara syncline. The present slope of the Timber Canyon fan is about 6° whereas that of the entrenched fan Qf5 in Orcutt Canyon is about 8°. Thus the critical slope that initiates entrenchment must be between 6° and 8° and it is expected that with continued steepening of the Timber Canyon fan it will eventually cross the threshold and entrench. That the entrenchment of the Orcutt Canyon fan is due for the most part to tectonic rather than climatic pertubations is supported by the observation that the adjacent Timber Canyon fan is out of phase with the Orcutt Canyon fill-entrench cycle. If climatic pertubations were primarily responsible for entrenchment the cycle of fill-entrenchment for the two canyons should be in phase.

The oldest alluvial fans associated with Front 4 occur at ridgetop positions surrounded, in some cases, by the Plio-Pleistocene mudstone indicating the relative ease that the mudstone is eroded with respect to the fan deposits (see Figs. 2 and 3). This also suggests that although the locus of deposition of fan alluvium has shifted through time, most if not all the intermontane piedmont has been covered by alluvial fans at one time or another.

The alluvial fans themselves are composed almost entirely of sub-angular to sub-rounded Matilija and Coldwater cobbles with minor amounts of reworked Saugus conglomerate clasts near the mouths of the canyons. When fans are not entrenched and thus con-

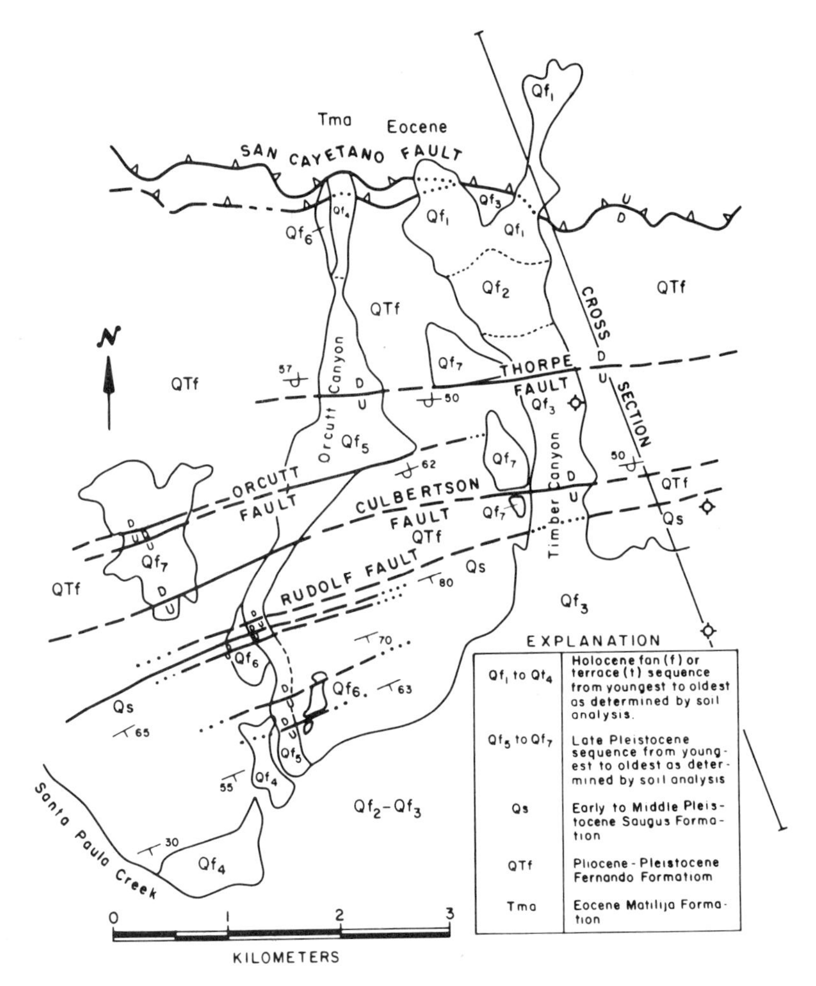


FIGURE 2. Geologic map of Orcutt and Timber Canyons, north flank of Santa Clara trough.

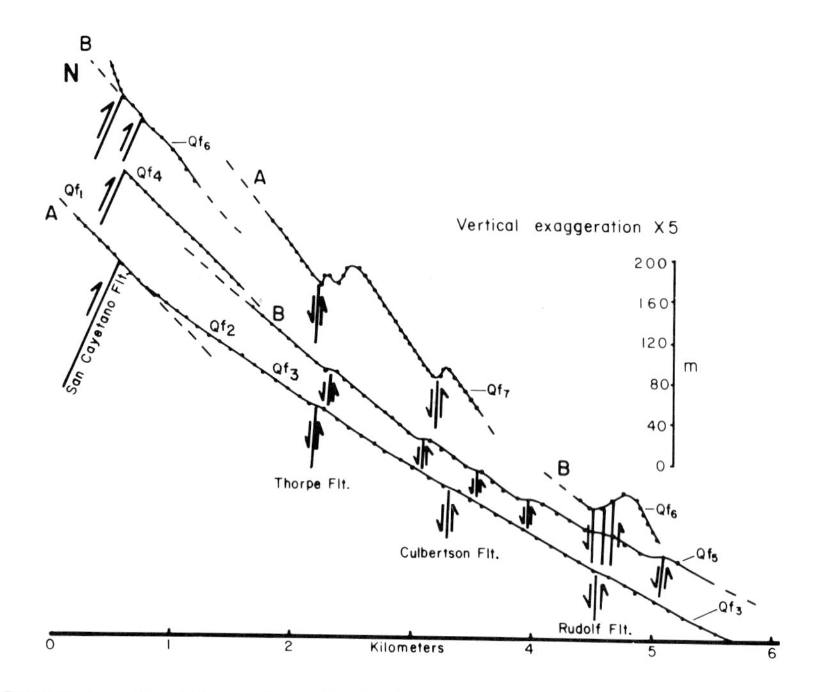


FIGURE 3. Profiles of faulted alluvial fan segments for Timber (A) and Orcutt (B) Canyons.

TABLE 2. Fault displacement and tilting of alluvial fans in Orcutt and Timber Canyons. Determinations made from USGS 7 1/2 minute topographic maps with 40 foot contour intervals and by hand level and tape measurements. Measurements are believed to be good in all cases to ± 3 m for the vertical displacements and ± 0.2° for the slope determinations.

| Geomorphic Surface | Displacement on Faults (Δ) | | | Tilting | | Est. Age (t) y.b.p. |
|---|---|---|---|---|---|---|
| | Thorpe | Culbertson | Rudolph | Present Slope(s) | Degrees Tilted (Δs) | |
| $QF_3$ | 4.5 m | 2 m | 6 m | 6° | 0° | 4000-5000 |
| $QF_5$ | 14 m | 4.5 m | 24-27 m | 8.2° | 2.2° | 25,000-30,000 |
| $QF_6$ | | | 61 m | 11.1° | 5.1° | 80,000-100,000 |
| $QF_7$ | 98 m | 37 m | | 17° | 11° | 160,000-200,000 |

FIGURE 4. Oblique aerial view of Orcutt Canyon.

FIGURE 5. Oblique aerial view of Timber Canyon.

tinuous to the San Cayetano fault, the fine textured Fernando formation contributes minor volumes of material to the fans as debris flows from the low, shallow canyon walls. Otherwise, the mudstone is quickly eroded by colluvial and minor landsliding processes, deposited in the entrenched channels and carried away by streams.

Because the fans are composed almost entirely of Eocene clasts, the drainage basin area was taken to be the area above the San Cayetano fault and the fan area as the composite of all the old and young alluvial fans originating from a source canyon. The eroded part of the intermontane piedmont, although at one time probably covered by fans, was not considered as drainage or fan area and was omitted from the calculations.

Front 5: San Cayetano Fault, Sisar to Santa Paula Creek

Front 5 coincides with the western section of the San Cayetano fault between Santa Paula Creek and Sisar Creek in the Upper Ojai Valley. Although, based on faulted alluvium, the fault is known to be active, only two fans (Sisar and Bear) are present due to the complex nature of the front, incision and capture of the Sisar and Bear drainages by Santa Paula Creek, and the convergence and constriction of the piedmont area between Fronts 5 and 8 (Fig. 1). Because of this constriction, the alluvial fans are relatively small in area relative to drainage basin area (see Fig. 1 and discussion below).

The Sisar Canyon fan has been repeatedly displaced by movement on the San Cayetano fault, and a 60 m composite fault scarp is present (Figs. 6 and 7). Several late Pleistocene fan segments are truncated at the fault, and the minimum vertical slip rate is estimated to be 0.6-0.9 mm/yr (Rockwell, 1983).

Front 6: North Side of the Ojai Valley

Front 6 is the northern boundary of the Ojai Valley between Wilsie Canyon and the Ventura River (Fig. 1). The northern bounding structure is the Matilija overturn composed primarily of Eocene sandstone and shale.

The fans of Senior, Gridley and west Gridley Canyons were grouped together in measuring fan and drainage areas because of the capture of the Gridley drainage area by the Senior fan, facilitated by headward erosion along the weak Cozy Dell shale (Rockwell, 1983).

FIGURE 6. Oblique aerial view of Sisar Canyon alluvial fan with 60 m fault scarp produced by displacement along the San Cayetano fault.

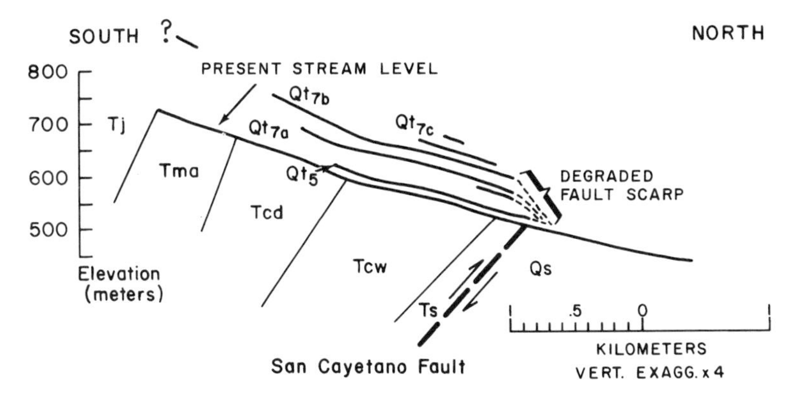


FIGURE 7. Alluvial fan and terrace profiles on hanging wall of San Cayetano fault at Sisar Canyon. Qs = Saugus Formation of the upper Ojai Valley; Ts = Sespe Formation; Tcw = Coldwater Formation; Tma = Matilija Formation; Tj = Juncal Formation.

Front 7: South Side of the Ojai Valley - Santa Ana Fault

Front 7 runs along the south side of Ojai Valley between Wilsie Canyon and the Ventura River and coincides with the Arroyo Parida-Santa Ana fault system. The alluvial fans are relatively small due to active fluvial erosion of the fans by Reeves Creek in the Wilsie Canyon to San Antonio Creek area. Because the drainage areas are also small, the fans were not studied but range front sinuosity and valley height ratios were calculated for comparison since this fault is active as indicated by faulted Ventura River terrace deposits.

Front 8: South Side of the Upper Ojai Valley - Lion Fault

Front 8 bounds the southern side of the Upper Ojai Valley and coincides with the active Lion fault (Fig. 1). Range front sinuosity and valley height ratios were computed but the fans were not studied.

FAN AND DRAINAGE BASIN AREA RELATIONSHIPS

The areas of drainage basins and alluvial fans were determined for most fans along Fronts 1, 2, 4, 5, and 6 (Table 3). The fans of front 3 are intrinsically related to and a part of those of front 4; the front 7 and 8 fans were not studied.

Relations between alluvial fan area and drainage basin area are shown on Figure 8. Two groups of fans exist, defined by $A_f = 3.84\ A_d^{0.55}$ and $A_f = 0.59\ A_d^{0.8}$ where $A_f$ is the fan area and $A_d$ is the drainage basin area in $km^2$.

Previous studies by Bull (1962, 1964, 1977a), Hooke (1967, 1972), and Denny (1965, 1967) indicate that the slope of the equation generally falls between 0.8 and 1.0 indicating that drainage basin area increases a little faster than fan area. Figure 9 shows the fans of this study in comparison to the fans of the previous studies listed. The fans of group 2 (Fronts 2, 5, and 6) fit an equation nearly identical to Denny's (1965) and are similar to the curves of the other workers with the difference probably explained by varying lithology, rainfall, and tectonic activity, (see Fig. 9).

The equation for the fans of group 1 (Fronts 1 and 4) have a lower exponent but higher coefficient than those of group 2 or for the other studies (Figs. 8 and 9). The fan areas for group 1 are

TABLE 3. Drainage basin and fan areas and fan slope. Some fans and drainage areas are grouped together due to closeness. In these cases, two fan slopes were measured, one for each fan.

| Front | Drainage | Drainage Basin Area (DBA) $km^2$ | Fan Area (FA) $km^2$ | Fan Slope(s) Degrees |
|---|---|---|---|---|
| | Hall Canyon* | 13.50 | 4.95 | 1.43 |
| 1 | Barlow Canyon* | 3.90 | 4.90 | 1.23 |
| | Sexton Canyon | 9.75 | 13.60 | .99 |
| | Harmon Canyon | 10.45 | 15.20 | 1.32 |
| | Long Canyon | 4.60 | 10.75 | 1.53 |
| | Aliso and Wheeler Canyons | 57.25 | 14.95 | .73, .74 |
| 2 | O'Hare and Adams Canyons | 31.85 | 9.20 | .73, .79 |
| | Fagan Canyon | 9.40 | 4.75 | .82 |
| | Orcutt Canyon | .90 | 4.60 | 4.65 |
| | Timber Canyon | 3.95 | 12.55 | 5.11 |
| | Canyon east of Timber Canyon | .25 | 1.55 | |
| 4 | Bear Canyon | .75 | 4.50 | 5.71 |
| | Boulder Canyon | 6.10 | 7.45 | 3.19 |
| | Snow Canyon | 1.10 | 2.40 | 5.71 |
| | Canyons east of Snow Canyon | 1.50 | 6.10 | |
| | Bear Creek | 3.90 | 2.70 | 4.90 |
| 5 | Sisar Creek | 16.70 | 5.40 | 3.71 |
| | Horn and Wilsie Canyons | 18.40 | 8.15 | 3.81 |
| 6 | Senior and Gridley Canyons | 31.35 | 8.05 | 2.92, 6.88 |
| | Stewart Canyon | 5.05 | 2.20 | 4.45 |
| | Cozy Dell Canyon** | 5.05 | .93 | 4.57 |

* Truncated by the coast-seacliff retreat

** Truncated by Ventura River

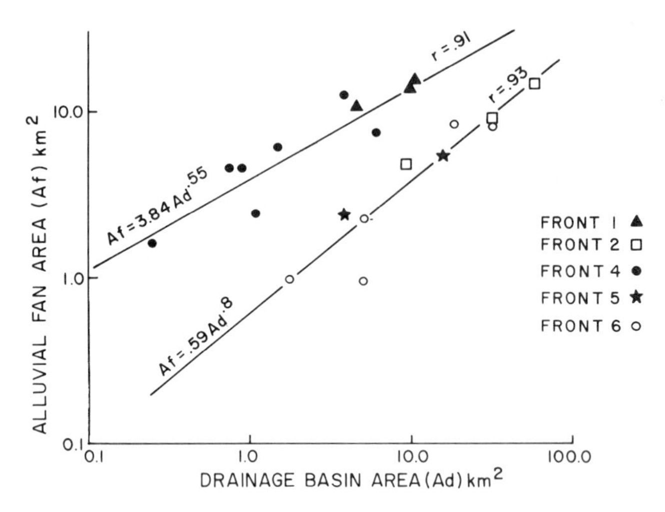


FIGURE 8. Alluvial fans versus drainage basin area for the Ventura Basin fans. Fans are divided into two groups based on fan/drainage area relations. Group 1 fans (upper curve) includes fans of fronts 1 and 4; group 2 fans (lower curve) include fans of fronts 2, 5, and 6.

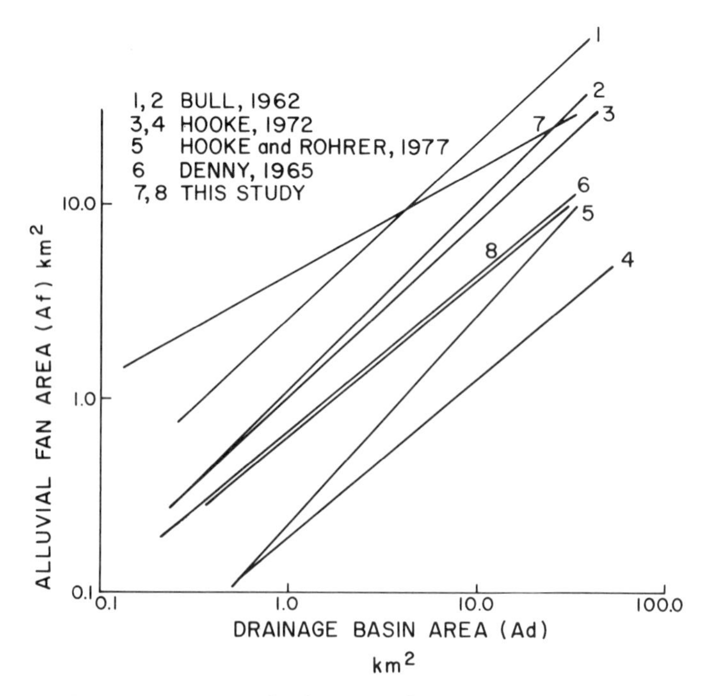


FIGURE 9. Comparison of alluvial fan versus drainage area relationships between Ventura fans and fans of other studies.

significantly larger (in area) for a given drainage basin area compared to the fans of group 2 and of other studies, suggesting that factors such as high rates of uplift and erosion in the drainage basin above the front, coupled with basinward tilting of the piedmont which eventually causes entrenchment and formation of new fan segments, are responsible for producing relatively large fans from small drainage basins.

Hooke (1972) found that the alluvial fans on the west side of Death Valley are much larger than those on the east because the entrenched western fans are being tilted eastward. The small eastern fans, all unentrenched, are forming along an active fault. This suggests that tilting is one cause of oversized fans. But in the Ventura Basin, Front 2 is being tilted as well as Fronts 1 and 4, although perhaps at a lesser rate.

High rates of uplift of a front may result in large quantities of debris being shed onto fans, facilitating development of relatively large alluvial fans providing that the fans are allowed to grow uninhibited to their maximum area. Fans of Fronts 1 and 4 are not now significantly restricted and their large area relative to the size of the drainage basins are probably due to uplift and tilting associated with the Ventura Avenue anticline and Ventura fault for Front 1 and the San Cayetano fault and Santa Clara syncline for Front 4. These fans are either presently receiving fanhead deposition or have received fanhead alluvium during the Holocene, suggesting that the tilting does not always keep the locus of deposition down fan. However, if the drainage basins of Timber and Orcutt Canyons increase in size in the future, the alluvial fans would have difficulty increasing in area because they are nearly at a maximum now, extending to the Santa Clara River which is presently against the Oak Ridge escarpment.

The relatively large fans of Fronts 1 and 4 are probably best explained by a combination of the above processes; tilting and high uplift rates - both of which collectively affect only these two fronts. The tilting periodically shifts the locus of deposition downfan causing entrenchment; high rates of uplift along the San Cayetano fault delivers eroded debris from the hanging wall of the fault to the fanhead causing fanhead deposition.

Front 5 is also along the San Cayetano fault and although

uplift is not as rapid as towards the central portion of the fault (~1 mm/yr compared to 4-8 mm/yr), it is certainly higher than Fronts 2 and 6. The fans of Front 5 drain to the Upper Ojai Valley which is a tectonically formed valley (Rockwell, 1983). The fans have reached a maximum size and now grow only in volume as the valley constrains the fan area. Thus, although associated with a tectonically active front, these fans plot on the lower group 2 curve (Fig. 8).

ALLUVIAL FAN SLOPE

Previous work suggests that fan slope is a function of lithology of the source area, particle size of the fan debris, drainage basin area, and discharge (Bull, 1977a, 1964; Hooke, 1972). Steeper fans are generally associated with coarser debris and smaller drainage basins (Bull, 1977a).

Slopes of the active fans are listed in Table 3. The drainage areas and fans of Fronts 3, 4, 5 and 6 are composed of coarse debris where as those of Fronts 1 and 2 are primarily fine grained. In general, the coarse alluvium (primarily sandstone cobbles) form fan slopes between 3° and 7° with no recognizable trend between fronts. The fan slope of Fronts 1 and 2, though, are significantly lower, between 0.7° and 1.7°, supporting the observation that coarser material forms steeper fans. Front 2 fans have larger drainage areas and lower fan slopes than those of Front 1, consistent with previous studies (Bull, 1969; Hooke, 1972) that suggest that larger drainages yield fans with lower slopes.

MOUNTAIN FRONT SINUOSITY

Mountain front sinuosity ($S_{mf}$) is defined as:

$$S_{mf} = \frac{L_{mf}}{L_s}$$

where $L_{mf}$ is the length of the mountain front along the mountain-piedmont junction and $L_s$ is the straight line length of the mountain front (Bull, 1977b and 1978). The $S_{mf}$ index according to these authors reflects the balance between uplift that tends to maintain a relatively straight front and erosion that tends to produce an irregular or sinuous front. Rapid uplift along range bounding faults produces a straight front and when uplift slows or stops, mountain front sinuosity increases with time as the front retreats due to erosional processes.

Tectonically active fronts near Ventura (Fronts 1, 4, 5, 7, 8) associated with known faults or active folds have low sinuosities between 1.01 and 1.34 with an average of 1.14. The less active fronts (2, 3, 6) have sinuosities ranging from 1.57 to 2.72 with an average of 2.04 (Table 4). Bull and McFadden (1977) suggest that if a threshold rate of uplift is exceeded, then mountain front sinuosity remains low. Examination of $S_{mf}$ values and rates of uplift of mountain fronts near Ventura (Table 4) suggests that for the tectonically active fronts, a rate of uplift of about 0.4 mm/yr is sufficient to keep $S_{mf}$ values low.

Mountain front sinuosity suggests that Front 6 is the least active of the fronts. This is suported by the presence of significantly older fan segments ($Qf_7$) with virtually the same slope only a few meters above presently active fans suggesting little or no tilting deformation in at least the last 160,000 yrs. No known active fault is associated with Front 6, thus geomorphic and geologic data both suggest the rate of tectonic deformation for this front is low.

VALLEY WIDTH TO HEIGHT RATIO

The ratio of valley floor width to valley height defines an index ($V_f$):

$$V_f = \frac{2V_{fw}}{(E_{ld} - E_{sc}) + (E_{rd} - E_{sc})}$$

where $V_{fw}$ is the width of the valley, $E_{ld}$ and $E_{rd}$ are the elevations of the left and right valley divides respectively, and $E_{sc}$ is the elevation of the valley floor or stream channel (Bull, 1977b and 1978). The index thus reflects the difference between "V" shaped valleys that are actively downcutting in response to active uplift (low values of $V_f$) and broad-floored valleys that are eroding laterally into adjacent hill slopes in response to base level stability (high values of $V_f$) (Bull, 1977b and 1978).

Data for the $V_f$ index was collected upstream from the front, at a distance of 0.1 of the drainage basin length as suggested by

TABLE 4. Tectonic geomorphic parameters for Fronts 1 through 8.

| Front Number | Location of Tectonic Mountain Front | Type of Front | Orientation of Front | Front Length (km) | Maximum Relief (m) | Alluvial Land Form | Mountain Front Sinuosity | $V_f$ | Tectonic Activity Class | Uplift Rate |
|---|---|---|---|---|---|---|---|---|---|---|
| 1 | Ventura River to Harmon Cyn. | Bounding folded (VAA) | E-W | 6.45 | 438 | entrenched alluvial fans | 1.09 | .7 | 1 | 4 mm/yr |
| 2 | Aliso Cyn. to Fagon Cyn. | Bounding folded | N60E | 8.88 | 670 | entrenched alluvial fans | 1.57 | 1.8 | 2 | unknown |
| 3 | Orcutt Cyn. to Snow Cyn. | Bounding folded | N60E | 10.1 | 1280 | entrenched and unentrenched fans | 1.83 | 1.91 | 2 | unknown |
| 4 | S.P. Creek to Sespe Creek | Internal faulted (SCF) | E-W | 6.45 | 915 | entrenched and unentrenched fans | 1.14 | .43 | 1 | 2-8 mm/yr |
| 5 | Sisar to Santa Paula Creek | Internal/bounding faulted (SCF) | N60W | 5.65 | 1639 | | 1.14 | .47 | 1 | 0.5-1.5 mm/yr |
| 6 | Wilsie to West Gridley Canyon | Bounding folded | Curving front NE-W | 5.25 | 1363 | entrenched and unentrenched fans | 2.72 | 1.89 | 2-3 | unknown |
| 7 | Wilsie to San Antonio Creek | Bounding (APF) faulted | N75E | 5.85 | 305 | unentrenched fans | 1.01 | <.73 | 1 | 0.4 mm/yr |
| 8 | South side Upper Ojai Valley | Bounding faulted (Lion F.) | N80E to EW | 4.23 | 451 | unentrenched fans | 1.34 | <1.09 (~.80) | 1 | unknown |

Bull (1978). $V_f$ ratios agree well with the sinuosity and geologic data for the Ventura Basin (Table 4). Active fronts all have $V_f$ ratios of less than 1.0 with an average of 0.63. The less active fronts have higher ratios (1.8 to 1.9).

TECTONIC ACTIVITY CLASS

Bull (1977b and 1978) assigns relative tectonic activity classes to mountain fronts. The most active fronts (class 1) have low values of $S_{mf}$ and $V_f$ as well as unentrenched fans or fans with only Holocene deposits at the fan head. The rate of uplift upslope of the front is greater that the sum of the rate of channel downcutting and piedmont deposition. Relative tectonic activity classes 2, 3, and 4 describe situations where stream channel downcutting is the dominant local base-level process relative to tectonic uplift and piedmont erosion. The characteristic landform for each of these classes is permanently entrenched alluvial fans with Pleistocene fan segments near the fan heads; $S_{mf}$ and $V_f$ are considerably higher than those of class 1 fronts (Bull, 1978).

The chief difference between classes 2, 3, and 4 concern erosion of the mountain front escarpment (Bull, 1978). Class 2 has a rate of channel downcutting sufficient to maintain a V-shaped cross-valley profile upstream from a mountain bounding fault or fold. Class 3 is associated with terrains of lesser tectonic activity; U-shaped cross-valley profiles are present but the mountain-piedmont junction is still fairly straight. Class 4 includes those situations of slight or minimal uplift at a mountain front. Both U-shaped cross-valley profiles and embayed mountain fronts are present. In such situations, erosional processes dominate over tectonic processes to the extent that initial development of pediments has begun. Class 5 consists of landform associations that describe pediments in tectonically inactive terrains (Bull, 1978).

Bull (1978) also delineated approximate uplift rates required to develop each of the tectonic activity class landforms and relations. Class 1 fronts are on the order of 1 to 5 m/1000 yrs, class 2 fronts are roughly 0.5 m/1000 yrs and class 3 rates may be approximately 0.05 m/1000 yrs. Class 4 and 5 fronts are almost inactive or are inactive tectonically with uplift rates on the order of 0.005 m/1000 yrs or less (Bull, 1978). In this study, five of the eight fronts are bounded by active faults and were designated as relative tectonic activity class 1. Vertical slip rates on some of the faults, however, were as low as about 0.5 mm/yr (0.5 m/1000 yrs), equivalent to those of class 2 fronts in Bull's (1978) study. Fronts 2 and 3, bounding the northern side of the Santa Clara River plain, are both class 2 fronts. Although an uplift rate is not known, Bull (1978) found that each tectonic activity class was approximately an order of magnitude apart in uplift rates. If this is true in the Ventura Basin and if class 1 fronts indicate uplifts of 0.5 to 5 mm/yr (0.5 to 5 m/1000 yrs) as is apparently the case, then class 2 fronts probably indicate rates on the order of 0.05 to 0.5 mm/yr (0.05 to 0.5 m/1000 yrs). Front 6 in Ojai Valley is apparently about a class 2-3 based on its higher sinuosity and more dominant U-shaped cross-valley profiles.

## CONCLUSIONS

Eight range fronts are delineated in the study area based upon geographic location, geomorphic expression and/or presence of known active faults and folds. Alluvial fans, discussed in groups by fronts, exhibit diagnostic characteristics associated with tilting and uplift or relative stability.

Fans associated with Front 4 along the San Cayetano fault are being actively tilted basinward resulting in a complex system of fans which have a recurring history of fanhead deposition, tilting, entrenchment, uplift and faulting, and then backfilling, examples of which are present in Orcutt and Timber Canyons. The fans of Fronts 1 and 2 show a similar but less complex history of fan development associated with tilting produced by active folding.

The drainage basin area versus fan area relations suggest two groupings of fans; those of Fronts 1 and 4 (group 1) and those of Fronts 2, 5 and 6 (group 2). Group 1 fans have highly active fronts associated with faults and/or folds and are being uplifted and tilted basinward. These two components apparently work together to develop relatively large alluvial fans for given drainage areas. Group 2 fans are associated with less active tectonism or are more constrained in their depositional areas resulting in alluvial fans which fall on a similar regression line to previous studies.

Fan slopes are steep for fans composed of relatively coarse-grained alluvium ranging between 3° and 7° as opposed to 0.7° to 1.7° for fans composed of relatively fine-grained alluvium. Although no trend between drainage basin area and fan slope is apparent between fronts with coarse grained fans, the fans composed of finer grained alluvium do show such a trend; larger basins produce alluvial fans with lower slopes.

Mountain front sinuosity ($S_{mf}$) and ratio of valley width to valley height ratios ($V_f$) are useful indicators of relative tectonic activity in the Ventura area. Highly active fronts consistently produced lower values of $S_{mf}$ and $V_f$, than less active fronts. A threshold uplift rate of 0.4 mm/yr is sufficient to keep $S_{mf}$ values low. Thus, Front 7 with an uplift rate of about 0.4 mm/yr has a similar sinuosity to those fronts with an order of magnitude higher uplift rates (4-8 mm/yr).

ACKNOWLEDGMENTS

Research was supported by contracts 14-08-0001-17678 and 14-08-0001-19781 as part of the Earthquake Hazards Redirection Program, U. S. Department of Interior Geological Survey.

REFERENCES CITED

Blackwelder, E., 1931, Desert Plains: Jour. of Geology, v. 39, p. 133-140.

Bull, W. B., 1962, Relations of alluvial-fan size and slope to drainage-basin size and lithology in western Fresno County, California: U. S. Geological Survey Prof. Paper 450-B, p. 51-53.

Bull, W. B., 1964, Geomorphology of segmented alluvial fans in western Fresno County, California: U. S. Geological Survey Prof. Paper 352-E, p. 89-129.

Bull, W. B., 1977a, The alluvial-fan environment: Progress in Physical Geography, v. 1, no. 2, 1977, p. 222-270.

Bull, W. B., 1977b, Tectonic geomorphology of the Mojave Desert: U. S. Geological Survey Contract Report 14-08-001-G-394; Office of Earthquakes, Volcanoes, and Engineering, Menlo Park, California, 188 p.

Bull, W. B., 1978, Geomorphic tectonic activity classes of the south front of the San Gabriel Mountains, California: U. S. Geological Survey Contract Report 14-08-001-G-394; Office of Earthquakes, Volcanoes, and Engineering, Menlo Park, California, 59 p.

Bull, W. B. and McFadden, L. D., 1977, Tectonic geomorphology north and south of the Garlock Fault, California: in Proceedings Vol. of 8th Annual Geomorph. Symp. at N. Y. State Univ. at Binghampton, Sept. 23- , 1977, Donald O. Doehring, ed., p. 116-138.

Clark, M. N., 1982, Tectonic geomorphology and neotectonics of the Ojai Valley and upper Ventura River: Unpublished M. A. thesis, University of California, Santa Barbara, 77 p.

Crowell, J. C., 1976, Implications of crustal stretching and shortening of coastal Ventura basin, California: Pacific Sect. Am. Assoc. Petrol. Geol. Misc. Publ. 24, p. 365-382.

Dembroff, G., 1983, Tectonic geomorphology and soil chronology of the Ventura Avenue anticline, Ventura, California: Unpublished M. S. thesis, Univ. of Calif., Santa Barbara, 122 p.

Denny, C. S., 1965, Alluvial fans of the Death Valley region, California and Nevada: U. S. Geol. Survey Prof. Paper 466, 62 p.

Denny, C. S., 1967, Fans and pediments: Am. Jour. of Sci., v. 265, p. 81-105.

Hooke, R. L., 1967, Processes on arid-region alluvial fans: Jour. of Geol., v. 75, p. 438-460.

Hooke, R. L., 1972, Geomorphic evidence for late Wisconsin and Holocene tectonic deformation, Death Valley, California: Geological Society of America Bulletin, v. 83, p. 2073-2098.

Hooke, R. L., and Rohrer, W. L., 1977, Relative erodibility of source-area rock types as determined from second-order variations in alluvial-fan size: Geological Society of America Bulletin, v. 88, p. 1177-1182.

Keller, E. A., Rockwell, T. K., Clark, M. N. and Dembroff, G. R., 1982, Tectonic geomorphology of the Ventura, Ojai and Santa Paula areas, western Transverse Ranges, California: In Cooper, J. D. (ed.), Neotectonics in Southern California: Geological Society of America Guidebook for 78th Annual Meeting, Cordilleran Section, p. 25-42.

Luyendyk, B. P., Kamerling, M. J. and Terres, R., 1980, Geometric model for Neogene crustal rotations in Southern California: Geological Society of America Bulletin, Part I, v. 91, p. 211-217.

Rockwell, T. K., 1983, Soil chronology, geology, and neotectonics on the north central Ventura basin: Unpublished Ph.D. thesis, Univ. of California, Santa Barbara, 288 p.

Sarna-Wojcicki, A., Williams, K. M., and Yerkes, R. F., 1976, Geology of the Ventura fault, Ventura County, California: U. S. Geol. Survey Map File 781, scale 1:6,000, 3 sheets.

Wehmiller, J. F., Lajoie, K. R., Sarna-Wojcicki, A. M., Yerkes, R. F., Kennedy, G. L., Stephens, T. A., and Kohl, R. F., 1978, Amino-acid racemization dating of Quaternary mollusks, Pacific coast, United States: In Zartman, R. E., ed., Short papers of the 14th Internat. Conf., Geochronology, Cosmochronology, Isotope Geology: U. S. Geol. Survey Open-File Report 78-701, p. 445-448.

Yeats, R. S., 1968, Rifting and rafting in the southern California borderland: In W. R. Dickinson and A. Grantz,(eds.), Conf. on Geol. Problems of San Andreas fault system: Proc. Stanford, Calif. 1967, Stanford Univ. Pubs., Geol. Sci., v. 11, p. 307-322.

Yeats, R. S., 1971, East Pacific Rise and Miocene Tectonics of the

Southern Ventura Basin: Geological Society of America, Abstracts with Programs, v. 3, no. 2, p. 222.

# 9

# Equilibrium tendency in piedmont scarp denudation, Wasatch Front, Utah

*James F. Petersen*

ABSTRACT

Piedmont scarps in the Wasatch fault zone of Utah exhibit a variety of hillslope forms. Slope form characteristics and relationships between fault scarps and geomorphic features of Lake Bonneville yield evidence of scarp age. Relationships among slope form attributes of fault scarps and shoreline cliffs were tested on an age-controlled set of profile data. Several form attributes are age-dependent: midslope angle, profile height/length ratio, crest and footslope curvature, and midslope length. These components are poorly correlated immediately after offset, but become adjusted as erosion modifies the scarp. Piedmont scarps rapidly grade into a convex-straight-concave profile that is a relatively stable form. Rejuvenated scarps also rapidly regrade into this form, obscuring evidence of the faulting history. Many multiple event scarps have similar topographic profiles, a result of rejuvenation and regrading. The degradation of piedmont scarps can be explained by a shift in erosional stresses to maladjusted parts of the slope system. A ternary diagram model of hillslope profiles illustrates the equilibrium tendency of piedmont scarps.

## INTRODUCTION

Piedmont scarps were defined by G. K. Gilbert (1928) as small fault scarps offsetting unconsolidated materials. The term avoids confusion between references to scarps in surficial deposits and the bedrock escarpments of fault-block mountains. Near the base of the Wasatch Range in Utah, piedmont scarps are well exposed in a discontinuous zone of normal faults. The scarps represent the most recent surface faulting and offset Quaternary deposits. No surface faulting has occurred along the Wasatch Front since the region was settled in 1847, but an estimate for the last event is about 300-500 years ago (Morrison, 1965).

The Wasatch Front could be considered the "type locality" of the piedmont scarp, where Gilbert made many important observations concerning their geomorphology.

1. Surface offset is generally less than topographic scarp height. Offset can be determined by the vertical distance between projections of the upper and lower parts of the pre-faulting surface (1890, 1928).

2. Geomorphic characteristics of piedmont scarps and wave-cut cliffs of Lake Bonneville are similar (1890).

3. Relative ages may be estimated by geomorphic criteria, particularly steepness of the scarp and sharpness of the crest (1928).

4. Relative age can be determined by relationships between scarps and other landforms such as Lake Bonneville shoreline features (1928). This is an early application of the principle of crosscutting relationships to geomorphology.

These insights, like many of Gilbert's ideas, have provided a basis for additional study (Slemmons, 1957; Wallace, 1977; Bucknam and Anderson, 1979; Nash, 1981). Based on profiles surveyed in Nevada, Wallace (1977) developed a denudation chronology (Figure 1) for piedmont scarps, assuming offset of unconsolidated materials followed by stillstand. His model outlines profile changes and the shift in dominant erosional processes over time. The height and slope of an initial scarp are controlled by the topographic displacement and the fault dip at the surface (Figure 1-A). Initially, the scarp face is steep, generally over 60 degrees. The free face is rapidly degraded by gravity-induced movement of materials (1-B and 1-C), and should be removed from a scarp in alluvium in less than 1000 years (Wallace, 1977). The free face is replaced by a debris slope, rounding of the crest begins, and a wash-controlled slope develops at the base (1-C and 1-D). Slope decline follows as the wash-controlled slope and crest merge (1-E).

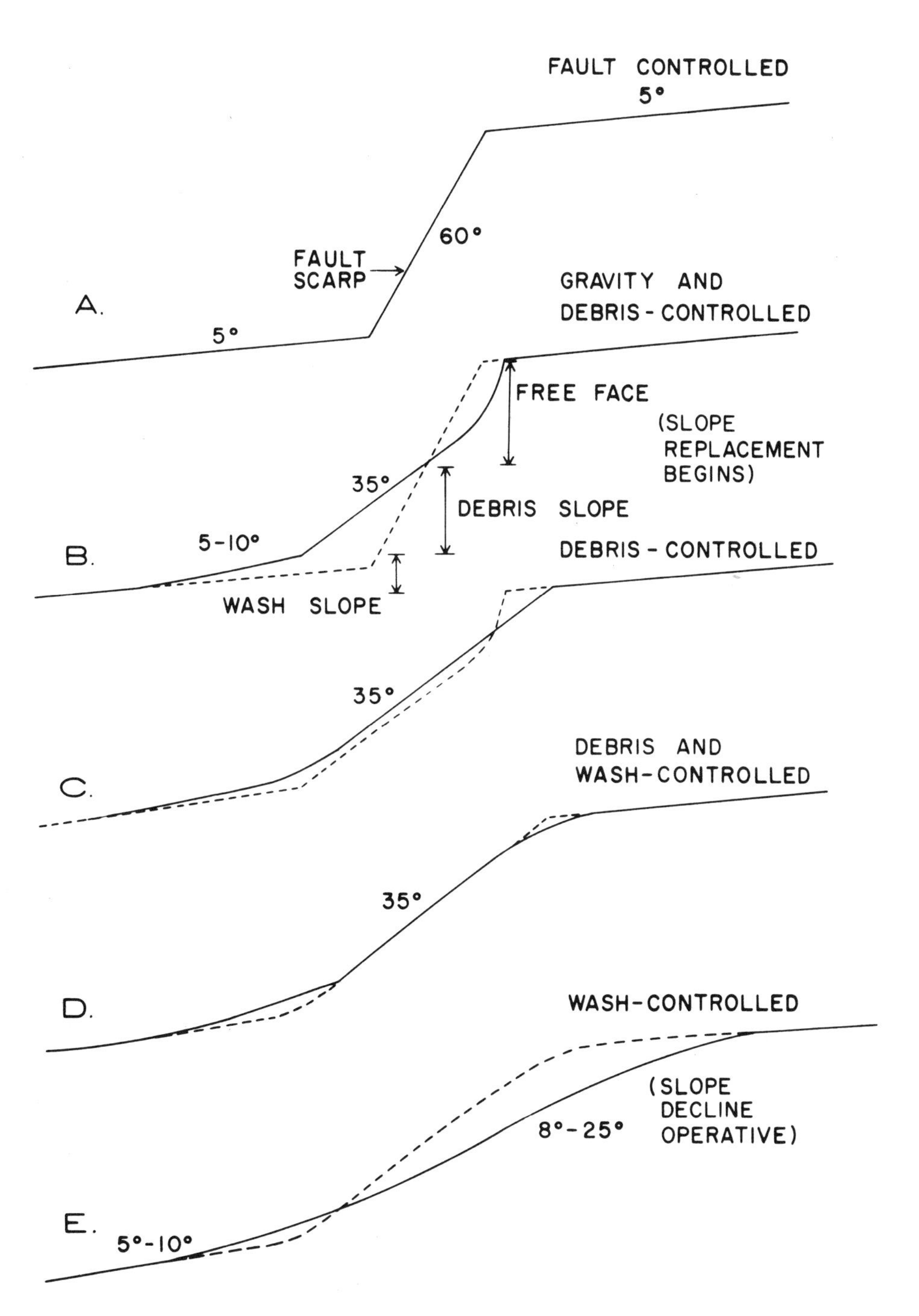


FIGURE 1. Wallace's (1977, p. 1269) model of fault scarp denudation based on measurement of piedmont scarp profiles in Nevada.

Recent models of hillslope evolution have stressed the need for considering many factors of slope form, rather than only one or two variables (Blong, 1975). While many qualitative assessments of scarp degradation have been made, few relationships have been quantitatively tested. Wallace's (1977) model suggests several quantifiable scarp form components that may indicate age, such as crest and footslope curvature, and height/length ratio.

Surface faulting produces a discontinuity (disequilibrium) in the piedmont slope system. The initial scarp is fault-controlled, and slope forms should be poorly correlated. As erosional modification progresses, relationships among the slope components of a scarp should become stronger. Young scarps should consist of poorly correlated forms and old scarps should be composed of strongly related forms. Disequilibrium is replaced by increasing adjustment. A well-correlated set of morphometric variables is indicative of equilibrium conditions (Chorley and Kennedy, 1971). The most recent piedmont scarps may not have been subjected to denudation long enough to attain equilibrium, even though the scarps are in easily erodible unconsolidated materials. This equilibrium tendency may apply to the interpretation of fault scarp form and age, and these relationships can be tested on data of known relative age.

## AN AGE CONTROLLED TEST OF FORM RELATIONSHIPS

Obtaining age control for a scarp study is a difficult problem. A general method is to use several lines of evidence. Wallace (1977) compared historic to prehistoric scarps and related scarps to Lake Lahontan shoreline features. Bucknam and Anderson (1979) measured wave-cut cliffs of the high stand of Lake Bonneville for comparison with piedmont scarps. In this study, time control was obtained by measuring slopes which have age evidence based on indicators other than slope form. Crosscutting relationships between scarps and landforms with a degree of age limitation are the most frequently available criteria. At some locations, an age sequence can be interpreted where fault scarps cut landforms related to episodes of Lake Bonneville. Along the Wasatch, shoreline cliffs for two levels of Lake Bonneville may be compared to piedmont scarps, the Bonneville (c. 16,000 B.P.) and Provo (c. 13,500 B.P.) shorelines are present in sizes and surficial materials comparable to the piedmont scarps. Age estimates for the Bonneville and Provo levels are from Scott (1982), and Currey (1984). A set of hillslope profiles was measured on piedmont scarps and shoreline cliffs which fit into four relative age categories: young scarps, Provo shoreline cliffs, Bonneville shoreline cliffs and pre-Bonneville scarps. All slopes studied are in unconsolidated deposits, in a narrow zone of latitude, altitude, climate, and vegetation. Age related changes should be more easily discriminated from these data than would be possible with a more heterogeneous profile sample.

## FORM ATTRIBUTES OF SLOPE PROFILES

Five categories of form attributes were measured or derived: lengths, form ratios, angles, areas, and indices of curvature (Figure 2). Wherever possible, a dimensionless variable was used, to permit direct comparison between scarps of different size.

Scarp height (SCPH) is measured by projecting the upper and lower original surfaces to intersect with a projection of the midslope. The vertical distance between the two intersections of these three lines is the scarp height (Figure 2-A).

Surface offset (SURO) is the vertical distance between projections of the upper and lower original surfaces, measured at the midpoint of the midslope (Gilbert, 1890; Bucknam and Anderson, 1979). Surface offset was not used in the correlations as it is of dubious value in describing shoreline cliffs.

Height/length ratio (HLR) is the ratio of profile height to profile length (Figure 2-C).

Slope component length percentages. The profiles were divided into convex, straight, and concave components by overlaying a straight edge on the drafted profile and marking deviations from a straight line (Figure 2-B). Length percentages for the overland length of the crest (PCC), midslope (PMS), and footslope (PBC) were used as dimensionless expressions of the importance of each slope component to the entire profile.

Midslope angle (MANG), or principal slope angle (Wallace, 1977), is the mean angle of the midslope segment (Figure 2-C).

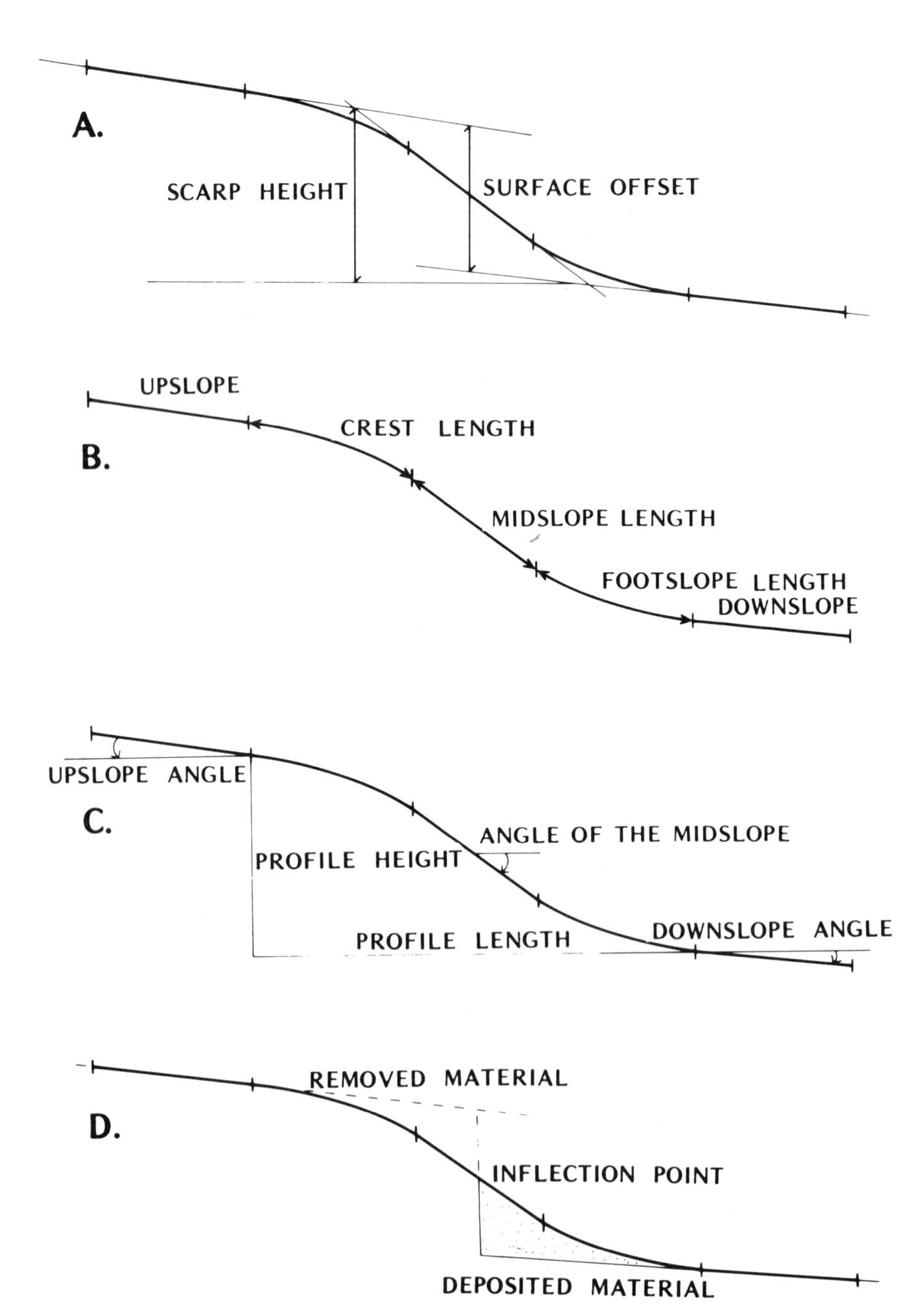


FIGURE 2. Diagrams illustrating the measurement of form attributes of hillslope profiles.

Material removed from the crest (ARM) estimates erosional modification, assuming an originally vertical scarp face and the fault located in the center of the midslope. This area is encompassed within a vertical line drawn through the center of the midslope, a projection of the upper original surface and a line defining the surface of the crest (Figure 2-D). ARM was measured with a polar planimeter and expressed in square meters.

Dimensionless indicators of curvature were calculated for both the crest (CCP) and footslope (ACBP). Relative curvature is the slope angle change over a curved element in degrees per one percent of overland length (Ahnert, 1970). The index is positive for convex and negative for concave slope elements, so the absolute value was used to compare the crest and footslope.

$$Cr = \frac{B-A}{L\%}$$

where:

Cr is relative curvature of a nonlinear slope component in degrees per one percent of element length.
A is the slope angle at the top of the curved element.
B is the slope angle at the bottom of the curved element.
L% is the length percentage of the segment as a part of the total overland slope profile length.

## RELATIONSHIPS AMONG ATTRIBUTES OF SCARP FORM

Five form attributes: midslope angle, height/length ratio, midslope length percentage, crestal curvature, and footslope curvature, yield a data set that describes the major parts of a three-component slope profile. Assumptions about the population distributions of these attributes were minimized by using Spearman's rank correlation, for all combinations of the form variables (Table 1). Values of Spearman's Rho $\pm.40$ are significant at the .01 level. Relationships among these five variables of hillslope form are illustrated with a diagram (Figure 3) linking significant inverse (-) and direct (+) correlations (after Towler, in Chorley and Kennedy, 1971). The variables are all directly related and each is inversely related to time. The form components are well-correlated even though the profile sample includes a variety of ages, suggesting that they change together as a morphological system of interdependent parts. For example, relative curvatures of the crest and footslope decrease as scarp height, midslope angle, and midslope length decline.

TABLE 1. Correlation Matrix: Form Attributes of Profiles of Known Relative Age.

| | HLR | PMS | CCP | ACBP | ARM | SCPH |
|---|---|---|---|---|---|---|
| MANG | .97 | .79 | .81 | .82 | -.27 | -.05 |
| HLR | | .80 | .79 | .81 | -.27 | .01 |
| PMS | | | .67 | .85 | .03 | .17 |
| CCP | | | | .53 | -.18 | .02 |
| ACBP | | | | | -.16 | -.01 |
| ARM | | | | | | .46 |

n = 30

All correlations ±.40 are significant at the .01 level.

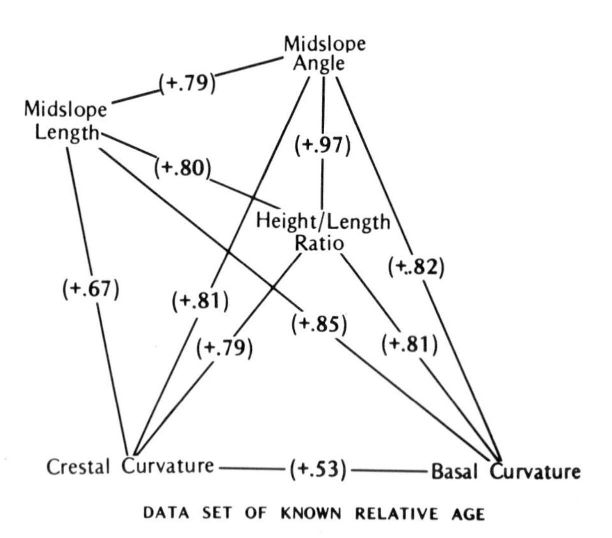


FIGURE 3. Correlation diagram of form attributes, based on the data set of four known relative ages: young scarps, Provo shoreline cliffs, Bonneville shoreline cliffs, and pre-Bonneville fault scarps.

Midslope angle (MANG) and height/length ratio (HLR) are geometrically constrained on a slope that is in a state of adjustment. As development of the crest and base expands the planimetric profile length (PLEN), the midslope angle diminishes. Profile length is a good indicator of scarp age, but HLR is substituted for PLEN as a dimensionless expression. The age dependency of MANG and HLR is suggested by their high correlation with relative curvatures of the crest and footslope. If the sample did not include slopes that have been subjected to recent tectonic disturbance (and thus are in a state of disequilibrium), MANG and HLR would be almost perfectly correlated. Residuals reveal a poor correlation of these two variables on young scarps, but they are strongly correlated on older scarps and shoreline cliffs.

Scarp height was not significantly related to any of the age-sensitive form variables. Correlations of midslope angle and scarp height (log) were not significant at the .05 level. Bucknam and Anderson (1979), however, found a predictable relationship between scarp height and midslope angle in scarps and shoreline cliffs of <u>similar</u> ages. More study is needed concerning the relationship between slope angle and hillslope relief. Bucknam and Anderson (1979), Melton (1957), and Schumm (1966) reported increasing slope angles with height. Chorley and Kennedy (1971) reported a non-significant relationship and Blong (1975) found a tendency toward decreasing slope angle with increasing relief.

Material removed from the crest (ARM) was also poorly related to the other variables, except those describing slope size. ARM is inversely related to the upslope angle for a given scarp. Where the upslope is steep, the area of removed material will be less than that measured where the upslope is flatter. Erosion of the crest is not only dependent on time, but also on the difference in angle between the upslope and the scarp face. Removal of crestal sharpness is really a reduction in the slope break between the upslope and the scarp face.

Relationships between form variables and ranked ages support the age dependency of the selected scarp form variables. The values of these components are a function of elapsed time since cessation of the process that produced the discontinuity, either wave erosion or tectonic offset. The two most highly correlated variables, midslope angle and height/length ratio, illustrate the inverse relationship of the form attributes to relative time (Figure 4).

MANG and HLR, however, are not necessarily well correlated throughout the time continuum. A plot of these two variables shows four groups of points ranked by age (Figure 5). The scatter of values representing younger scarps is partially a function of age differences among individual sites. Slight age differences may result in considerable form variation among young scarps, but very little in older scarps. As the slope is altered by erosion, tectonic-offset control diminishes. Although the data representing younger scarps probably include a variety of scarp ages, the variance also results from a lack of adjustment among the slope components of a young scarp. As time passes, HLR and MANG become better correlated, and increasing strength of correlation with time is a general tendency among most other variable pairings. An overlap of values exists between point groups in Figures 4 and 5, a problem in determining an index of scarp age based on form attributes. The overlap particularly affects attempts to resolve small increments of time based on form characteristics. The shoreline cliff observations each represent a synchronous point in time, yet the scatter of values suggests the influence of factors other than time. In comparison, the high variance is expected in the young scarps group, which probably includes a wide variation in scarp ages. This is also true of the pre-Bonneville scarps, but variance tends to diminish with time.

## SCARP FORM: TIME RELATIONSHIPS AND GEOMORPHOLOGY

Example profiles of young scarps (Figure 6-A), Provo and Bonneville age wave-cut cliffs (6-B and 6-C) and scarps older than the Bonneville shoreline (6-D) illustrate hillslope degradation and the temporal succession of profile forms. While these profiles were selected as good examples, they are relatively typical of measured traverses for each age group. The values of form attributes (MANG, HLR, PMS, CCP, and ACBP) are shown for each profile. The general decline in these variables with age is illustrated along with the slope profile changes. These relationships support the general applicability of Wallace's (1977) denudation model to the Wasatch

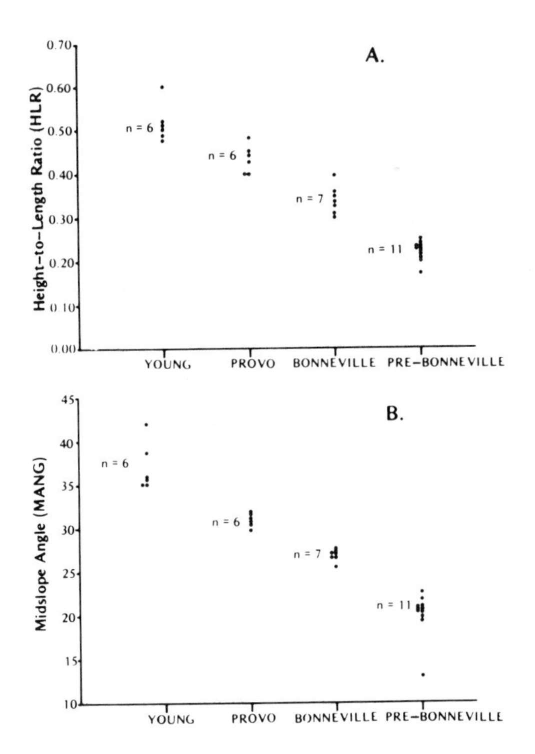


FIGURE 4. Plots of profile height/length ratio (HLR) against relative age (A.) and midslope angle (MANG) against relative age (B.).

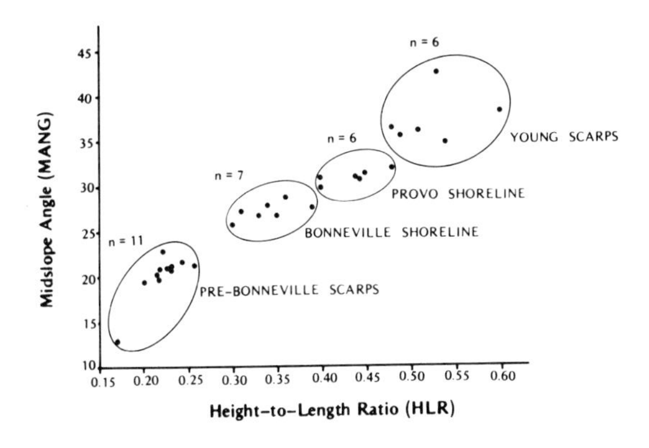


FIGURE 5. Plot of profile height to profile length ratio (HLR) against midslope angle (MANG). Data set of known relative age.

Front. Mean values of the midslope angle, height/length ratio, curvature of the crest and footslope, and midslope length, decline with age (Table 2). An exception is midslope length for observations on the Provo shoreline cliff. These cliffs are higher (mean height = 31 meters) than most other profiles and height may influence the midslope length percentage.

## Young Scarps

The profile form of very recent scarps is relatively well known, based on measurement and observation of scarps that have developed in historic time. Erosional changes in historic scarps have been photographically documented by Morisawa (1975) and Wallace (1980). Although historic offset has not occurred along the Wasatch Front, several scarps measured display forms similar to historic scarps. Gilbert (1884) compared piedmont scarps near Salt Lake City to those produced by the 1872 earthquake on the east side of the Sierra Nevada, near Lone Pine, California. He reported little visual difference between the freshest appearing Wasatch scarps and the Lone Pine scarps. Scarps at both locations exceeded the angle of repose and had not yet been completely colonized by vegetation.

Figure 6-A is the profile of a young scarp that postdates the Provo level. Scarp age is indicated by its location. It cuts a landslide on valley walls incised into the Lake Bonneville delta of the Odgen River. Downcutting of the delta coincided with a drop in baselevel as the lake receeded from the Provo shoreline, at about 11,000 B.P. (Scott, 1982; Currey, 1984). The appearance of this scarp suggests that it is much younger than the limiting age (c. 11,000 B.P.). The midslope angle is steep (35 degrees), the midslope dominates the profile (53 percent), curved elements are narrow and sharp, and the height/length ratio is high (.48).

## Lake Bonneville Shoreline Cliffs

The prominent shorelines of Lake Bonneville were named by Gilbert (1890). The Bonneville shoreline was produced by a transgression of the lake to its highest level prior to its overflow through Red Rock Pass, Idaho. Downcutting and outlet failure during the Bonneville Flood caused the lake level to drop about 105 meters. Subsequent outlet threshold control developed the Provo shoreline,

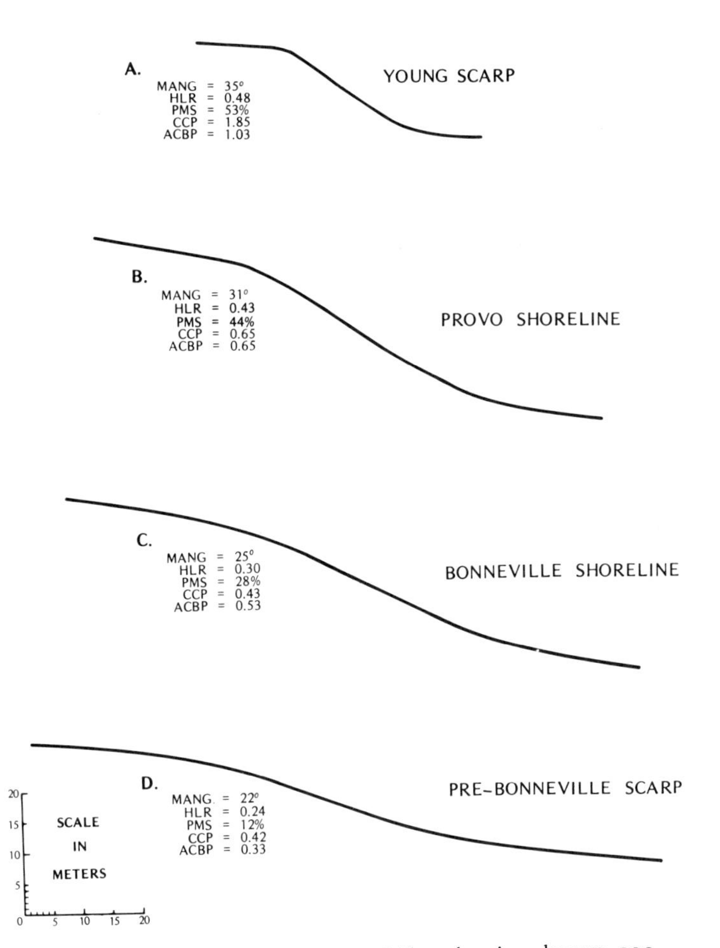


FIGURE 6. Examples of hillslope profiles having known age relationships. Note the sequential changes in the selected form variables: midslope angle (MANG), profile height to profile length ratio (HLR), percentage length of the midslope (PMS), relative curvature of the crest (CCP) and the footslope (ACBP).

TABLE 2. Mean Values for Selected Form Attributes: Profiles of Known Relative Age.

| Relative Age | MANG | HLR | CCP | ACBP | PMS |
|---|---|---|---|---|---|
| Young Scarps (n = 6) | 37.2 | .50 | 1.54 | 0.78 | 41.9 |
| Provo Shoreline (n = 6) | 30.8 | .43 | 0.75 | 0.73 | 48.5 |
| Bonneville Shoreline (n = 7) | 27.2 | .34 | 0.65 | 0.51 | 26.2 |
| Pre-Bonneville Scarps (n = 11) | 20.1 | .22 | 0.46 | 0.37 | 17.5 |

which was abandoned as the lake dropped due to climatic change. Each shoreline age is the time of abandonment by wave erosion. These two relict beach cliffs, separated by a few thousand years, provide additional evidence of the degradation of slopes. Figures 6-B and 6-C are Provo (c. 13,500 B.P.) and Bonneville (c. 16,000 B.P.) wave-cut cliffs. The shoreline cliff profiles reveal several apparently time-related changes which distinguish these older slopes from the young scarps. The crest and footslope have become more gently curved, and widened at the expense of the midslope, and the midslope angle has declined. These relationships can be seen in the values adjacent to the profiles.

### Pre-Bonneville Scarps

Proof of age for old piedmont scarps has been limited in most scarp studies. Wallace (1977), however, noted a scarp with a muted form that was truncated by wave erosion and pre-dated the high stand of Lake Lahontan. Few piedmont scarps of pre-Bonneville age (>16,000 B.P.) exist along the Wasatch Front. The Bonneville level in most areas of the Wasatch was at or near the bedrock escarpment, and little of the pre-Bonneville piedmont remains intact. Although most pre-Bonneville scarps were obscured by transgressions of the lake, a few remain above the highest shoreline. Figure 6-D is a pre-Bonneville fault scarp that was truncated by waves at the Bonneville level. The example Bonneville shoreline cliff (6-C) is the wave-cut edge of this pre-Bonneville horst (6-D). Profile comparison suggests that the scarp is considerably older than the Bonneville shoreline, perhaps by thousands of years. The scarp is in a swarm of horsts and grabens that have broken the piedmont and thus, little overland flow occurs over these scarps in comparison to the other sites. Degradation at this site has probably been a slow process, but the scarps have a muted form. The crest and footslope of the pre-Bonneville fault scarp are gently curved, and dominate the profile length, the midslope segment is almost an inflection point (12 percent of the profile), midslope angle is low (22 degrees), and height/length ratio is low (.24).

### Multiple Event Scarps

A problem in the study of piedmont scarps is a lack of evidence for the number of surface faulting events that produced a scarp.

Thirteen profiles of unknown age, which comprise the highest fault scarps observed, were examined separately. It is assumed that these scarps are too high (mean = 30 m.) to represent a single event and have resulted from repeated offset on the same fault trace. Although profiles of these scarps exhibit a convex-straight-concave form as do the other scarps, relationships among form components differ from those of smaller scarps and shorelines. A correlation diagram (Figure 7) and correlation matrix (Table 3) for the set of high multiple scarps illustrate these differences. Several relationships that are strong among the other scarps are not significant for these multiple event scarps. None of these scarps has a low midslope angle. Renewed offset has progressed more rapidly than degradation of the midslope angle. The correlation between HLR and MANG is poor, but attributes which describe curved elements are strongly related. This indicates that a well-adjusted slope system existed prior to renewal of offset. The curved elements persist as relicts of the pre-rejuvenation scarp profile. The midslope has been rejuvenated, but the curved elements retain much of their previous form because prior adjustment had graded the crest and footslope to a stable and persistent, gently curved shape.

Height/length ratio is smaller than expected for multiple event scarps considering the recency of faulting as evidenced by steep midslope angles. HLR is an indicator of how long faulting has been active on a multiple scarp. Multiple event scarps have a HLR that is indicative of older scarps, and a midslope angle that represents younger events. The poor correlation between HLR and MANG suggests that the slope system is not adjusted. Each variable conveys a different piece of evidence about surface faulting. Curved elements indicate the length of faulting history on a scarp, and midslope steepness indicates relative recency of surface faulting. None of the high multiple event scarps has a sharp crest.

Midslope angle and height/length ratio of single event scarps and shorelines are strongly correlated because those slope components developed synchronously. The components of most scarps have formed together and the slope system is adjusted. However, this is not true of the high multiple event scarps. The curved elements of these scarps developed prior to the rejuvenation that reshaped the

TABLE 3. Correlation Matrix: Multiple Event Scarps.

| | HLR | PMS | CCP | ACBP | ARM | SCPH |
|---|---|---|---|---|---|---|
| MANG | .03 | -.26 | .04 | -.06 | -.21 | -.33 |
| HLR | | -.26 | .72 | .69 | .05 | .17 |
| PMS | | | .79 | .82 | -.12 | .08 |
| CCP | | | | .69 | -.20 | -.14 |
| ACBP | | | | | -.08 | .09 |
| ARM | | | | | | .91 |

n = 13

All correlations ±.40 are significant at the .01 level.

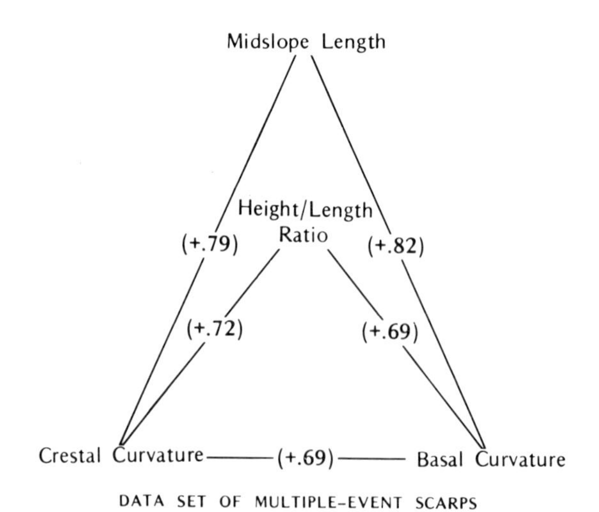


FIGURE 7. Correlation diagram of form attributes: high multiple event fault scarps. Note that a strong correlation with midslope angle is not present in these scarps, as in the set of known relative age.

midslope. Thus, MANG and HLR are poorly correlated on multiple event scarps which have been recently rejuvenated, unless the scarp is so young that it displays little or no development of the curved slope forms. On long-inactive multiple scarps, the relationship between MANG and HLR will redevelop.

Many multiple offset scarps have very similar profiles (Figure 8-A, B, C). The resemblance is striking, because they have resulted from repeated offset in different unconsolidated materials and piedmont settings. These scarps have a much greater form similarity than the pre-Bonneville scarps, which have been subjected to a longer term of degradation. Profile 8-A is a scarp that crosses an alluvial fan and profile 8-B is a site 50 miles south, that offsets fine-grained lake sediments. When these two profiles are overlain, they are almost identical. Along the Wasatch, scarps that display this form resemblance are generally the highest multiple-offset scarps. Evidently their geomorphic likeness results from an approximately equal amount of offset within the same time span, and a similar history of denudation.

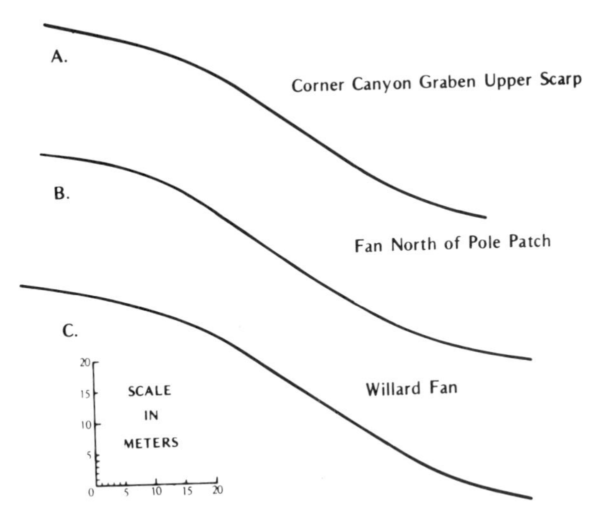


FIGURE 8. Profiles illustrating the form similarity of high multiple event scarps.

## SCARP FORM AND EQUILIBRIUM TENDENCY

Dynamic equilibrium in landforms was proposed by G. K. Gilbert (1877), who made many important observations concerning a tendency toward balance between processes of uplift and degradation.

1. Slope angle is a major control of erosional power and the relationship between slope and erosion is greater than a simple ratio (p. 115).
2. A tendency exists toward equality of denudation rates over the parts of a slope (p. 123).
3. A change in one slope component may influence another because parts of a slope system are interdependent (p.124).
4. The assemblage of slope forms present in an area has resulted from a balance between the erodibility of the rock, amount of downwasting and uplift (p. 115).

These concepts have been developed by many workers (e.g. Strahler, 1950; Hack, 1960; Scheidegger, 1961; Chorley, 1962; Schumm and Lichty, 1965). The processes involved are often difficult to observe or measure, but they have been important in the development of scarps along the Wasatch.

Wasatch piedmont scarps exhibit two phases of form evolution: rapid convergence into a convex-straight-concave profile, followed by gradual slope decline and a progressive sequence of degradational forms. Scarps rapidly grade to a state of adjustment, or near-equilibrium, characterized by strong correlation among the components of scarp form. Controlled by negative exponential rates of decay, rapid initial change is followed by slower long term change. Further change will not affect the correlation as it is proportional among the parts of the slope system. Temporal change in piedmont scarps is characterized by a tendency toward interadjustment among slope components. Prior to adjustment, slope components are weakly correlated as shown by a wide scatter of points for young scarps in Figure 5. Afterward, scarp forms remain interadjusted but continue to undergo degradational slope change. This paradoxical relationship, of being in a near-equilibrium state, and yet continuing to change through downwasting has been discussed by Howard (1965). The difference is the time span being considered; observation of a scarp at a point in time (graded time or one profile) indicates interadjustment and stability. Longer observation (cyclic time or a time-transgressive set of profiles) shows cumulative erosion of forms through time. A system may never reach perfect equilibrium, but within a finite period of time it will approximate equilibrium (Howard, 1965). Hack (1960) noted that the time required for most slopes to reach equilibrium is usually short. Few Wasatch scarps have an angular form, and the presence of a free face is rare.

The rapid gradation of a piedmont scarp to a state of adjustment can be explained by the degradational energy on different parts of the slope. The formation of a fault scarp creates a maximum slope angle, with great energy for degradation. It is also a state of disturbance in the slope system. A newly formed piedmont scarp has an inequality of erosion rates on various parts of the slope. This process affects all scarps, and multiple event scarps provide an excellent illustration. Renewed offset rejuvenates the midslope, which is rapidly graded because of the high energy potential provided by the oversteepened scarp face. The previously adjusted crest and footslope have a lower energy potential, and are much less affected by rejuvenation and subsequent degradation than the midslope. The renewed free face rapidly regrades to a straight debris slope and the scarp again has a convex-straight-concave form. On these multiple scarps, rejuvenation has apparently operated faster than degradation. The profile is smoothed, but the midslope remains steep.

Once adjustment is attained, further scarp modification occurs more slowly. Erosion of the crest and midslope combine with deposition at the base to lower the slope angle. Available energy decreases with slope decline, and erosional power becomes more evenly distributed. The rate of degradation lessens and erosion potential diminishes. On slopes where debris accumulates at the foot, erosion of the crest and deposition at the base operate jointly (Scheidegger, 1961). A piedmont scarp profile has a balance point between degradational and depositional slope components. Subsystems on either side of this axis are interdependent, and a change on one side will accompany a change in the other (Figure 9). This relationship has been the basis for computer modeling of scarp degradation (Nash, 1981). The erosional/depositional balance for most scarps is located where the fault intersects the surface. In piedmont scarps the fault is located in midslope, rather than in front of the scarp as is common to eroded bedrock scarps. The upper part of the footwall is eroded to form the crest. Debris accumulates at the base and preserves the fault trace beneath the lower midslope. The upper midslope is erosional and the lower midslope is depositional. The initial scarp (t1) has a large area of slope discontinuity (midslope) and erosional energy is concentrated on this zone. Erosion and deposition diminish the area of disequilibrium (t2, t3, t4). The crest and footslope expand at the expense of the midslope, and the area of the slope that is poorly adjusted is reduced.

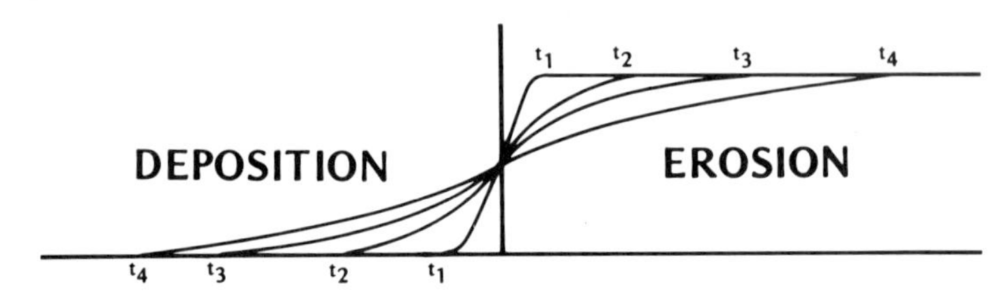


FIGURE 9. The degradation of piedmont scarps relative to an erosional/depositional balance point (after Scheidegger, 1961). The example profiles illustrate temporal slope changes found in Wasatch Front piedmont scarps and Lake Bonneville shorelines.

## A TERNARY DIAGRAM MODEL

A model of piedmont scarp degradation illustrates the characteristic denudation sequence of hillslope profiles. Length percentages of the crest, midslope, and base (footslope) may be represented on a ternary diagram (Figure 10). While the ternary diagram represents only length percentage distributions, the example profiles also integrate other time related changes in the slope system, particularly rounding of the curved elements and the decline in midslope angle. The ternary diagram shows slope forms that represent a typical sequence of long-term scarp degradation (Figure 10-A, 10-B, 10-C, and 10-D). Young scarps are represented by the end of the triangular graph that indicates a midslope dominated profile (10-A). Degradation will shift the slope toward a more even length percentage distribution among the three components (10-C). Apparently the near-equilibrium state is one of roughly equal length percentages of these components. The high multiple event scarps also converge on the center of the graph, although rejuvenation temporarily disrupts the even distribution of the component lengths. Continued degradation shortens the midslope relative to the curved slope elements. Pre-Bonneville scarps, characterized by short midslope lengths, are represented by the lower center of the graph (10-D), indicating profile domination by the crest and footslope.

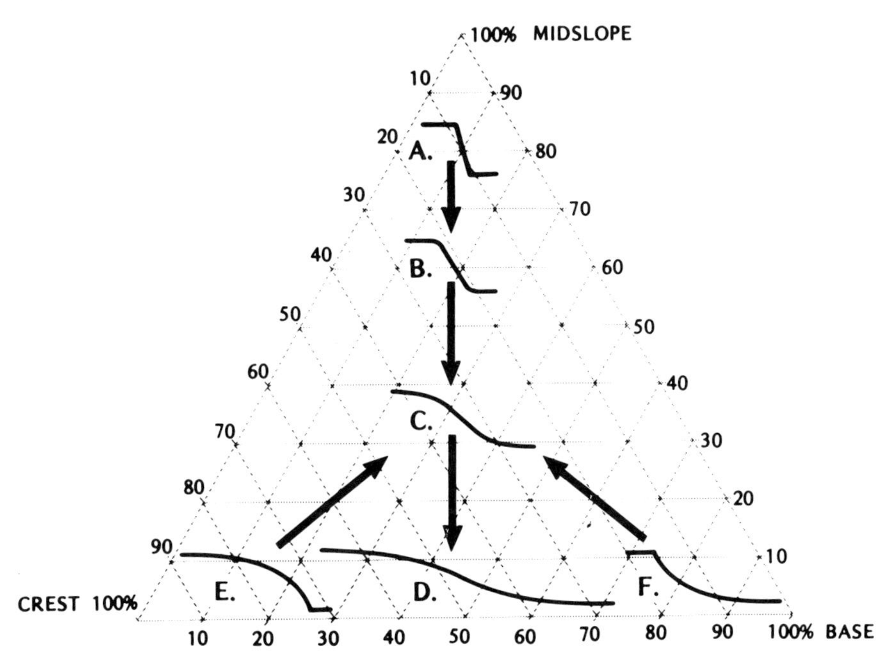


FIGURE 10. Ternary diagram model of piedmont scarp degradation, illustrating the tendency toward equilibrium and the convergence of hillslope profile forms. The arrows indicate the sequence of forms resulting from degradation.

Profiles representing ternary diagram end members are examples of unstable forms. Wasatch scarp profiles display three variations of slope disequilibrium that approach these hypothetical forms. The creation of a single event scarp produces a profile form that is dominated by the midslope. This imbalance will be altered by denudation unless the scarp undergoes rejuvenation before the curved elements develop. On scarps that have not yet been rounded, rejuvenation may produce a multiple event scarp with a form that is indistinguishable from those of single event scarps. Two or more events of surface faulting, closely spaced in time, may produce the same scarp form as one larger event. Length of recurrence interval relative to degradation rates is the controlling factor in this process. Dependent on the length of the recurrence interval, a scarp will progress through all or part of the form sequence represented on the center of the diagram (10-A, B, C, and D) prior to rejuvenation. Previously graded multiple event scarps may undergo several variations of slope change, dependent on the nature of the rejuvenation. Renewed faulting may not coincide exactly with the previous fault trace. While this temporarily complicates the slope form, degradation will shift the profile back toward a balanced three component hillslope. Renewed offset near the scarp base tends to produce forms that are dominated by the crest (10-E), and offset near the crest produces slopes that are mainly concave (10-F). Variations 10-E and 10-F represent imbalanced slopes that quickly regrade to a convex-straight-concave form (10-C). Rejuvenation on the previous fault trace lengthens the midslope. On most multiple scarps, midslope lengthening caused by renewed offset reinforces the tendency toward an even distribution of slope length percentages. If a long time elapses between the last offset and rejuvenation, degradation will reduce midslope length relative to the curved profile elements. Few scarps have forms that are greatly imbalanced among the crest, midslope, and footslope components. Apparently, regrading operates rapidly (faster than the recurrence interval for surface faulting) which tends to restore a balance among parts of the three-component slope form.

The regrading of multiple event scarps is a geomorphological illustration of Le Chatelier's principle; a system responds to an external change with an action that tends to minimize the disturbance. Furthermore, "...the initial rate of change back toward equilibrium is proportional to the degree of disequilibrium induced, and thereafter it decreases exponentially," (Ruxton, 1968, p. 31). Degradational energy focuses on the slope component that is most out of adjustment with the rest of the slope system. If rejuvenation oversteepens the base, gravity and slope wash processes will deposit material over the new break, extending the midslope and footslope at the expense of the crest. If the crest is oversteepened by faulting near the top of the scarp, re-rounding of the crest will proceed rapidly. The convex-straight-concave slope is restored as energy is shifted to the most disturbed portion of the slope profile. This process maintains the convex-straight-concave shape as a very persistent slope form. Form convergence greatly complicates attempts to derive indices of scarp age based on form measurement. Based on hillslope form it is often difficult or impossible to separate young single event scarps and young but rapidly rejuvenated multiple event scarps. Rejuvenation that occurs before rounding develops, will obscure the evidence for multiple events, once the slope break created by renewed faulting is regraded. If surface faulting ceases and no longer rejuvenates the slope, a scarp will progress through the normal degradational sequence. Single and multiple event scarps which have long been inactive also tend to have a similar profile form. Multiple event scarps which represent numerous events of surface faulting, particularly those with a well-rounded crest and footslope, can be discriminated on the basis of poor correlation between midslope angle and either height/length ratio, or the indices of curvature.

## CONCLUSION

Topographic profiles of piedmont scarps and wave-cut cliffs of Lake Bonneville suggest a degradation chronology characterized by a decline in midslope angle, extension of profile length, midslope shortening, and increased rounding of the curved slope elements. Immediately after offset, young scarps are in disequilibrium as indicated by a poor interadjustment of hillslope profile components. During this early stage, intensity of erosional stresses is great and grading to state of interadjustment among form attributes is rapid. Rapid grading of maladjusted parts creates a tendency toward the development, maintenance, or restoration of a convex-straight-concave slope profile. This tendency toward convergence into a balanced three-component slope strongly affects scarps that have developed through repeated offset. Correlation of form components indicates a lack of complete interadjustment in multiple event scarps. Nevertheless, the similarity in profile form of many high multiple scarps is striking. They appear to have converged in form to a convex-straight-concave profile in a relatively short time in comparison to the recurrence interval of surface faulting. Several episodes of rejuvenation and the resulting increase in scarp height provide an erosional environment favorable to the intensive regrading that these scarps have apparently undergone. Three factors greatly complicate attempts to derive indices of scarp ages based on hillslope form attributes: the convergence of different scarp types on a similar hillslope form, the overlap of slope attribute values among scarps of differing ages, and the difficulty in discriminating the number and size of events that produced a multiple event scarp. A more complete understanding of the denudation chronology and rates of change will require additional age-controlled study, and a better knowledge of the number of events responsible for a scarp.

The ternary diagram model describes profile form and depicts how piedmont scarp profiles change through time. Representation of three-component profiles on a ternary diagram may be a useful method of classifying slope profiles in other applications.

## REFERENCES

Ahnert, F., 1970, An approach toward a descriptive classification of slopes: Zeitschr. Geomorphologie, Supp. 9, p. 71-84.

Blong, R. J., 1975, Hillslope morphometry and classification: a New Zealand example: Zeitschr. Geomorphologie, v. 19, p. 409-529.

Bucknam, R. C., and Anderson, R. E., 1979. Estimation of fault-scarp ages from a scarp height-slope angle relationship: Geology, v. 7, p. 11-14.

Chorley, R. J., 1962, Geomorphology and general systems theory: U. S. Geol. Survey Prof. Paper 500-B, p. B1-B10.

Chorley, R. J., and Kennedy, B., 1971. Physical geography-- a systems approach: London, Prentice-Hall International, 370 p.

Currey, D. R., Atwood, G., Mabey, D., 1984, Levels of the Great Salt Lake and its predecessors: Utah Geological and Mineral Survey, Miscellaneous Map (in press).

Gilbert, G. K., 1877, Report on the geology of the Henry Mountains: U. S. Geol. Survey, 160 p.

______, 1884, A theory of the earthquakes of the Great Basin, with a practical application: Am. Jour. Sci., ser. 3, v. 27, p. 49-53.

______, 1890, Lake Bonneville: U. S. Geol. Survey Monograph 1, 438 p.

______, 1928, Studies of Basin-Range structure: U. S. Geol. Survey Prof. Paper 153, 92 p.

Hack, J. T., 1960, Interpretation of erosional topography in humid temperate regions: Am. Jour. Sci., v. 258-A, p. 80-97.

Howard, A. D., 1965, Geomorphological systems-- equilibrium and dynamics: Am. Jour. Sci., v. 263, p. 302-312.

Melton, M. A., 1957, An analysis of the relations among elements of climate, surface properties, and geomorphology: Columbia Univ., Office of Naval Research, Tech. Report 11, 102 p.

Morisawa, M., 1975, Tectonics and geomorphic models, in W. N. Melhorn and R. C. Flemal, eds., Theories of landform development: Binghamton, State University of New York Publications in Geomorphology, p. 199-216.

Morrison, R. B., 1965, Lake Bonneville: Quaternary stratigraphy of eastern Jordan Valley, south of Salt Lake City, Utah: U. S. Geol. Survey Prof. Paper 477, 80 p.

Nash, D. B., 1981, Fault: A FORTRAN program for modeling the degradation of active normal fault scarps: Comput. Geosci., v. 7, p. 249-260.

Ruxton, B. P., 1968, Order and disorder in land, in G. A. Stewart, ed., Land evaluation: Melbourne, Macmillan, p. 29-39.

Scheidegger, A. E., 1961, Mathematical models of slope development: Geol. Soc. Am. Bull., v. 72, p. 37-50.

Schumm, S. A., 1966, The development and evolution of hillslopes: Jour. Geological Education, v. 14, p. 98-104.

Schumm S. A., and Lichty, R. W., 1965, Time, space, and causality in geomorphology: Am. Jour. Sci., v. 263, p. 110-119.

Scott, W. E., 1982, Guidebook for the 1982 Friends of the Pleistocene, Rocky Mountain Cell, field trip to Little Valley and Jordan Valley, Utah: U.S. Geological Survey Open-File Report 82-845.

Slemmons, D. B., 1957, Geological effects of the Dixie Valley-Fairview Peak, Nevada, earthquakes of December 16, 1954: Seis. Soc. Am. Bull., v. 47, p. 353-375.

Strahler, A. N., 1950, Equilibrium theory of erosional slopes approached by frequency distribution analysis: Am. Jour. Sci. v. 248, p. 673-696 and 800-814.

Wallace, R. E., 1977, Profiles and ages of young fault scarps, north-central Nevada: Geol. Soc. Am. Bull., v. 88, p. 1267-1281.

______, 1980, Degradation of the Hebgen Lake fault scarps of 1959: Geology, v. 8, p. 225-229.

Young, A., 1972, Slopes: Edinburgh, Oliver and Boyd, 228 p.

# 10
# Tectonic geomorphology of the basin and range Colorado Plateau boundary in Arizona

*Larry Mayer*

**ABSTRACT**

The boundary between the Colorado Plateau and Basin and Range provinces in Arizona developed diachronously from south to north in Tertiary time. Topographic differentiation was certainly complete by 12 million years ago and was followed by the cutting of the Grand Canyon about 5 million years ago. Major faulting appears to be pre-Quaternary, though some Pleistocene faulting is recorded by fault scarps. Documented Holocene faulting is absent in the area between the Grand Wash fault and the Virgin River and also in the Transition Zone. The timing of these events is based on tectonic geomorphologic estimates that are derived from models describing the changes in landform morphology that result from erosion. These models include fault scarp degradation, stream profile adjustment, and escarpment retreat.

**INTRODUCTION**

Arizona is subdivided into three physiographic provinces (Fig. 1): the Basin and Range, Mountain Highlands and Colorado Plateau (Ransome, 1904; 1923; Fenneman, 1931). The Mountain Highlands Province lying immediately adjacent to the Colorado Plateau is transitional in its topographic and structural character between the Basin and Range and Colorado Plateau and has been referred to the the Transition Zone (Wilson and Moore, 1959). The boundary between the Basin and Range and the Transition Zone is characterized by a westward decrease in base-level as one approaches the Colorado River and also a southward decrease in level following the course of the Colorado River (Fig. 2). The edge of the Colorado Plateau in central Arizona is placed at the Mogollon Rim, an escarpment with a half kilometer of local relief. In northwestern Arizona the Colorado Plateau edge is located at the Grand Wash Cliffs, an escarpment generated by a fault that has perhaps a total accumulated displacement of 5 km since about 17 MA (Lucchitta, 1972). This paper briefly describes the development of the boundary between the Basin and Range province - Colorado Plateau province but concentrates on tectonic geomorphological methods useful in studying landforms within in the region.

## Tectonic Setting

The physiographic boundary separating the Colorado Plateau from the Basin and Range parallels the ancient margin of the Cordilleran miogeocline which formed during the late Precambrian (Stewart and Suczek, 1977; Armin and Mayer; 1983). The location of this rifted margin is recorded as a rapid thickening of miogeoclinal strata from the Grand Wash Cliffs in northwestern Arizona, westward into Nevada (Stewart and Suczek, 1977). The coincidence of the Basin and Range - Colorado Plateau boundary with the Precambrian margin of North America may suggest that fundamental Precambrian rift structures in the lithosphere strongly influenced the geometry of subsequent tectonic episodes.

The Basin and Range province is characterized by lithospheric thinning and extension. The structural, geophysical, and volcanic characteristics of Basin and Range extension have been the subject of many recent studies (Best and Hamblin, 1978; Dickinson and Snyder, 1978; Eaton et al., 1979; Eaton, 1979, 1982; Lipman, 1980; Smith, 1978; Snyder et al., 1976; Stewart, 1978; Thompson and Zoback, 1979; Zoback et al. 1981). Briefly, proposed models for Basin and Range extension include oblique extension within a wide transform boundary (Atwater, 1970) and volcanic arc related extension (Scholz et al., 1971; Coney and Reynolds, 1977; Eaton, 1983).

Both the Basin and Range and Colorado Plateau are anomalously high topographically. The regional topography may be related to the migration of the North American Plate over hot spots (Suppe et al.,

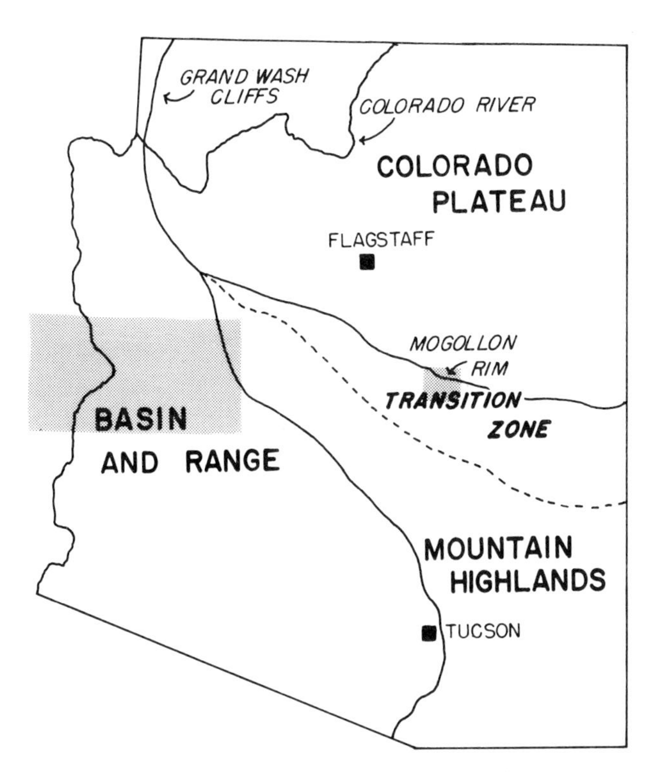


FIGURE 1. Location map showing physiographic provinces of Arizona and place names. Large shaded area in west central Arizona is the location of Figure 2. Small shaded area along the Mogollon Rim in central Arizona is the location of Figure 11.

1975) or by non-uniform extension of the lithosphere and thermal buoyancy (Mayer, 1982). The lithosphere under the Basin and Range is as thin as 25 km in contrast to the Colorado Plateau which is 70 km thick in places (Smith, 1978; Thompson and Zoback, 1979). "Normal" continental lithosphere is in excess of 100 km thick.

## Stratigraphic Constraints on the Development of the Plateau Edge

The Basin and Range of southern and central Arizona was the location of a persistant topographic high during the Mesozoic. It was the source for sediments deposited on the Colorado Plateau (Chamberlin and Salisbury, 1906; Schuchert and Dunbar, 1941; McKee et al., 1956) and was referred to by Harshbarger et al. (1957) and

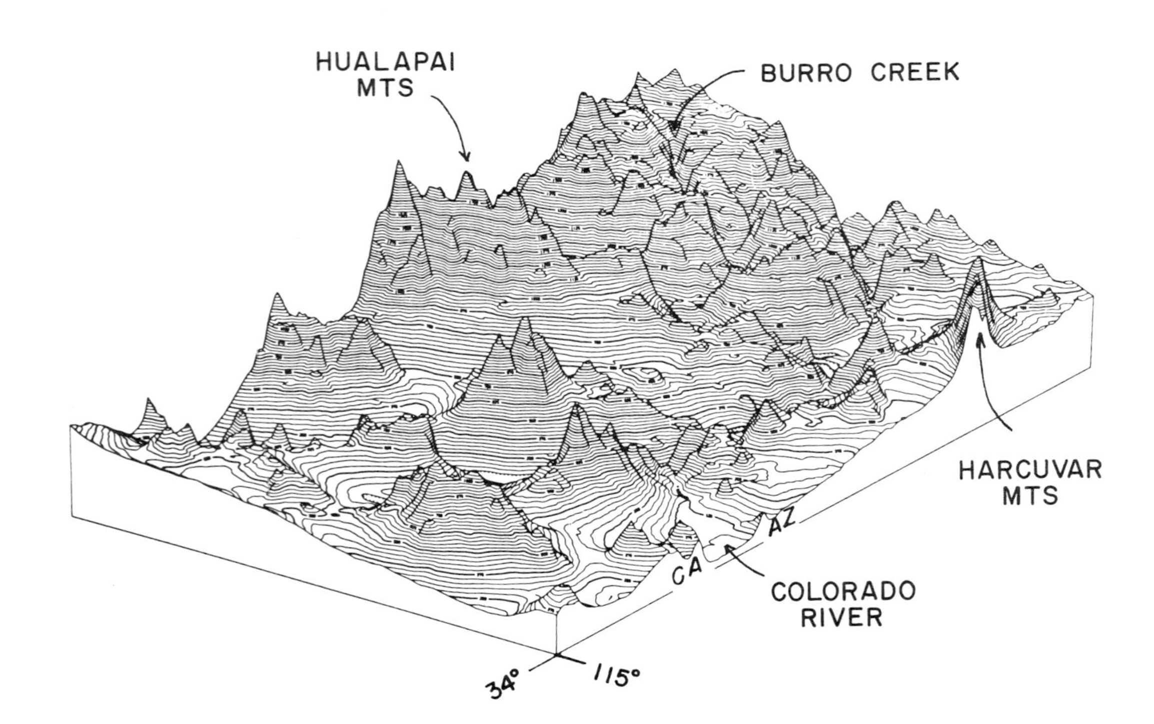


FIGURE 2. Three dimensional contour perspective diagram showing the transition between the Mountain Highlands province and the Basin and Range province in western Arizona. The topographic profiles along the edges of the block figure illustrate the decrease in elevation westward toward the Colorado River.

Cooley and Davidson (1963) as the Mogollon Highlands (Fig. 3). The Mogollon Highlands of central Arizona remained higher than the adjacent Colorado Plateau into the early Tertiary (Lindgren, 1926; Anderson and Creasey, 1958) and by middle Tertiary time the Paleozoic miogeoclinal strata had been stripped away exposing Precambrian basement (McKee, 1951).

Gravels from the Mogollon Highlands were deposited on the Colorado Plateau and their age estimates are constrained by the ages of volcanic clasts contained within them or overlying volcanic flows. These gravel deposits document a period of time during which drainage flowed northward across the present Colorado Plateau - Basin and Range boundary. The northward flow directions is based upon provenance of rock-types found in the deeply eroded Transition Zone and imbricated gravel deposits. The youngest of these gravels in west-central Arizona, the Rim Gravels (Price, 1950) are probably Oligocene in age, ca. 28 MA (Peirce et al., 1979). Imbricated gravel deposits preserved beneath the Buckhead Mesa basalts which are dated at 12 MA (Peirce et al., 1979) and are situated well below the present elevation of the Colorado Plateau, show flow directions to the east-southeast (Mayer, 1979). Thus for this section of the Colorado Plateau edge, drainage reversal occurred prior to 12 MA.

Farther west, a latite cobble in the gravels of Oak Creek Canyon was dated as 23.1 MA (McKee and McKee, 1972). Younger basalt cobbles dated at ca. 12 MA were also reported (McKee and McKee, 1972) but the source of the gravels that contained them is in question (Peirce et al., 1979).

The Peach Springs Tuff in northwestern Arizona, dated at 17 MA, is the youngest age for the Mogollon Higlands in that area (Young and Brennen, 1974). Imbricated gravels preserved beneath this ash flow tuff indicates a northeasterly drainage direction, though there is some evidence of pre-tuff drainage divide trending northwest that presumably separated northeasterly flowing from southerly flowing streams (Young and Brennen, 1974).

These data indicate that, in Arizona, the major structural and related topographic differentiation between the Colorado Plateau and Basin and Range is younger than about 20 MA, though the establishment of drainage divides representing concomitant drainage reversal may have occurred earlier (Young and Brennen, 1974).

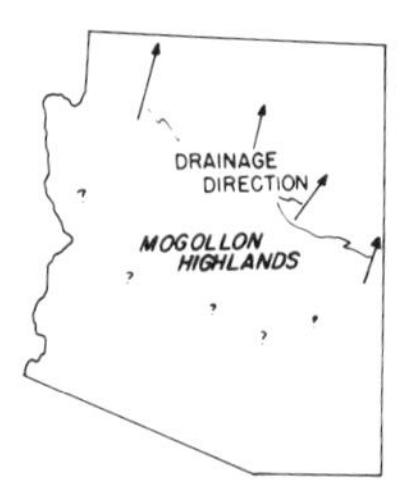


FIGURE 3. Location of the Mogollon Highlands in central Arizona during early Cenozoic time (after Cooley and Davidson, 1963). The Mogollon Highlands was the source area for sediments that were shed northward across what is now the Colorado Plateau edge. The southern extent of the Mogollon Highlands is not known.

## TECTONIC GEOMORPHOLOGICAL METHODS

Geomorphological data help further constrain the tectonic history of the Basin and Range - Colorado Plateau transition. These data are based on landforms that change with time either in a predictable fashion or in a way that allows comparison and ranking with landforms of known age. Landforms with short survival times such as fault scarps cutting alluvium, permit relatively accurate determinations of recent faulting history. In contrast, landforms with long survival times such as high bedrock escarpments, permit very approximate determinations of fault history but describe much longer time spans.

### Escarpments

Escarpments are a spectacular result of normal faulting or combinations of faulting and erosion along the margins of the Colorado Plateau in Arizona. The major escarpments bounding the Colorado Plateau are the Grand Wash Cliffs and the Mogollon Rim, whose ages or history are known only in a general way. The purpose of the following discussion is to introduce methods that may be useful in quantifying the relative tectonic activity of the escarpments or their evolution.

The morphologic change of escarpments depends on many factors including the location of major drainage divides, their drainage areas, rock-types, and escarpment height. In a general fashion, escarpments resemble fault generated mountain fronts. Relative tectonic activity of mountain fronts based on morphology and the balances between various piedmont processes have been successfully applied (Bull, 1970, 1973; Bull and McFadden, 1977) in the Basin and Range. Escarpments however, tend to be more diverse with respect to piedmont processes and their morphology, eg. sinuosity, is dependent on stream size in addition to time since faulting. A tectonic classification of escarpments analogous to Bull's tectonic activity classes of mountain fronts is desirable but not yet at hand.

Escarpments tend to follow either a slope retreat or slope replacement type of evolution (Bakker and LeHeux, 1950, in Scheidegger, 1961; Koons, 1955; Carson and Kirkby, 1972). In either case the escarpment crest migrates back due to erosion. The Grand Wash Cliffs, an escarpment caused by faulting along the Grand Wash Fault, appears to follow a retreat model.

Age estimates of escarpments derived by morphology are complicated by the fact that escarpments form over a period of time which is controlled by uplift rates. Therefore the "age" of an escarpment implies something quite different from the age of a single rupture fault scarp. In addition, the processes acting on escarpments are difficult to model in a way that can be used for estimating their age. In the case of perfectly parallel retreat, the escarpment form is esentially time independent and thus not useful in dating. From the outset it is clear that morphology-derived ages will not have high resolution. Ancillary data, such as those provided by stratigraphy and faulted volcanic units, are of great value for placing bounds on escarpment ages. Despite these complications, two features of escarpments that seem very useful in determining their age are the morphology of longitudinal stream profiles and the morphology of embayments.

**Models**. Stream channels are vertically offset by faults that generate escarpments resulting in a break in slope of the bed. The shape of the break in slope can be treated as a transient perturbation (analogous to a thermal pulse in the heat flow context) that is modified by stream downcutting. If the tectonically caused slope breaks of the longitudinal stream profile varies systematically with time, then these systematic changes in the longitudinal profile can be used to estimate the time between some initial condition and the present one.

The longitudinal profile of streams has been the object of many analytical investigations. Several workers have tried to explain this feature from the perspective of theoretical physics or statistics (Culling, 1960; Scheidegger, 1961; Leopold and Langbein, 1962). For example, Culling (1960) used Fourier theory of heat flow to develop differential equations describing stream profiles and valley side slopes. Sprunt (1972) developed a computer algorithm to simulate the evolution of drainage basins, their networks and topography.

Leopold and Langbein (1962) modeled longitudinal stream profiles using a random walk method (Fig. 4). These profiles represented the tendency to maximize entropy and minimize work along the channel. Their procedure was as follows. A deck of cards, initially consisting of five white cards and one black card, was prepared. Each white card represented a unit of elevation above base level. A card was selected at random and replaced by a black card. If a white card was selected, the profile decreased one unit of elevation but if a black card was selected, then the profile remained at the same elevation. The experiment is over when all the cards are black, that is when the profile has reached base-level. The experimental sample space for this type of experiment is shown on Figure 4.

Originally, Leopold and Langbein (1962) concluded that the the longitudinal profile of a stream is dependent on elevation above base-level. Langbein (1964), related the previous results to discharge and stream length. The key, linking the Leopold and Langbein results to geomorphic processes is that for any monotonically sloping surface, vertical distances can be expressed as horizontal distances. This is easy to visualize for a retreating scarp where vertical lowering of a scarp by erosion results in horizontal translation of the scarp face. By mapping, in a mathematical sense, vertical changes onto the horizontal, a simple mathematical model can be used to describe channel slope change related to distance from the headwaters.

To apply such notions to dating the age of an escarpment uplift, several assumptions are needed. We need to assume that the uplift is instantaneous, that the profile will adjust to some equilibrium form, and that the rate of stream downcutting following an uplift is proportional to the curvature of the profile. A

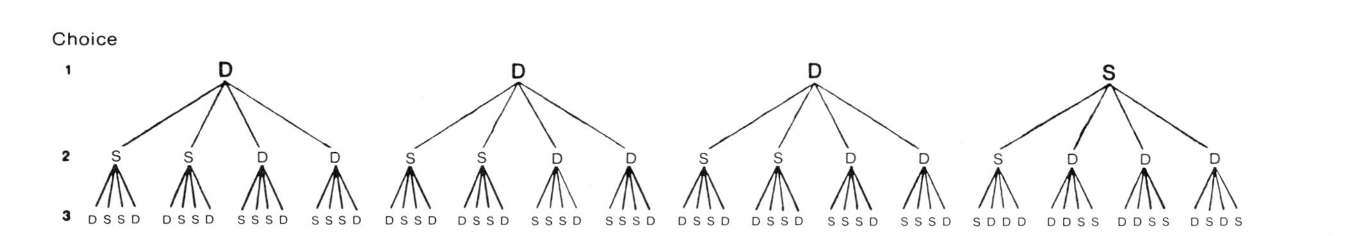


FIGURE 4. Tree diagram illustrating the sample space of the random walk experiment of Leopold and Langbein (1962). In the case shown here there are four cards initially represented by choice number one. The probabilities for three card selections can be calculated. D represents a downward step and S is no change in elevation. The probability of a downward step at each card choice is the number of D's divided by the total number of possible outcomes. For choice number one the probability of a downward step is 3/4 or 0.75.

diagram of these assumptions and their symbolic form is shown on Figure 5, where the profile is plotted on semi-log axis in order to represent the profile as a straight line (see for example Hack, 1973).

This system can be described by the heat conduction equation or diffusion equation (B. Begin, written communication, has independently applied the diffusion equation using different boundary conditions to the migration of knickpoints in alluvial floored channels) and states

$$\frac{\partial D}{\partial t} = k \frac{\partial^2 D}{\partial H^2}$$

where the constant k is related to all factors influencing the erodibility of the channel floor.

The boundary conditions can be stated as

$$D_0=0, \quad H=H_1$$
$$D=D_1, \quad H=0$$

In words, at the headwaters $D_0$, the height above base-level is $H_1$ and at the fault located at $D_1$, the height above base-level is zero. The solution for D is (Carslaw and Jaeger, 1959)

$$D=D_1\{1- \frac{H}{H_1} + \frac{2}{\pi}\sum_{n=1}^{\infty} \frac{(-1)^{n+1}}{n} \frac{\mathrm{Sin}\, n\pi(\frac{1}{P})\,\mathrm{Sin}\frac{n\pi H}{H_1}}{n\pi(\frac{1}{P})} \exp(-n^2\pi^2 Kt/H_1^{\,2})\}$$

where D is a function of H and time for a given P and $H_1$.

There are several constraints on the application of this method. First, to date, there are too few calibrated examples to allow for an estimate of the error involved. Second, there are cases where geologic controls on a stream channel do not permit rounding of the knickpoint, rather it is maintained to some degree independent of time. For noncohesive materials, there is some experimental evidence to support knickpoint rounding (Brush and Wolman, 1960).

To determine the rate constant, or diffusivity, the solution to the diffusion equation was fitted by eye (Fig. 6) to a stream profile, the Virgin River, with known P and t. The Virgin River channel was displaced by faulting across the Hurricane fault repeatedly during the Quaternary. Basalts that flowed into the pre-faulted channel are dated as 0.3 MA (Hamblin et al., 1981) and have been offset vertically about 90 meters. Recall that the model assumes all the uplift to have occurred at the same time. Bearing this in mind, the results of dating profile changes in tributary streams of the Grand Canyon (Fig. 7) indicate that the fall in

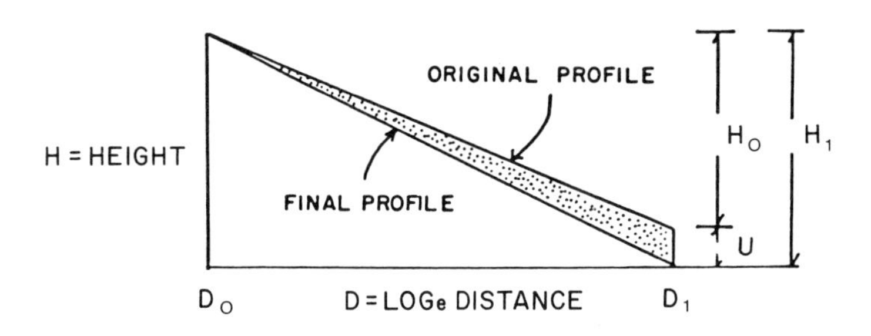


t = TIME SINCE UPLIFT
U = MAGNITUDE OF UPLIFT
P = TECTONIC PERTURBATION

$$P = \left( \frac{1}{1 - U/H_1} \right)$$

FIGURE 5. Diagram showing the relationships between the parameters used in the diffusion model of stream profile adjustment to instantaneous tectonic uplift. Symbols are defined in the text. P, the tectonic perturbation is a function of the amount of uplift relative to the basin relief. The stippled pattern indicates the area of the stream channel that is eroded after complete adjustment to the uplift. Equations in the text predict how much of the adjustment has occurred in a given period of time.

base-level recorded by the Colorado River's downcutting east of the Grand Wash Cliffs occurred about 5 million years ago. The age of the lower Colorado River is constrained by the ages of the Muddy Creek Formation and Colorado River gravels. A basalt flow located at Fortification Hill and resting on Muddy Creek deposits is dated at 5.9 MA (Shafiquallah et al., 1980). In another location a 3.8 MA basalt rests on Colorado River gravels (Shafiquallah et al., 1980).

The reasonable results of the Grand Canyon tributary "dates" are permissive evidence in support of analytical models of profile change. However, several additional studies are needed to establish the validity of the model.

A more straightforward approach to dating escarpments is based on information about slope retreat rates. Fault-generated escarpments retreat from the fault with time assuming instantaneous uplift. The width of the resulting pediment is a relative age indicator and if the rate of retreat is known, then the pediment width divided by the rate of retreat is an age estimate. Related phenomena are deep embayments cut in the escarpment face by large rivers that drain the Plateau. The geometry of these embayments often resemble a triangular slice cut into an escarpment (Fig. 8).

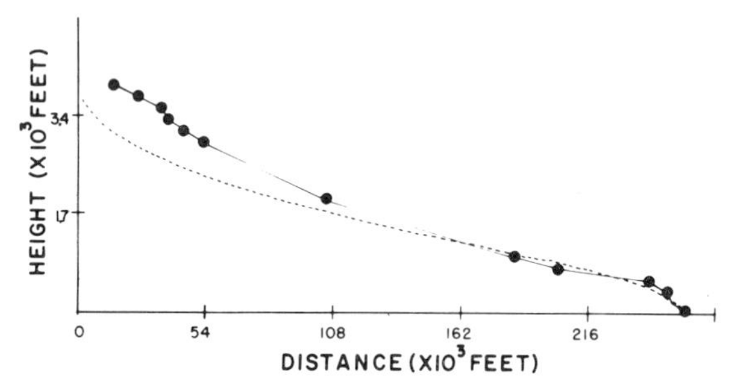


FIGURE 6. Fit of the diffusion model to the stream profile of the Virgin River at the Hurricane Fault. Solid line is the Virgin River profile measured from 1:250 000 scale topographic maps while the dashed line is the theoretical curve for a 0.3 MA old uplift.

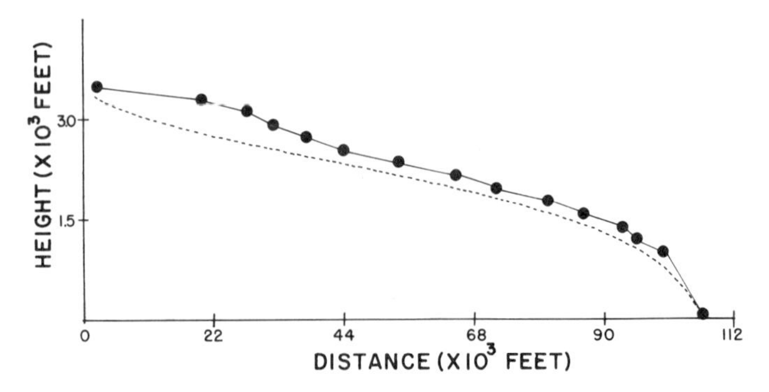


FIGURE 7. Fit of the diffusion model to the stream profile of Whitmore Wash, a tributary to the Colorado River in the Grand Canyon. Solid line is the actual profile while the dashed line is the model profile for a 5 MA uplift.

FIGURE 8. Photograph of an embayment into the Hurricane Cliffs south of the town of Hurricane, Utah. Often, embayments resemble triangular slices into the escarpment. Note the pronounced knickpoint that has migrated upstream from the fault.

The relative dimensions of the embayment for an escarpment of a given age are determined by the size of the stream that cut the embayment, that is the larger the stream, the longer the length of the embayment (Fig. 9). For tributaries to the Grand Canyon, the relationship between embayment length, defined by the predominant knickpoint location, and stream length is linear as expected (Fig. 10). The width of the embayment is controlled by the rate of escarpment retreat and therefore for a given age escarpment, the embayments would be expected to have similar widths. The sinuosity of the escarpment is then largely a function of the size of the streams flowing across it and the time since faulting. An advantage of using embayment width rather than pediment width is that the location of the fault is not needed. However all these methods are clearly crude approximations. Based on published estimates of cliff retreat (Yair and Gerson, 1974; Young, 1974; Cole and Mayer, 1983) a reasonable age estimate can be obtained using a cliff retreat rate of 0.5 0.4 meters per thousand years.

The Mogollon Rim is at least 12 km away from the Diamond Rim fault at Buckhead Mesa (Fig. 11) which may have taken 24 MA of scarp retreat. The Upper Grand Wash Cliffs lie 6-9 km away from the Grand Wash Fault. Assuming cliff retreat from the fault implies an age of about 12-18 MA. The older age is close to that derived from the Peach Springs Tuff. The embayment at Pigeon Canyon on the Upper

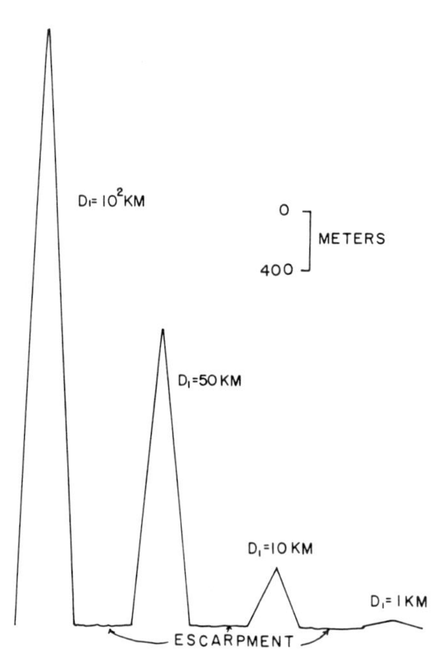


FIGURE 9. Idealized relationship between the geometry of a triangular embayment and the size of the stream.This is a map view. Note that where the stream cutting the escarpment is very small the overall morphology is controlled by the rate of scarp retreat.

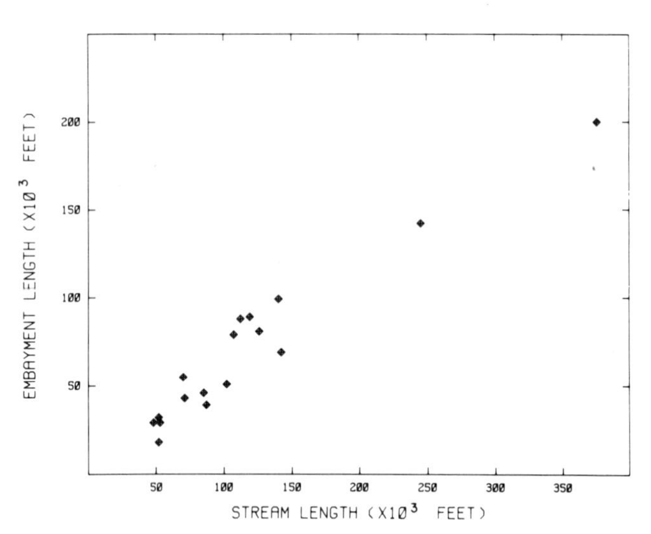


FIGURE 10. Plot of embayment length versus stream length for tributaries of the Colorado River in the Grand Canyon. Embayment length is defined by the major knickpoint on stream profiles measured from 1:250 000 scale topographic maps.

Grand Wash Cliffs is about 6km wide which also implies an age of about 12 MA. The crest of the Lower Grand Wash Cliffs lie about 3 km from the Grand Wash Fault implying an age of about 6 MA. This latter figure is consistent with the age of the Grand Canyon derived from stream profiles and stratigraphy.

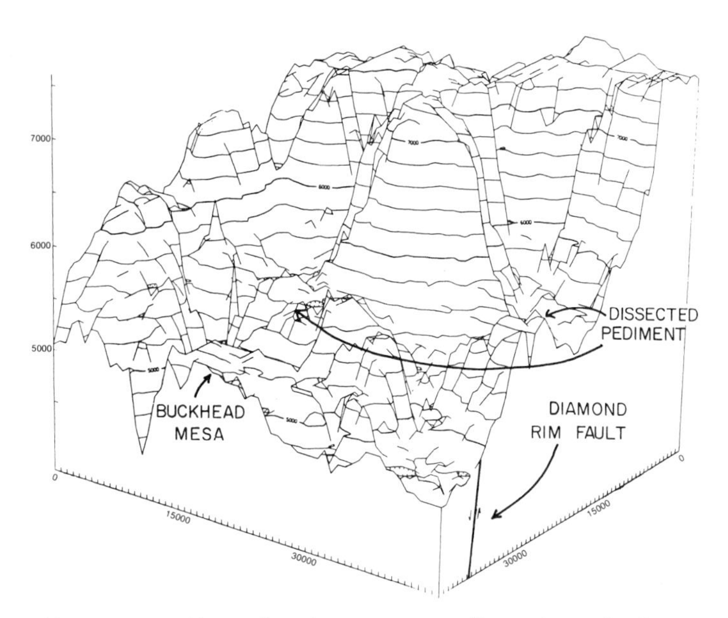


FIGURE 11. Three dimensional contour perspective of the Mogollon Rim in central Arizona illustrating the relationship between the escarpment bounding fault and the present location of the Plateau edge. Vertical scale is elevation in feet above sea level. Horizontal scales are in feet.

## Fault Scarps

The relative freshness of a fault scarp has long been used as a qualitative way to determine scarp age. Wallace (1977) surveyed fault scarp profiles and used their form and the dominant geomorphic process operating on the fault scarp as an age indicator (Fig. 12). Bucknam and Anderson (1979) demonstrated for a scarp of a given age, that the relationship between the logarithm of scarp height and the maximum scarp slope is linear. Further, they established a reference scarp, the Lake Bonneville shoreline scarp, that has been subsequently used to calibrate models and allow relative age comparisons. Assuming that the initial fault scarp slope in cohesionless or slightly cohesive material is controlled by the angle of internal friction of the material, then the initial scarp slope is relatively constant and independent of scarp height. With time, scarp slope becomes dependent on scarp height, or in other words, low scarps degrade more rapidly than high scarps.

**Processes Operating on Scarps.** Scarp degradation is accomplished by a combination of channelized flow in channels and rills, overland flow, gravity movements and raindrop impact. Gravity movements begin immediately upon faulting when loose material falls off the free face and is deposited on the debris slope. Gravity movement predominates on free faced scarps. Gravity movements can also occur as mass movements such as rotational slumping during or soon after faulting. Raindrop impact also move material downslope. The potential of raindrop erosion is related to the diameter of the drop and its velocity. This process may be related to the convex form of hillslopes in transport-limited settings (Gilbert, 1909). Raindrop size is believed to be a function of storm intensity and under present climatic conditions, intense summer storms take place in the southwestern United States.

Overland flow occurs as a thin sheet of braids of water that flow across an unrilled surface. Overland flow can also be quite effective in transporting sediment (Emmett, 1970).

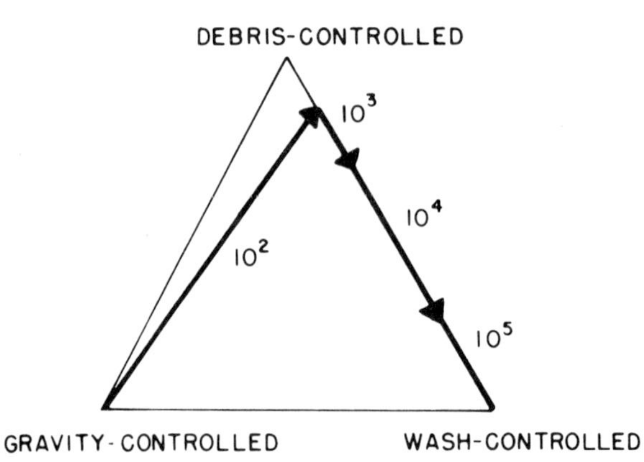


FIGURE 12. Triangular diagram showing the time spans of slope processes acting on fault scarps (concept from Wallace, 1977). The numbers along each time vector is the period of time, in years, during which the geomorphic process is most important.

**Rate Controlling Factors.** Factors that affect the rate of scarp degradation can be grouped under the general headings of lithologic and climatic controls. The physical characteristics of the faulted material determine its erodiblity and hence the degradation rates (Wallace, 1977; Dodge and Gross, 1980; Mayer, 1982). These physical characteristics include particle size, particle roughness, particle films, pedogenic induced cementation, and moisture content. Collectively, these physical parameters are related to the angle of internal friction (Carson and Kirkby, 1972).

Climatic conditions that favor ground covering vegetation such as grasses result in increased cohesion, restricted overland flow, and perhaps decreased degradation rates. Climatic changes may also affect degradation rates as perturbed vegetation assemblages readjust to new conditions.

**Scarp Models.** Fault scarp degradation models attempt to describe the evolution of a scarp with time. A relatively successful model is based on the diffusion equation describing one dimensional heat flow (Culling,1960; Nash ,1980; Colman and Watson, 1983). This model assumes that the cohesionless material eroded from the upper convex portion of the scarp will be deposited on the lower concave portion of the scarp and follow a slope rounding type of slope evolution. Slope rounding describes hillslope evolution where erosion and deposition is proportional to the surface curvature. The success of scarp evolution models depends on their application to areas and settings where basic model assumptions are valid. Thus, in an area that is actively aggrading, a fault scarp may ultimately be buried while in an area that is actively degrading, dissection may overwhelm other forms of hillslope evolution. Under either of these conditions the diffusion model is inadequate.

There are also many factors that influence the rate of scarp degradation not described in erosion models, including topography upslope from the scarp. Ridges can disperse overland flow while a topographic swale concentrates flow causing differences in erosion rates.

**Model Application.** Application of a model involves a rigorous evaluation of the sources of variation. While the model itself is important, how well a dating tool works and the confidence we have in its application, ultimately determines the usefulness of the method. Therefore too, in the case of model based scarp morphology derived dates, it is not sound practice to **assume** that all variation in scarp morphology is age related. The effects of such factors such as measurement error, scarp material, aspect (K. Peirce, oral communication, 1982), and climate require careful attention in the application of a model. Many of these factors have been examined elsewhere (Mayer, 1982; in press) and therefore I will only summarize the statistical characteristics of scarp measurements.

A fundamental concern in the application of morphology derived dating models is the matter of measurement error. Simply stated, one needs to know whether two or three people measuring the same scarp, will obtain the same measurements. Also in question are the statistical properties of these measurements and do these statistical characteristics differ from scarp to scarp. To evaluate this, three persons surveyed the identical young fault scarp profile a total of twenty times using two different techniques. The results indicate that both the maximum scarp angle and scarp height generally vary less than ten percent around their respective means.

The distribution of maximum scarp angle measurements appear to be Gaussian (Table 1). Comparison of the variances of maximum scarp angles between two scarps of very different age, one a historic rupture the other pre-Holocene, indicate that the variances are not equal at the 0.10 level but equality of variances can be accepted at the 0.15 significance level (Table 2). These data collectively indicate that standard statistical tests of equality between the means of morphological variables, such as t-tests or analysis of covariance, can be used without severe violations of assumptions of normality and homoscedasticity. This result establishes the validity of applying linear discriminant analysis and allows bounds to be placed on diffusion model age estimates.

Fault scarps cutting alluvium are found on the west side of the Virgin Mountains and in Chino Valley. The existence of these scarps indicate relatively young tectonism related to continued extension in the Basin and Range but are not related to continued faulting along any of the major Colorado Plateau bounding faults in

Table 1. Kolmogorov-Smirnov test of fit between observed maximum scarp angle distribution and a normal distribution.

| X=Angle ($^{o}$) | FREQUENCY | $S_N(X)$ | (X-26.92)/1.58 | F(X) | $\lvert S_N(X-1)-F(X)\rvert$ | $\lvert S_N(X)-F(X)\rvert$ $\lvert\Delta\rvert$ |
|---|---|---|---|---|---|---|
| 24.5 | 1 | .05 | -1.53 | .063 | .063 | .013 |
| 25.0 | 3 | .20 | -1.22 | .111 | .061 | .089 |
| 25.5 | 2 | .30 | -0.90 | .184 | .016 | .116 |
| 26.0 | 2 | .40 | -0.58 | .281 | .019 | .121 |
| 26.5 | 2 | .50 | -0.27 | .394 | .006 | .106 |
| 27.0 | 1 | .55 | 0.05 | .480 | .020 | .020 |
| 27.5 | 2 | .65 | 0.37 | .644 | .094 | .006 |
| 28.0 | 4 | .85 | 0.68 | .752 | .102 | .098 |
| 29.0 | 1 | .90 | 1.32 | .907 | .057 | .007 |
| 29.5 | 1 | .95 | 1.63 | .948 | .048 | .002 |
| 30.0 | 1 | 1.0 | 1.95 | .974 | .024 | .026 |

$D = 0.121$ $p = .20$

$H_O$:the distribution of the observed variable is normally distributed with mean 26.92 and standard deviation 1.58.

$H_A$:the distribution of the observed variable is different than hypothesized.

Decision: Accept $H_O$.

Table 2. Data for a comparison between the variances of the maximum scarp angles of the Pitaycachi and Mesquite scarps using the nonparametric Siegal-Tukey test. -- P is the probability that the null hypothesis is true for the particular sample data.

| X | $X-M_X$ | S | Y | $Y-M_X$ | S |
|---|---|---|---|---|---|
| 12.0 | 1.0 | 14.5 | 29.5 | 4.0 | 19.0 |
| 12.5 | 1.5 | 16.0 | 27.5 | 2.0 | 17.5 |
| 11.5 | 0.5 | 12.5 | 27.5 | 2.0 | 17.5 |
| 12.0 | 1.0 | 14.5 | 24.5 | -1.0 | 2.5 |
| 10.5 | -0.5 | 7.0 | 25.0 | -0.5 | 7.0 |
| 10.0 | -1.0 | 2.5 | 25.0 | -0.5 | 7.0 |
| 10.5 | -0.5 | 7.0 | 25.0 | -0.5 | 7.0 |
| 10.0 | -1.0 | 2.5 | 30.0 | 4.5 | 20.0 |
| 10.0 | -1.0 | 2.5 | 25.5 | 0 | 10.5 |
| 11.0 | 0 | 10.5 | 26.0 | 0.5 | 12.5 |

$$T_X = \sum_{i=1}^{10} S_i \text{ for all } X = 89.5$$

$H_0: \sigma_X = \sigma_Y$ $\quad$ $P \simeq 0.13$

$H_A: \sigma_X < \sigma_Y$ $\quad$ Decision: Reject $H_0$

Arizona. The Mesquite fault on the west side of the Virgin Mountains appears to be 40,000 years old based upon morphology-derived age estimation described above and other faults in the Virgin Mountains area also appear to be latest Pleistocene (Mayer, 1982). The same is true for the Chino Fault (Soule, 1978) but like the Mesquite fault, a much longer fault history is implied by the very large scarp heights which are in excess of 20 meters. The lack of fault scarps along the Grand Wash Fault (with one local exception) indicates that this fault has been inactive at least as long as the survival time of an alluvial fault scarp, or on the order of $10^5$ years.

## CONCLUDING REMARKS

Tectonic geomorphology of escarpments along the Colorado Plateau - Basin and Range in Arizona suggests that major topographic differentiation occurred about 25 MA ago in central Arizona and about 18 MA ago in northwestern Arizona. These estimates are consistent with previously published estimates based largely on stratigraphy and therefore support the usefulness of geomorphic studies for time periods extending as far back as 25 MA. The survival time of a particular landform determines the period of geologic time that can be studied. Escarpment retreat from a known initial location covers the longest period of time. Knickpoints in streams resulting from large base-level changes are useful for time periods on the order of $10^6$ years while knickpoints resulting from smaller tectonic perturbations survive considerably less time. Fault scarps cutting alluvium survive the shortest period of time but offer the best resolution of the methods discussed.

## ACKNOWLEDGMENTS

I am grateful to Bill Bull for his support during the course of my research. Also, I wish to thank R. Bucknam, S. Calvo, P. Heller, P. Kneufer, and P. Pearthree for assistance or helpful discussions. This paper has benefitted from a careful review by John Hack.

REFERENCES

Anderson, C.A., and Creasey, S.C., 1958, Geology and ore deposits of the Jerome area, Yavapai county, Arizona: U.S. Geol. Survey Prof. Paper 308, 185 p.

Armin, R.A. and Mayer, L., 1983, Subsidence analysis of the Cordilleran miogeocline - Implications for the timing of Proterozoic rifting and amount of extension: Geology, v. 11, p. 702-705.

Atwater, T., 1970, Implications of plate tectonics for the Cenozoic evolution of western North America: Geol. Soc. America Bull., v. 81, p. 3513-3536.

Bakker, J.P., and LeHeux, J.W.N., 1950, Proc. Konikle. Akad. Wetenschap. Amsterdam, v. 50, p. 959.

Best, M.G., and Hamblin, W.K., 1978, Origin of the northern Basin and range province - Implications from the geology of its eastern boundary, in, Smith, R.B., and Eaton, G.P., eds., Cenozoic tectonics and regional geophysics of the western Cordillera: Geol. Soc. America Mem. 152, p. 313-340.

Brush, L.M., and Wolman, M.G., 1960, Knickpoint behaviour in non-cohesive material - A laboratory study: Geol. Soc. America Bull., v. 71, p. 59-74.

Bucknam, R.C., and Anderson, R.E., 1979, Estimation of fault-scarp ages from a scarp-height -- slope angle relationship: Geology, v. 7, p. 11-14.

Bull, W.B., 1970, Effect of climatic and tectonic changes on denudation rates in part of the Diablo Range, California (abs): Geol. Soc. America Abs. Progs., v. 2, p. 77.

__________, 1973, Local base-level processes in arid fluvial systems (abs): Geol. Soc. America Abs. Progs., v. 5, p. 562.

__________, and McFadden, L.D., 1977, Tectonic geomorphology north and south of the Garlock Fault, California, in, Doehring, D.O., ed., Geomorphology of Arid Regions: Ann. Bing. Conf., SUNY Binghamton, London, Allen and Unwin, p. 115-138.

Carslaw, H.S., and Jaeger, J.C., 1959, Conduction of Heat in Solids: Oxford University Press, Oxford, 510 p.

Carson, M.A., and Kirkby, M.J., 1972, Hillslope Form and Process: Cambridge University Press, Cambridge, 475 p.

Chamberlin, T.C., and Salisbury, R.D., 1906, Geology, v. 3, Earth History - Mesozoic, Cenozoic: New York, 692 p.

Cole, K., and Mayer, L., 1982, The use of packrat middens to determine rates of cliff retreat in the eastern Grand Canyon: Geology, p. 597-599.

Colman, S.M., and Watson, K., 1983, Ages estimated from a diffusion-equation model for scarp degradation: Science, v. 221, p. 263-265.

Coney, P.J., and Reynolds, S.J., 1977, Cordilleran Benioff zones: Nature, v. 270, p. 403-406.

Cooley, M.E., and Davidson, E.S., 1963, The Mogollon Highlands-their influence on Mesozoic and Cenozoic sedimentation: Ariz. Geol. Soc. Digest, v. 6, p. 7-35.

Culling, W.E.H., 1960, Analytical theory of erosion: Jour. Geology, v. 68, p. 336-344.

Dickinson, W.R., and Snyder, W.S., 1979, Geometry of subducted slabs related to San Andreas transform: Jour. Geology, v. 87, p. 609-627.

Dodge, R.L., and Grose, L.T., 1980, Tectonic and geomorphic evolution of the Black Rock Fault, northwestern Nevada: U.S. Geol. Survey Open File Rept., 80-801.

Eaton, G.P., 1979, A plate tectonic model for late Cenozoic crustal spreading in the western United States, in, Reicker, R.E., ed., Rio Grande Rift - Tectonics and Magmatism: American Geophys. Union, Washington, D.C., p. 7-32.

__________, 1982, The Basin and Range Province - origin and tectonic significance: Ann. Rev. Earth Planet. Sci., v. 10, p. 409-440.

__________, 1983, The Miocene Great Basin of western North America as an extending back-arc region: Tectonophysics, v. 102, p. 275-295.

__________, Wahl, R.R., Prostka, H.J., Mabey, D.R., and Kleinkopf, M.D., 1978, Regional gravity and tectonic patterns - their relation to late Cenozoic epeirogeny and lateral spreading in the western Cordillera, in, Smith, R.B., and Eaton, G.P., eds., Cenozoic tectonics and regional geophysics of the western Cordillera: Geol. Soc. America Mem. 152, p. 51-92.

Emmett, W.W., 1970, The hydraulics of overland flow on hillslopes: U.S. Geol. Survey Prof. Paper 662-A, 68 p.

Fenneman, N.M., 1931, Physiography of the western United States: McGraw-Hill, New York, 534 p.

Gilbert, G.K., 1909, The convexity of hilltops: Jour. Geology, v. 17, p. 344-351.

Hack, J.T., 1973, Stream profile analysis and stream gradient index: U.S. Geol. Survey Jour. Res., v. 1, p. 421-429.

Hamblin, W.K., Damon, P.E., and Bull, W.B., 1981, Estimates of vertical crustal strain rates along the western margins of the Colorado Plateau: Geology, v. 9, p. 293-298.

Harshbarger, J.W., Repenning, C.A., and Irwin, J.H., 1957, Stratigraphy of the uppermost Triassic and the Jurassic rocks of the Navajo country: U.S. Geol. Survey Prof. Paper 291, 74 p.

Koons, D., 1955, Cliff retreat in the southwestern United States: Am. Jour. Sci., v. 253, p. 44-52.

Langbein, W.B., 1964, Geometry of river channels: Proc. Amer. Soc. Civ. Eng. Hydrol. Div., p. 301-312.

Leopold, L.B., and Langbein, W.B., 1962, The concept of entropy in landscape evolution: U.S. Geol. Survey Prof. Paper 500-A, 20 p.

Lindgren, W., 1926, Ore deposits of the Jerome and Bradshaw Mountains quadrangle, Arizona: U.S. Geol. Survey Bull. 782, 192 p.

Lipman, P.W., 1980, Cenozoic volcanism in the western United States - implications for continental tectonics, in, Burchfiel, B.C., Oliver, J.E., and Silver, L.T., eds., Continental Tectonics, National Research Council, Washington, D.C., p. 161-174.

Lucchitta, I., 1972, Early history of the Colorado River in the Basin and Range Province: Geol. Soc. America Bull., v. 83, p. 1933-1948.

Mayer, L., 1979, Evolution of the Mogollon Rim in central Arizona, in, McGetchin, T.R., and Merrill, R.B., eds., Plateau Uplift - Mode and Mechanism: Tectonophysics, v. 61, p. 49-62.

Mayer, L., 1982, Constraints on morphologic ages of fault scarps based on statistical analysis in the Basin and Range province of Arizona and northeastern Sonora, Mexico (abs): Geol. Soc. America Abs. Progs., v. 14, p. 213.

________, 1983, Constraints on non-uniform stretching models in the Basin and Range province of Arizona and adjacent areas (abs): Geol. Soc. America Abs. Progs., v. 15, p. 319.

________, in press, Dating fault scarps formed in alluvium using morphologic parameters: Quaternary Research.

McKee, E.D., 1951, Sedimentary basins of Arizona and adjoining areas: Geol. Soc. America Bull., v. 62, p. 481-506.

__________, and McKee, E.H., 1972, Pliocene uplift of the Grand Canyon region - Time of drainage adjustment: Geol. Soc. America Bull., v. 83, p. 1923-1932.

__________, et al., 1956, Paleotectonic maps, Jurassic system: U.S. Geol. Survey Misc. Geol. Inv. I-300.

Nash, D., 1980, Morphologic dating of degraded normal fault scarps: Jour. Geology, v. 88, p. 353-360.

Peirce, H.W., Damon, P.E., and Shafiqullah, M., 1979, An Oligocene (?) Colorado Plateau edge in Arizona: Tectonophysics, v. 61, p. 1-24.

Price, R.E., 1950, Cenozoic gravels on the rim of Sycamore Canyon, Arizona: Geol. Soc. America Bull., v. 61, p. 501-508.

Ransome, F.L., 1904, Description of the Globe quadrangle, Arizona: U.S. Geol. Survey Geological Atlas, Folo 111.

Scholz, C.H., Barazangi, M., and Sbar, M.L., 1971, Late Cenozoic evolution of the Great Basin, western United States, as an ensialic interarc basin: Geol. Soc. America Bull., v. 82, p. 2979-2990.

Schuchert, C., and Dunbar, C.O., 1941, Historical Geology: John Wiley, New York.

Shafiqullah, M., et al., 1980, K-Ar geochronology and geologic history of southwestern Arizona and adjacent areas: Ariz. Geol. Soc. Digest, v. 12, p. 201-260.

Scheidegger, A.E., 1961, Theoretical Geomorphology: Springer-Verlag, Germany, 333 p.

Smith, R.B., 1978, Seismicity, crustal structure and intraplate tectonics of the interior of the western Cordillera, in, Smith, R.B., and Eaton, G.P., eds., Cenozoic tectonics and regional geophysics of the western Cordillera: Geol. Soc. America Mem. 152, p. 111-144.

Snyder, W.S., Dickinson, W.R., and Silberman, M.L., 1976, Tectonic implications of space-time patterns of Cenozoic magmatism in the western United States: Earth Planet. Sci. Lett., v. 32, p. 91-106.

Sprunt, B., 1973, Digital simulation of drainage basin development, in, Chorley, R.J., ed., Spatial Analysis in Geomorphology: Methuen, London, p. 371-389.

Stewart, J.H., 1978, Basin and Range structure in western North America, in, Smith, R.B., and Eaton, G.P., eds., Cenozoic tectonics and regional geophysics of the western Cordillera: Geol. Soc. America Mem. 152, p. 1-17.

____________, and Suczek, C.A., 1977, Cambrian and latest Precambrian paleogeography and tectonics in the western United States, in, Stewart, et al., eds., Paleozoic paleogeography of the western United States: Soc. Econ. Paleont. and

Mineralogists, Pacific Section, Pacific coast Paleogeography Symposium 1, p. 1-17.

Suppe, J., Powell, C., and Berr, R., 1975, Regional topography, seismicity, Quaternary volcanism and the present-day tectonics of the western United States: Am. Jour. Sci., v. 275-A, p. 397-436.

Thompson, G.A., and Zoback, M.L., 1979, Regional geophysics of the Colorado Plateau, in, McGetchin, T.R., and Merrill, R.B., eds., Plateau Uplift- Mode and Mechanism: Tectonophysics, v. 61.

Wallace, R.E., 1977, Profiles and ages of young fault scarps in north-central Nevada: Geol. Soc. America Bull., v. 88, p. 1267-1281.

Wilson, E.D., and Moore, R.T., 1959, Structure of the Basin and Range province in Arizona: Ariz. Geol. Soc. Guidebook 2, p. 89-97.

Yair, A., and Gerson, R., 1974, Mode and rate of escarpment retreat in an extremely arid environment, in, Slope processes, Sec. 5 of Schick, A., Yaalon, D.H., and Yair, A., eds., Geomorphic processes in arid environments: Zeitschr. Geomorph., Supp. v. 21, p. 202-265.

Young, A., 1974, The rate of slope retreat, in, Brown, E.M., ed., Progress in Geomorphology: Inst. British Geog. Spec. Pub. 7, p. 63-78.

Young, R.A., and Brennen, W.J., 1974, Peach Springs Tuff - Its bearing on the structural evolution of the Colorado Plateau and development of Cenozoic drainage in Mohave county, Arizona: Geol. Soc. America Bull., v. 85, p. 83-90.

Zoback, M.L., Anderson, R.E., and Thompson, G.A., 1981, Cainozoic evolution of the state of stress and style of tectonism of the Basin and Range province of the western United States: Philos. Trans. Royal Soc. London, Ser. A., p. 407-434.

# 11

# Geomorphic evolution of the Colorado Plateau margin in west-central Arizona: A tectonic model to distinguish between the causes of rapid, symmetrical scarp retreat and scarp dissection

*Richard A. Young*

ABSTRACT

Early Eocene viviparid gastropods have established the Laramide age of the oldest, widespread, Tertiary erosion surface along the southwest margin of the Colorado Plateau, permitting the major Tertiary landscape-forming episodes to be separated by age and tectonic setting.

Scarp retreat in Paleozoic rocks during Laramide tectonism was relatively rapid (1500 to 3800 m/$10^6$ years) when base levels were stable or rising. Much slower rates of scarp recession and dissection (160 to 170 m/$10^6$ years) occurred between the end of the Laramide Orogeny and the present.

Both tectonic stability from Late Eocene through Oligocene time and regional drainage incision from Late Miocene through Pliocene time (Grand Canyon) are associated with these order-of-magnitude slower rates of erosion. Active tectonic uplift appears to promote rapid, uniform scarp retreat when strike valley tributaries flow into stable or aggrading master streams. Conversely, when base level is lowered and regional drainage incision becomes dominant, major stream courses become fixed, and individual scarp-face tributaries erode headward and dissect scarps. Depending upon which of these two tectonic settings is dominant, the rate of landscape evolution may vary by an order of magnitude or more.

Both overall scarp retreat and scarp dissection by streams in similar stratigraphic settings have proceeded at comparable rates since Laramide time, whether directly adjacent to the Grand Canyon or in basins graded to valleys off the plateau. Erosion by the Colorado River does not appear to be associated with a conspicuous change in scarp recession rates, although it has caused scarp dissection.

The distinction between rapid scarp recession and scarp dissection may help elucidate the tectonic histories of relict or compound landscapes, especially the persistence of relatively featureless erosion surfaces in proximity to landforms assumed to be indicative of rapid or dynamic change.

## INTRODUCTION

Geomorphologists have long attempted to understand the processes and measure the rates involved in scarp recession in sequences of flat-lying to gently-dipping sedimentary rocks. This interest in scarp retreat as a regional, landscape-forming phenomenon results from the observation that many landscapes developed on sedimentary rocks consist of a series of cliffs formed by the erosion of shales underlying more competent sandstones or limestones. In many instances the scarp at the highest stratigraphic level has migrated tens of kilometers from its inferred original position.

The different processes that contribute to slope retreat have been discussed at great length by numerous authors. Recent models have begun to deal quantitatively with the variables that control the dynamics of existing scarps and hillslope elements (Oberlander, 1977; Crickmay, 1975).

This paper focuses on changes in *regional* scarp retreat rates and changes in scarp aspect that can be related to specific geologic events along the southwestern margin of the Colorado Plateau since Late Cretaceous time. This regional analysis has been made possible by the recent discovery of age-diagnostic fossils in early Tertiary fluvial sediments south of Grand Canyon, Arizona (Young and Hartman, 1984). The relevant regional geology is described by Young (1979, 1982).

Throughout the discussion a distinction is made between the phenomena of regional scarp retreat and scarp dissection. Regional scarp retreat is the condition whereby a line of cliffs maintains a relatively uniform crest or divide as it migrates, except locally where it is breached by consequent master streams. Scarp dissection, in contrast, occurs when a number of scarpface (obsequent) tributaries independently erode headward, producing a markedly embayed escarpment with evidence of numerous stream captures and isolated features such as buttes. Under these circumstances irregular scarp dissection is more conspicuous than overall recession.

Although both uniform scarp retreat and irregular scarp dissection both involve headward erosion and associated processes, it will be argued that the two conditions result from distinctly different tectonic settings. The contrasting responses to erosion suggest important differences in erosional settings, which are generally dependent upon the tectonic environment or history. Uniform scarp retreat can progress at a rate which is at least an order of magnitude greater than scarp dissection, implying a basic difference in the fundamental controls that govern erosion in each case.

## GEOLOGIC SETTING

The Tertiary geology of the southwestern Colorado Plateau from the Colorado River southward to the Aquarius Mountains has been summarized by Young (1966, 1979, 1982). Recently, age-diagnostic, viviparid gastropods of probable Early Eocene age have been collected and identified from limestones within a predominantly fluvial section on the Coconino Plateau (Figure 1) 48 km south of Grand Canyon (Young and Hartman, 1984). Related species of gastropods all had become extinct throughout North America by Middle Eocene time (Hartman, 1984).

The Early Eocene rocks south of Grand Canyon are part of the discontinuous, but widely distributed "rim gravels" that rest unconformably on the eroded Paleozoic rocks around much of the southern and southwestern plateau margin in Arizona (Peirce and others, 1979). They represent an Early Tertiary north- to northeast-flowing drainage system that clearly predates the development of the Colorado River. The rim gravels appear to have blanketed the plateau over much of northern Arizona with 200 feet or more of conglomeratic sandstones presently found at elevations

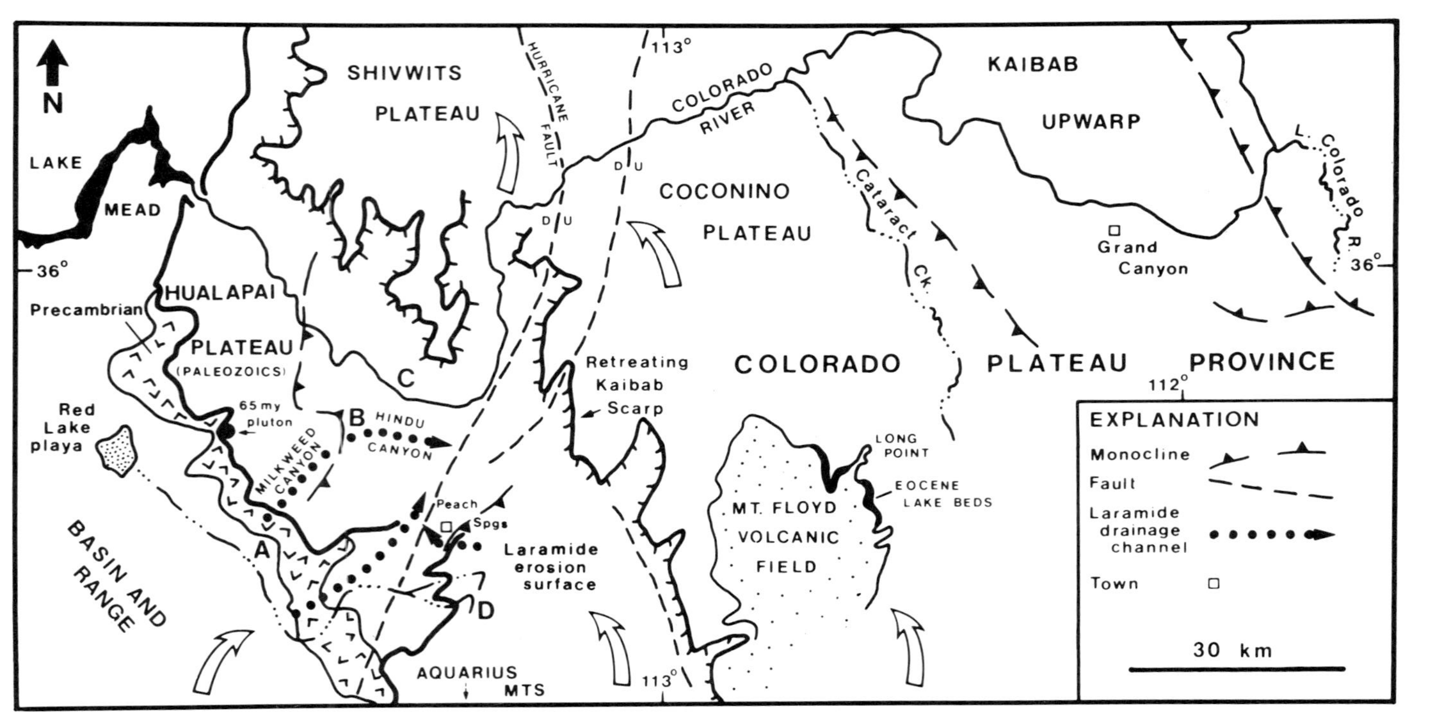


Figure 1. General location diagram. A-B is minimum distance of pre-Early Eocene scarp recession. B-C is distance of post-Laramide to Grand Canyon scarp recession (see text). D is area of post-Laramide scarp dissection for westward drainage to Red Lake playa. Large open arrows are Early Tertiary drainage directions (see Young, 1982, Figures 2, 4, 5, for details).

up to or about 6000 feet. Weathering and local reworking of these sediments have created numerous distinctive lag gravel deposits that have hindered the correct age assignment for the original fluvial sequences (Peirce and others, 1979). Although the oldest sediments may actually range in age from Late Cretaceous through Eocene time across various parts of Arizona, it now can be confidently stated that deposition was ongoing during Early Eocene time and that it probably gradually terminated later during the Eocene as the intensity of Laramide events decreased.

The fossiliferous Early Eocene lacustrine facies in the upper part of the fluvial facies south of Grand Canyon probably record a late Laramide structural disruption of the regional drainage. This possibility was suggested by Young (1979, 1982) based on stratigraphic and structural relationships preserved further west on the Hualapai Plateau (Figure 1). Oligocene volcanic rocks (24-28 m.y.) rest on the same regional erosion surface in at least three places in western and central Arizona, thereby demonstrating that the extensive erosion and associated fluvial deposition were completed prior to the onset of Middle Tertiary volcanism (Young, 1982).

The present edge of the Colorado Plateau in central Arizona (Figure 1) was first uplifted during the Late Createaceous (Miller, 1962) at the beginning of Laramide time. It is reasonable to conclude that the widespread erosion, which was followed by deposition of exotic fluvial clasts derived from terranes west and south of the plateau, was basically a Laramide event. Adjacent to the uplift, drainage flowing onto the plateau incised deep canyons completely through the tilted Paleozoic rocks at the plateau edge (Young, 1966, 1979). However, on the plateau proper a less dissected, stripped surface (Kaibab Fm.), created by cliff recession in Mesozoic strata, was gradually buried by a growing apron of alluvial debris spreading outward from the Laramide uplifts. This sequence of events has preserved a partially buried, relict landscape of Early Tertiary age that can be contrasted with the landscape being produced by the much younger Colorado River drainage. Essentially, the basic landscape elements of northern Arizona were created within the tectonic framework of this Laramide interval. The modern Grand Canyon drainage system has only modified this older landscape slightly since Late Miocene time in the western Grand Canyon region.

## EARLY EOCENE LANDSCAPE

Figures 1 and 2 illustrate the major physiographic and stratigraphic relations critical to this discussion. Details that document the field relations can be found in Young (1979, 1982). The most critical relationships involve reconstruction of the positions of the Eocene drainages and associated scarps on the western plateau as they relate to the known Laramide chronology. A more precise delineation of the timing of key Laramide events is necessary to develop a plausible Early Tertiary history for the plateau margin in the western Grand Canyon region.

From a plate tectonics perspective Coney (1973) places the start of the Laramide at 80 m.y. Damon (Shafiqullah et al., 1980) has suggested restricting the Laramide in Arizona to the 75 to 50 million-year-old cluster of magmatic ages in the Basin and Range province. Drewes (1978) documented two peaks of Laramide activity at 73 to 75 m.y. (Late Cretaceous) and 55 m.y. (Early Eocene) ago. Haxel and others (1984) argue for synchronous thrusting, metamorphism, and granitic plutonism in southern Arizona peaking 58 to 60 m.y. ago. Rocks west of the Colorado Plateau along the lower Colorado River give clusters of ages between 70 to 50 m.y. for major thermal resetting (Martin et al., 1982). Two distinctive volcanic clasts from the gravels discussed in this article gave ages near 80 m.y., requiring a somewht younger age for their erosion and transport from the southwest onto the Colorado Plateau south of Grand Canyon (E.H. McKee, written communication, 1983).

The Laramide Orogeny is characterized by compressional deformation, crustal thickening, uplift, volcanism, and granitic intrusives adjacent to the Colorado Plateau (Drewes, 1978; Haxel and others, 1984). In response to the compressional stresses, monoclines formed on the plateau and reversed movement reactivated some west-dipping normal faults (Huntoon, 1981). If the synchronicity of deformation and igneous activity seen regionally is assumed to hold for the plateau margin, several plutons south of Lake Mead, including one at the very edge of the plateau

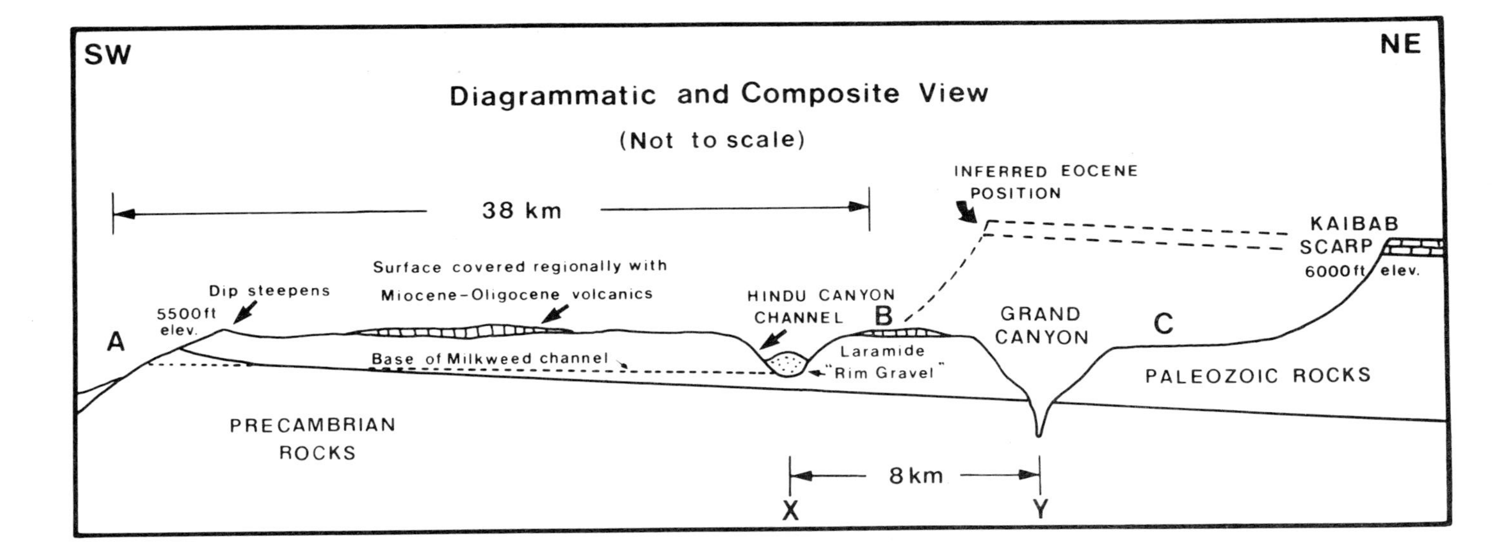


Figure 2. Generalized profile through area of A, B, C on Figure 1. A–B is minimum pre-Early Eocene scarp recession distance. Channel separation (X–Y) is assumed to be a more accurate measure of change in scarp position suggested by B–C (assuming both canyons formed in similar positions at the foot of the scarp). Laramide sediments in Hindu Canyon are equivalent (rim gravels) in age to sediments near Long Point (Figure 1) containing Early Eocene gastropods.

(Figure 1), imply that deformation occurred near the Cretaceous-Tertiary boundary (Young, 1979).

Overall, it appears reasonable to bracket the Laramide Orogeny in southwestern Arizona between 80 and 50 m.y. ago, with magmatic activity becoming pronounced beginning 70 to 75 m.y. ago. Subduction models imply that initial magmatic activity might be somewhat younger toward the eastern edge of the magmatic arc (Martin et al., 1982), and field evidence documents that intrusive activity on the plateau margin occurred 65 m.y. ago (Young, 1979). A composite overview of the published literature suggests that peaks of deformational and magmatic activity occurred in Late Cretaceous time and also near the Paleocene-Eocene boundary (55 to 60 m.y. ago).

Young (1982) has argued that Late Laramide monoclinal deformation accounts for the presence of lacustrine facies filling the incised Early Tertiary canyons on the Hualapai Plateau. The newly determined Early Eocene ages for equivalent limestones south of Grand Canyon can be presumed to strengthen this argument. These Early Eocene sediments may have formed in lakes produced by movement on the Kaibab Upwarp itself, the only significant structure in the path of the north- to northeast- flowing drainage on the Coconino Plateau.

Currently available chronologic evidence suggests that Laramide uplift adjacent to the southwestern plateau margin began between 65 and 80 m.y. ago, initiating erosional retreat of the scarps in the Mesozoic and Paleozoic rocks, such as the northern extension of the Mogollon Rim (Kaibab Fm.). In the region of Figure 2, $5^{o}$ to $15^{o}$ dips at the plateau margin controlled the development of consequent master streams flowing from the Laramide Uplifts onto the plateau. The abrupt eastward bend in the Milkweed-Hindu channel (Figure 1) is considered to mark the place where the Early Eocene (minimum age) channel encountered the retreating Kaibab scarp. The "deflected" drainage paralleled the scarp and finally joined the well-developed valley already eroded along the Hurricane fault trend. The existence of a much broader erosion surface beveled across the Early and Middle Paleozoic rocks is documented by the widespread distribution and elevations of somewhat younger Tertiary sediments containing Oligocene and

Miocene volcanic rocks south of the Colorado River (Young, 1979, 1982).

Thus by Early Eocene time the drainage consisted of Laramide-generated north- to northeast-trending consequent streams superposed across southwest-facing scarps, with some adjustment to major preexisting structures. Further to the northeast at higher elevations, the drainage flowed across successively higher stratigraphic units probably terminating in the Eocene basins of southern Utah (Young and McKee, 1978). The physiographic setting might have been similar to that seen along the eastern edge of the Rocky Mountains, except for the more prominent west-facing escarpments that had developed in the thick Paleozoic limestones on the plateau.

## TERTIARY SCARP RECESSION RATES

From the relations on Figures 1 and 2, it can be noted that the minimum amount of pre-Early Eocene scarp recession needed to create the topography near Hindu Canyon was 38 km. Given the Laramide chronology, this erosion must have occurred over a time interval 10 to 25 m.y. in length, depending on when Laramide events are assumed to have initiated cliff recession near the plateau margin in western Arizona. This retreat translates to a rate of recession of 1500 to 3800 m/$10^6$ years. If one assumed that a 75 m.y. age represents a reasonable estimate for the initiation of scarp retreat near the plateau margin, then the rate is close to 1900 m/$10^6$ years. However, it is also possible that scarp retreat occurred over a considerably greater distance in a slightly longer time period. Thus the 1900 m/$10^6$ years rate is a conservative value.

The point to be emphasized is that the major regional scarps had retreated several tens of kilometers, essentially during Paleocene time. An estimate of 10 to 15 m.y. for the total retreat is realistic if the beginning of Kaibab scarp recession is assumed to lag behind initial uplift, due to the requirement that the overlying Mesozoic section must first be stripped away. Also, it is necessary to allow some time for the deep canyons containing the Early Eocene sediments to be eroded after the Kaibab scarp had retreated to a position north of Hindu Canyon. Certainly it seems likely that the 1900 m/$10^6$ years recession rate is near the lower end of the permissible range.

SCARP RECESSION SINCE EARLY EOCENE TIME

The Grand Canyon lies about 8 km north of, and parallel to, the early Tertiary drainage channel in Hindu Canyon (Figure 2) at the foot of the same retreating scarp. It can be assumed that the Colorado River developed at the base of this scarp in a manner analogous to that which earlier formed the Eocene channel, although flowing in the opposite direction. It is not necessary to infer the distance of each channel from the base of the scarp, but only to assume that each channel formed in a similar position relative to the scarp. Based on this simple assumption, the scarp receded only 8 km between the initiation of the Early Eocene(?) channel and the development of the Colorado River, an interval of approximately 50 m.y. Lucchitta (1975) and Young (1979) have discussed the time constraints relating to the Late Miocene-Pliocene origin of the western Grand Canyon.

This slower, post-Early Eocene rate of retreat translates into 160 m/$10^6$ years or 9.5 to 24 times slower than the estimates for the Laramide. If the conservative assumption of 1900 m/$10^6$ years is used, the post-Early Eocene rate must have been about 12 times slower. This slower rate is, conservatively, an order of magnitude slower at the very least, regardless of whether the larger or smaller estimates are used. Thus, despite some uncertainty, the Laramide rates of recession are significantly greater than Middle to Late Tertiary rates, regardless of the extreme relief being created by modern Colorado River erosion.

A TECTONIC MODEL FOR VARIATIONS IN SCARP RECESSION RATES

The majority of Colorado Plateau geomorphologists no longer believe that the Colorado River formed gradually over tens of millions of years during a period of prolonged plateau uplift. However, the rather sudden appearance of the Colorado River gorge, starting in very latest Miocene time, raises the fundamental question of why this erosion interval appears to have had such a minor effect on local scarp retreat, as documented on the Hualapai Plateau. The answer appears to lie in the relationship between the regional drainage pattern and the tectonic framework.

During Laramide time the plateau margin was more strongly upwarped (Young, 1979, 1982) near the marginal uplifts but the master consequent streams flowing onto the plateau crossed onto successively younger stratigraphic rock units in northern Arizona and southern Utah. The ultimate base level for these streams was probably gradually rising as these Early Tertiary basins filled with sediment (Young and McKee, 1978) and as the Laramide uplift rates slowed. Under these conditions, drainage incision occurred only adjacent to the maximum uplifts at the plateau margin. Further to the north and east the main scarps (Kaibab Fm., Shinarump Congl.) would be breached only by the master streams. Subsequent (strike valley) tributaries would dominate the landscape development between master streams (Figure 3). These strike valley streams would not be actively incising while the master consequent drainages were flowing into aggrading Tertiary basins to the north.

Except for incised reaches adjacent to the Laramide Uplifts, the major streams should occupy broad open valleys in the shaley stratigraphic intervals, creating an open, cuestaform landscape. The strike valley drainages would be capable of lateral migration down the regional (northeast) dip slope while their obsequent tributaries eroded headwardly into the adjacent scarps (Figure 3) (Meyerhoff, 1975, p. 60).

Under these conditions of relatively little incision along master streams, a major portion of the energy available could be used to efficiently transport the sediment derived from these scarps out of the strike valleys and to undercut the cuestas by downdip valley migration. Scarps would maintain relative uniformity because obsequent valley spurs would be trimmed in the process of lateral stream migration down the regional dip. In this setting headward erosion on the scarp face would be a relatively slower process, and scarp dissection would be minimized by comparison.

In contrast to the case for aggrading master streams, if vertical incision of trunk streams and main tributaries becomes dominant, the positions of the strike valley tributaries become fixed and downdip migration effectively ceases (although some

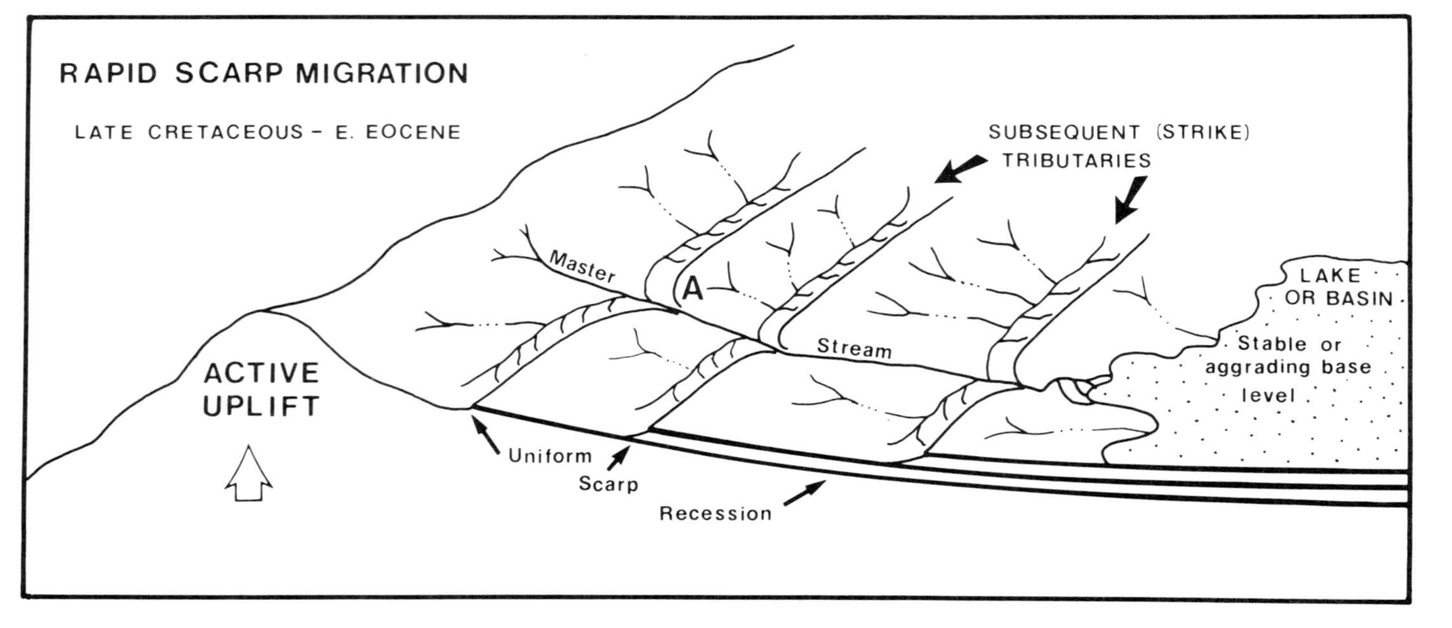


Figure 3. Downdip lateral migration of subsequent tributaries produces scarp recession at rate that exceeds headward dissection by scarp face (obsequent) streams. An open, cuestaform landscape is produced with deep incision of master streams only near A.

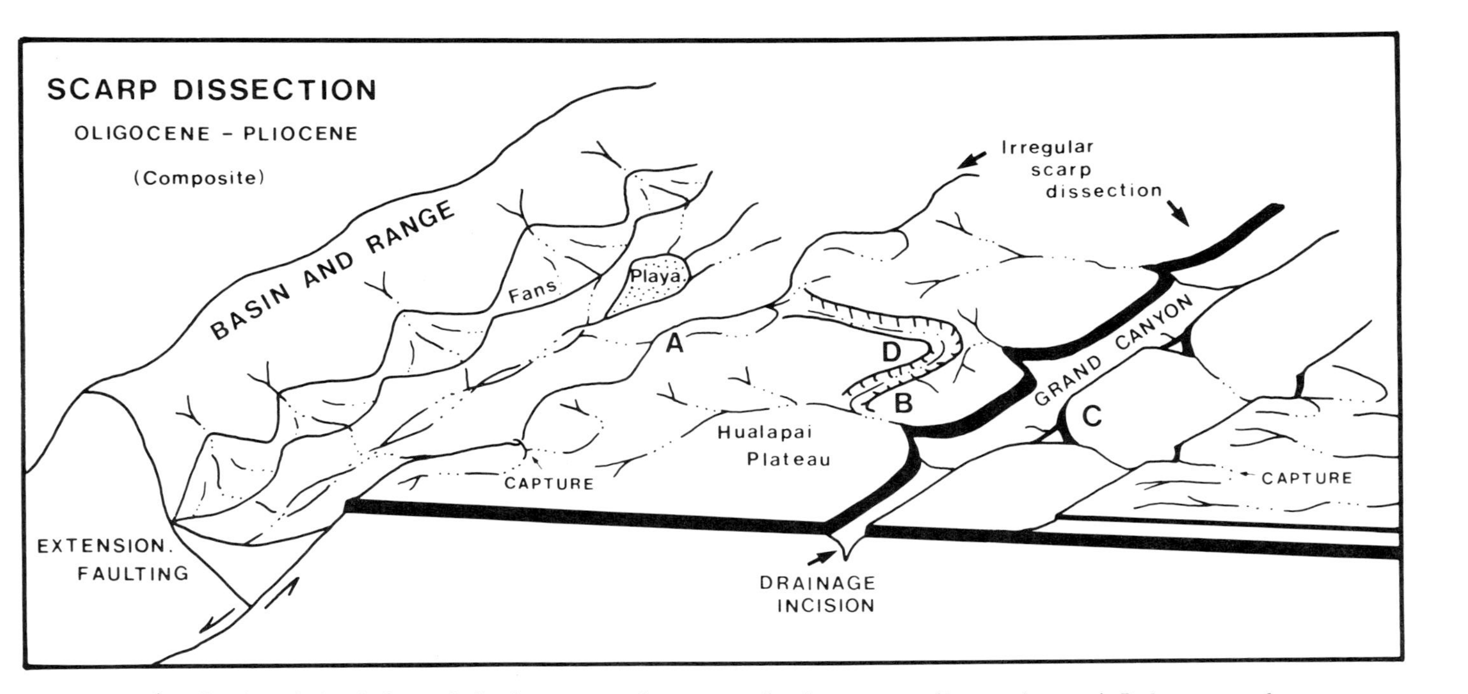


Figure 4. Regional incision of drainage accelerates relative scarp dissection. A–B is assumed scarp recession prior to regional incision. B–C is approximate scarp recession from post-Laramide to Grand Canyon time. D is relict of older drainage being reexumed (equivalent to Hindu Canyon on Figure 1). Stream captures increase scarp dissection at specific points relative to normal scarp erosion.

valley asymmetry may continue to develop). Once streams become regionally incised, minor tectonism is less able to cause significant changes in stream positions, especially by downdip lateral migration (Figure 4).

Following the episode of the Laramide (Late Creatceous through Middle Eocene) erosion and fluvial deposition under semitropical conditions, Late Eocene and Oligocene times produced deep weathering profiles (Telluride Erosion Surface) throughout much of western North America, (Gresans, 1981). Little evidence of concurrent tectonic activity on the plateau is apparent, but some normal faulting accompanied or closely followed the relaxation of compressional Laramide stresses (Rehrig and Heidrick, 1976; Loring, 1976). A period of magmatic quiescence corresponding to Late Eocene time has been noted adjacent to the plateau (Martin and others, 1982).

Middle Tertiary extension, detachment faulting, and volcanism began in the Basin and Range province in Late Oligocene time, but the period of most intense extensional faulting and contemporaneous volcanism in western Arizona occurred during Middle Miocene time (12 to 18 m.y. ago).

From Middle Eocene time until the cessation of widespread volcanism in the Late Miocene, regional scarp retreat was extremely slow before the Colorado River began to incise its tributaries on the Hualapai Plateau. Erosion of the Grand Canyon seems to have been concentrated in a relatively short period between 3 and 6 m.y. ago (Lucchitta, 1975, 1979). This canyon cutting event has been accompanied by scarp dissection due to the random headward erosion of selective tributaries, some aided by stream capture.

## EROSION OF OTHER SCARPS

South of the Colorado River near Peach Springs (Figure 1, near D), post-Early Eocene scarp dissection (Redwall Limestone) away from the immediate influence of the Grand Canyon has progressed approximately the same distance as the post-Early Eocene scarp retreat at the Colorado River near Hindu Canyon. Both areas (Figure 1) have comparable scarp settings, consisting of thick Paleozoic limestones over more shaley rock units. It is interesting, if not coincidental, that post-Early Eocene scarp retreat near the Kaibab scarp at Hindu Canyon is almost exactly the same as the minimum amount of post-Early Eocene scarp dissection south of Peach Springs (170 m/$10^6$ years) as measured by the amount of headward erosion along major tributaries flowing westward off the plateau (into a closed basin). The only assumption underlying this comparison of rates is that scarp dissection along the plateau margin could not have begun until the period of uplift and Eocene deposition ended, especially for small drainages flowing westward off the plateau (in the opposite direction from the Laramide paleoslope).

In other words, both scarp dissection and overall scarp retreat have proceeded at similar rates (160 to 170 m/$10^6$ years) since the cessation of Laramide tectonism. However, both these rates are significantly less than those of overall cliff recession. The evidence suggests that the relative effectiveness of scarp dissection by headward erosion depends on the relative rates of lateral stream migration and headward erosion between main strike valley drainages and their obsequent tributaries. Conversely, under similar conditions, average scarp retreat accompanying drainage incision proceeds at a rate comparable to random headward erosion (scarp dissection). This seems logical because both processes are caused by headward erosion without significant lateral migration of strike valleys.

Cole and Mayer (1982) tabulated rates of cliff recession from eight different studies in Utah, Arizona, and the Sinai Penninsula. The range of cliff recession rates in these studies was from 100 to 2000 m/$10^6$ years compared with 160 to 1900 m/$10^6$ years (best Laramide estimate) for the present study. However, the results of the present study demonstrate significant time-variable rates along an individual scarp.

## SUMMARY AND CONCLUSION

The most rapid scarp recession rates calculated in this study must be bracketed between 1500 and 3800 m/$10^6$ years for the values associated with active Laramide tectonism and lateral strike-valley migration. Under these circumstances regional base levels were either relatively stable or slowly rising.

In contrast, the long-term averages for either scarp recession or scarp dissection since Laramide time are very similar (160 to 170 m/$10^6$ years). These much slower rates appear to represent conditions of regional drainage incision, which precludes significant lateral stream migration in sequences of alternating weak and resistant sedimentary rocks.

The wide range of recession rates indicated for a single scarp under varying tectonic conditions suggest caution be used in inferring the causes of rate changes. Specifically, the implications of this study, if correct, indicate that tectonic controls of regional base levels can be of much greater significance than local relief in landscape evolution. Although a lowered base level may increase dissection locally, it may not be the most important cause of rapid regional landscape evolution. The implications may also be extended to the study of fault scarp evolution, where advanced dissection has sometimes occurred without significant scarp retreat being obvious.

Other implications relate to studies of relict erosion surfaces. Accelerated vertical uplift may not be the most effective means of destroying regional erosion surfaces, which might explain why so many have been partially preserved. Although localized tectonic uplift creates dramatic local relief, it does not necessarily increase headward erosion rates significantly throughout the entire area affected. Those processes or conditions which enhance lateral stream migration may be much more effective in shaping regional landscapes. This may be reflected in the observation that relict landscapes are generally identified by characteristics such as accordant summits, straths, pediments, and "peneplains." Such planar features may form relatively quickly in geologic time, given the right tectonic setting, and uplifting of such surfaces may only destroy them at rates that are an order of magnitude slower.

Applications of these general observations to analyses of relict or compound landscapes may help to unravel the complexities of their tectonic histories.

REFERENCES

Cole, K.L. and Mayer, L., 1982, Use of packrat middens to determine rates of cliff retreat in the eatern Grand Canyon, Arizona: Geology, v. 10, no. 11, p. 597-599.

Coney, P.J., 1973, Non-collision tectogenesis in western North America: in D.H. Tarling and S.K. Runcorn, eds., Implication of Continental Drift to the Earth Sciences, v. 2, London, Academic Press, p. 713-727.

Crickmay, C.H., 1975, The hypothesis of unequal activity: in W.N. Melhorn and R.C. Flemal, eds., Theories of Landform Development: London, Allen and Unwin, p. 103-109.

Drewes, H., 1978, The Cordilleran orogenic belt between Nevada and Chihuahua: Geol. Soc. Am. Bull., v. 89, p. 641-657.

Gresans, R.L., 1981, Extension of the Telluride erosion surface to Washington State and its regional and tectonic significance: Tectonophysics, v. 79, p. 145-164.

Hartman, J.H., 1984, Systematics, Biostratigraphy, and Biogeography of Latest Cretaceous and Early Tertiary Viviparidae (Mollusca, Gastropoda) of Southern Saskatchewan, Western North Dakota, Eastern Montana, and Northern Wyoming [Ph.D. dissert.]: Minneapolis, Minnesota, University of Minnesota, 919 p.

Haxel, G.B., Tosdal, R.M., May, D.J., and Wright, J.E., 1984, Latest Cretaceous and Early Tertiary orogenesis in south-central Arizona: Thrust faulting, regional metamorphism and granitic plutonism: Geol. Soc. Am. Bull., v. 95, no. 6, p. 631-653.

Huntoon, P.W., 1981, Grand Canyon monoclines: Vertical uplift or horizontal compression? in D.W. Boyd and J.A. Lillegraven, eds., Rocky Mountain Foreland Basement Tectonics, University of Wyo. Contributions to Geology, v. 19, no. 2, p. 127-134.

Loring, A.K., 1976, The age of Basin-Range faulting in Arizona: Arizona Geological Society Digest, v. 10, p. 229-258.

Lucchitta, I., 1975, The Shivwits Plateau: In Applications of ERTS Images and Image Processing to regional geologic problems and geologic mapping in northern Arizona, Jet Propulsion Lab Rept. 32-1597, p. 41-72.

____________, 1979, Late Cenozoic uplift of the southwestern Colorado Plateau and adjacent lower Colorado River region: Tectonophysics, v. 61, p. 63-95.

Martin, D.L., Krummenacher, D., and Frost, E.G., 1982, K-Ar Geochronologic Record of Mesozoic and Tertiary Tectonics of the Big Maria--Little Maria--Riverside Mountains terrane: in E.G. Frost and D.L. Martin, eds., Mesozoic-Cenozoic Tectonic Evolution of the Colorado River, California, Arizona, and Nevada: San Diego, Calif., Cordilleran Publishers, p. 518-549.

Miller, H.W., 1962, Cretaceous rocks of the Mogollon Rim area in Arizona: in R.H. Weber and H.W. Peirce, eds., Mogollon Rim Region, New Mexico Geol. Soc. Field Conf. Guidebook 13, p. 93.

Meyerhoff, H.A., 1975, The Penckian Model--With Modifications: in W.N. Melhorn and R.C. Flemal, eds., Theories of Landform Development, Longon, Allen and Unwin, p. 45-68.

Oberlander, T.M., 1977, Origin of segmented cliffs in massive sandstones of southeastern Utah: in D.O. Doehring, ed., Geomorphology in Arid Regions, Fort Collins, Colo., Citizens Printing Company, p. 79-114. (Available, Allen and Unwin, Winchester, MA).

Peirce, H.W., Damon, P.E., and Shafiqullah, M., 1979, An Oligocene (?) Colorado Plateau edge in Arizona: Tectonophysics, v. 61, p. 1-24.

Rehrig, W.A. and Heidrick, T.L., 1976, Regional tectonic stress during the Laramide and Late Tertiary intrusive periods, Basin and Range province, Arizona: Arizona Geol. Soc. Digest, v. 10, p. 205-228.

Shafiqullah, M., Damon, P.E., Lynch, D.J., Reynolds, S.J., Rehrig, W.A., and Raymond, R.H., 1980, K-Ar Geochronology and geologic history of southwestern Arizona and adjacent areas: Arizona Geological Society Digest, v. 12, p. 201-259.

Young, R.A., 1966, Cenozoic Geology along the edgte of the Colorado Plateau in northwestern Arizona [Ph.D. dissert.]: St. Louis: Washington University, 167 p.

__________, 1979, Laramide deformation, erosion, and plutonism along the southwestern margin of the Colorado Plateau: Tectonophysics, v. 61, p. 25-47.

__________, 1982, Paleogeomorphologic evidence for the structural history of the Colorado Plateau margin in western Arizona: in E.G. Frost and D.L. Martin, eds., Mesozoic-Cenozoic Tectonic Evolution of the Colorado River Region, California, Arizona, and Nevada, San Diego, Calif., Cordilleran Publishers, p. 29-39.

__________ and Hartman, J.H., 1984, Early Eocene fluviolacustrine sediments near Grand Canyon, Arizona: evidence for Laramide drainage across northern Arizona into southern Utah: Geological Society of America Abstracts with Programs, v. 16, no. 6, p. (in press).

__________ and McKee, E.H., 1978, Early and Middle Cenozoic drainage and erosion in west-central Arizona: Geol. Soc. Am. Bull., v. 89, p. 1745-1750.

# 12
# Geomorphic evidence for Pliocence–Pleistocene uplift in the area of the Cape Fear Arch, North Carolina

*Helaine Walsh Markewich*

ABSTRACT

The Cape Fear River of North Carolina flows just south of, and approximately along, the axis of the Cape Fear arch. Geomorphic, paleontologic and isotopic data indicate that two, and possibly three, styles of deformation have been active in the area of the arch for at least the past 3 Myr. Map patterns of Pliocene and Pleistocene terraces along the river suggest that the center (or axis) of uplift has been east-northeast of the present Cape Fear River, possibly along the Cape Fear River-Neuse River divide, since regression of the early Pliocene sea. Deformation associated with the Cape Fear arch is considered responsible for the progessive southwestward migration of the Cape Fear River during the past 3 Myr. Palynological data suggest that at least some of the small faults that have been reported along the inner edge of the Coastal Plain are post-4 Myr in age and are probably associated with one, or everal, major structures that are coincident with, or subparallel to the Fall Line. Regional geomorphic and paleontologic data from the southern limb of the Cape Fear arch and from the Charleston area of the Southeast Georgia Embayment indicate accelerated rates of regional uplift during the Pleistocene, particularly the late Pleistocene.

## INTRODUCTION

The Cape Fear arch of southeastern North Carolina and northeastern South Carolina has long been recognized as a major positive feature. The arch trends approximately northwest-southeast and is considered to be a distinct basement ridge that extends out onto the continental shelf (fig.1). The basement drops southwestward from approximately -300 m near Wilmington, North Carolina to to over -1000 m near Charleston, South Carolina and northeastward to about -3000 m near Cape Hatteras, North Carolina. On the Tectonic Map of the United States (Cohee and others, 1962) the arch is reflected in structure contours on the top of basement. J. P. Owens and D. C. Prowell (U. S. Geological Survey, verbal commun., 1983) have shown that the arch is reflected by structure contours drawn on the top of any formation of Late Cretaceous or Paleogene age.

The Cape Fear arch has been considered by some authors to be a horst between two northwest-southeast trending faults that are named for, and are approximately coincident with, the Cape Fear and Neuse Rivers (Harris and others, 1979). These faults, and the northeast-southwest trending Carolina fault postulated to be near the junction of the Cape Fear and Black Rivers (fig.1), were considered by Harris and others (1979) to have been episodically active throughout the late Mesozoic and Cenozoic and to have had direct effect on the distribution of Upper Cretaceous and younger sediments. Data presented by Sohl and Christopher (1983) and J. P. Owens (U. S. Geological Survey, written commun., 1984) substantially refute the existence of the Cape Fear, Neuse, and Carolina faults, but they do indicate that the area of the Cape Fear arch has been an active feature since the Cretaceous, and has undergone both positive and negative (oscillatory) vertical movement. The pattern of distribution of Cretaceous and Paleogene strata also suggest that the axis of the arch has not been stationary throughout the Cenozoic, and that subsidary folds (flexures) characterize the arch.

Neogene strata strike across the arch rather than wrap around it. Limited data prohibits reconstruction of early Neogene depositional patterns in the middle and inner Coastal Plain area of the Cape Fear arch. However, the lower Pliocene York-

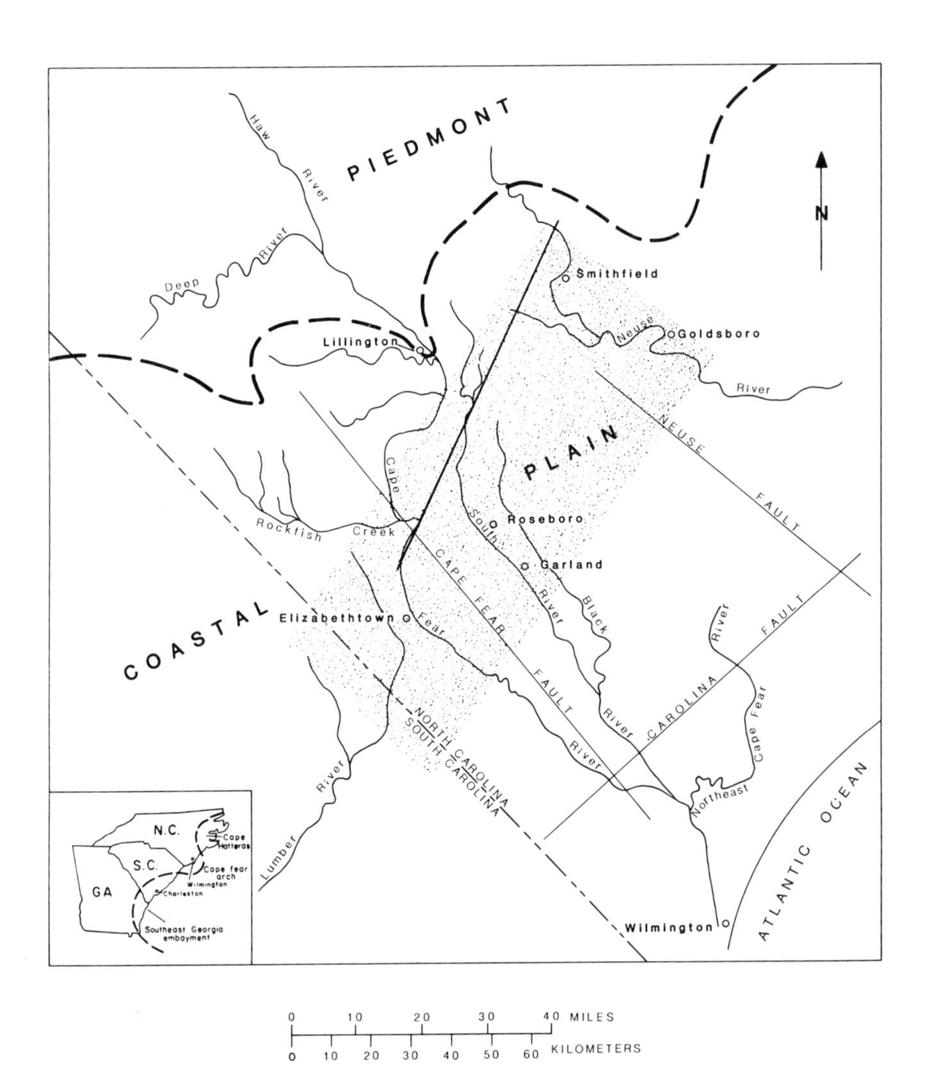


Figure 1. Generalized map of the Cape Fear Arch, which shows the major rivers and the three faults that have been proposed by Harris and others (1979) for the area. The stiple pattern represents the area of flexure and/or faulting proposed by Soller (1983) and is corroborated by data presented herein. The heavy northeast trending line from Rockfish Creek to near Smithfield represents either the top of the flexure or of a fault associated with the flexure.

town Formation forms an extensive east-southeastward sloping plane, that ranges in altitude from 80 to 30 m and represents a major marine transgression that created a shore line more than 140 km inland of the present coast. Below 30 m the Yorktown Formation is either overlain by younger marine sediments or have been removed by erosion.

## PURPOSE AND SCOPE

Data from the Cape Fear River terraces support the conclusion of Markewich and Soller (1983), Markewich and others (written commun., 1983) and R. E. Weems (U. S. Geological Survey, written commun., 1984) that there has been an acceleration of the rate of regional uplift during the late Pleistocene, and the conclusion of Soller (1983), that there is a flexure and/or zone of faulting across the Cape Fear River valley. This paper presents geomorphic data from the upper Cape Fear River valley that are indicative of late Cenozoic tectonism, and compares the data to that presented by Soller (1983) and Markewich and others (written commun., 1983).

## GEOLOGIC SETTING

Along the Cape Fear River, the Upper Cretaceous Cape Fear, Black Creek, and Pee Dee Formations underlie younger deposits and form high bluffs with nearly vertical faces. In the upper Cape Fear River valley the high cliffs along the river provide excellent exposures of the fluvial Cape Fear and overlying Black Creek Formations. The Cape Fear Formation is comprised of semi-consolidated, green to black, gravelly sands and interbedded clays. There are localized zones of chaotic bedding structures. The Black Creek Formation is composed of well cross-bedded, micaceous sands with a high percentage of finely divided carbonaceous matter. The thinness of the Cretaceous strata over the arch, a maximum of 300 m, compared to thickness of the same age strata in the adjacent Southeast Georgia embayment, 600 m (Gohn and others, 1980), and the pattern of structure contours on both the upper and lower contacts of the upper Cretaceous strata, indicate that the arch was active during upper Cretaceous time (Cohee and others, 1962; J. P. Owens, verbal commun., 1983).

Table I: Correlation of Cape Fear River Terraces

| Owens (see text) Wilmington to Rockfish Creek Elizabethtown (approx. ages) | Soller (1983) Wilmington to 10km NW of Elizabethtown | Markewich (this paper) gradients m/km b/w Fayette-ville and Elizabeth-town | Markewich (this paper) terraces and gradients m/km b/w Lillington and Fayette-ville |
|---|---|---|---|
| alluvium (no marine equivalent) | (Terrace I) (no marine equivalent) | approx. 0.26 | Terrace D approx. 0.10 |
| Wando 100,000 years | (Terrace II) Wando | ---- | ---- |
| Socastee 200,000 years or Canepatch 500,000 years | (Terrace III) Socastee | ---- | ---- |
| Waccamaw 1.6 million years | (Terrace III) Socastee | approx. 0.015 | Terrace C approx. 0.21 |
| Bear Bluff 3.0 million years | Terrace IV Waccamaw or Bear Bluff | approx. 0.01 | Terrace B approx. 0.21 |
| ---- | ---- | | Terrace A approx. 0.17 |

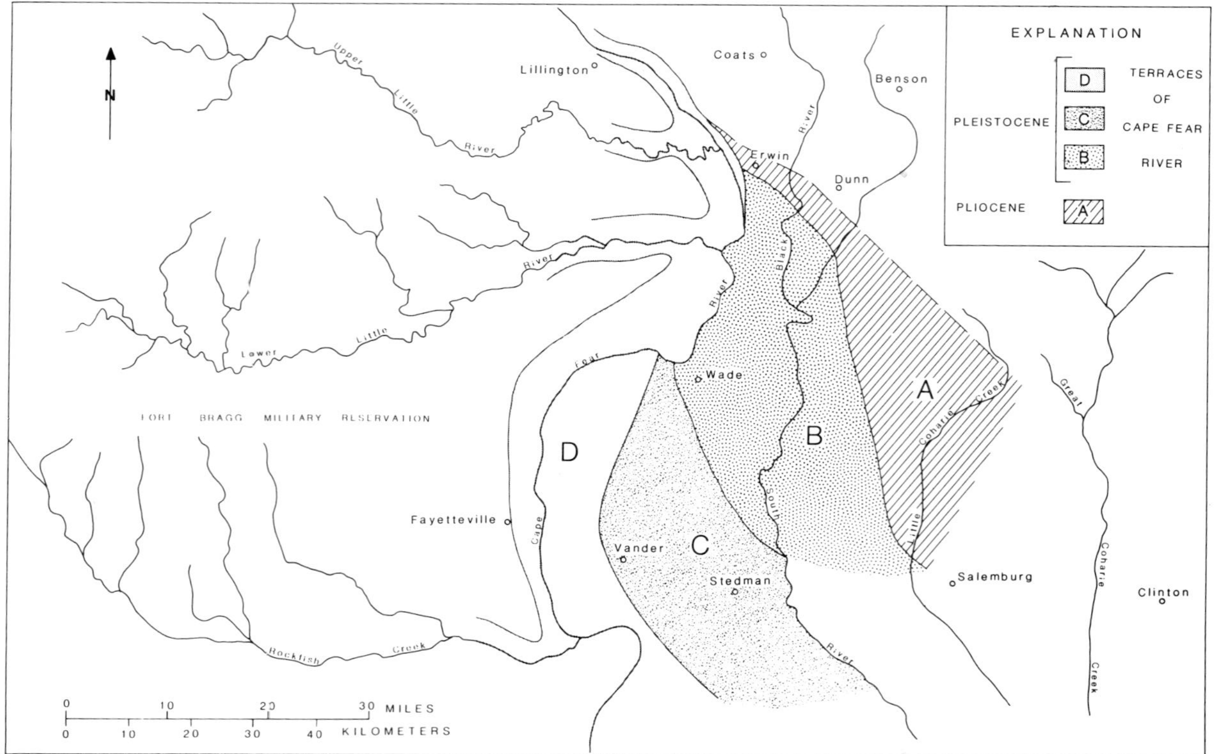


Figure 2. Drainage map of the upper Cape Fear River valley with the approximate boundaries of the four terraces associated with the Cape Fear River. Terrace D extends upstream into the three major tributaries, Rockfish Creek and the Upper and Lower Little Rivers.

Isolated remnants of Paleogene and lower Neogene strata crop out in the upper Cape Fear River valley and in the upper Coastal Plain between Lillington and Raleigh. The outcrops are discontinuous remnants of marine and marginal marine fluvial deposits that have been correlated with other Coastal Plain units of Eocene, Oligocene, and Miocene age. These remnants provide evidence that the inner Coastal Plain region of the Cape Fear arch has been periodically at or near sea level throughout the Cenozoic, suggesting oscillatory epeirogenic movements and/or major variations in eustatic sea level.

The Cape Fear arch is blanketed by Pliocene and Pleistocene age marine sediments that are thin, lithologically and paleontologically similar, and, in places, areally discontinuous. In the topographically highest part of the arch, along the inner margin of the Coastal Plain, the lower Pliocene Yorktown Formation (Blackwelder and Ward, 1979) and the Brandywine Formation (Cooke, 1930; Daniels and others, 1972) Formations form an only moderately dissected eastward-sloping plane, which has a maximum altitude of 83 m and forms the valley wall of the upper Cape Fear River valley. Younger marine units that crop out in the lower Coastal Plain of North and South Carolina (Table I) may be represented by the fluvial terraces of the Cape Fear River, as suggested by J. P. Owens (U. S. Geological Survey, written commun., 1984) and Soller (1983). It is, however, beyond the scope of this paper to make such correlations.

## THE UPPER CAPE FEAR RIVER AND ITS TERRACES

Figure 2 shows a generalized map configuration of four distinct fluvial terraces (A, B, C, and D,) in the Cape Fear River valley upstream from Rockfish Creek. The oldest terrace, A, is represented only· by isolated remnants on the valley side or as islands within younger fluvial sediments. Other terraces are relatively continuous. Terraces A, B, and C trend east-southeast; terrace D trends almost due south, as does the present river.

Yorktown age (Brandywine) Gravels and Terrace A

Terrace A is the oldest and most discontinuous of the four terraces. Remnants of terrace A are not extensive enough or numerous enough to allow determination of original terrace width, nor are they distinct enough to be definitely distinquished from fluvial facies of Yorktown strata (the Brandywine Formation?). The same constraints limit estimates of terrace gradients. From Erwin to a few kilometers downstream of Dunn (fig. 2), terrace A has an estimated gradient between 0.1 and 0.25 m/km. Remnants of terrace A gravel near Erwin are inset approximately 7 m into gravels of Duplin age that form the adjacent valley wall. Terrace A gravels appear to be somewhat less weathered than the Duplin age gravels, but are similar in composition and internal structure. Both are comprised of a $7^{+}$-m-thick bed of clayey, silty, medium- to coarse-grained sand with cobble-pebble gravel at the base.

Terrace B

Terrace B has a maximum width of 15 km. Its eastern boundary is a distinct 10-14 m scarp that separates terraces A and B. A 16 km long island of terrace A material is the most prominent morphologic form on terrace B. For a distance of about 20 km south of Erwin, terrace B has a gradient between 0.14 and 0.3 m/km. Downstream of the study area, from north of Roseboro to the town of Garland, terrace B has a gradient of about 0.01 m/km. No mineralogical data are available on the fluvial deposits of terrace B upstream of Roseboro.

Terrace C

Terrace C has a width of 2.5-15 km and is characterized by the most well developed Carolina Bays. The scarp separating terrace C from terrace B is indistinct, varies in height from 3 to 7 m, and is difficult to recognize in the field. For these reasons the boundary between terraces C and B has been only tentatively placed as seen on figure 2. Terrace C, from near Eastover to Stedman, has a gradient of about 0.21 m/km. Downstream of the study area, to an area around the junction of the Cape Fear and Black Rivers, terrace C has a gradient about 0.015

m/km. The surface of terrace C has been modified by aeolian activity. No data are available on depth of oxidation.

Terrace D

Terrace D is the youngest of the four terraces, is directly adjacent to the present Cape Fear River, and is the only fluvial or marine terrace in the Cape Fear River valley that does not have Carolina Bays. Between Erwin and Elizabethtown (figs.1, 2), the terrace is divided into two parts. The inner part is characterized by beautifully preserved primary topographic features including scroll work, ridges and swales, and small dune fields. The outer part is not topographically separated from the inner part but does not have distinct primary morphology. Terrace D ranges in width from 10 km to 2.5 km. The area of maximum width is associated with the confluence of the Cape Fear River with the Upper and Lower Little Rivers. During the time of formation of terrace D, the Upper and Lower Little Rivers were supplying a large part of the sediment load of the Cape Fear River. The large meander bend at the mouth of Rockfish Creek also formed during terrace D time when there was increased sediment load from Rockfish Creek to the Cape Fear River. The primary topography associated with terrace D along the Cape Fear River also characterizes the extension of terrace D upstream along these three tributaries.

The scarp between terrace D and C is unusually straight, continuous and distinct. The scarp ranges from 7 to 10 m in height. Terrace D has a gradient of 0.01-0.07 m/km from just north of Erwin to Fayetteville. From a point just north of Rockfish Creek, downstream for about 21 km, the gradient of terrace D increases to 0.2 to 0.4 m/km (fig.3).

The thickness of terrace D deposits ranges from 1 to 10 m; 6- to 7-m-thick exposures are common. Terrace-D deposits consist of a fining-upward sequence of pebble gravel to a silty, clayey sand. From Lillington to Erwin, cobbles and small boulders occur sporadically through the sands and gravels. Locally, individual dunes and small sand sheets overlie the fluvial deposits.

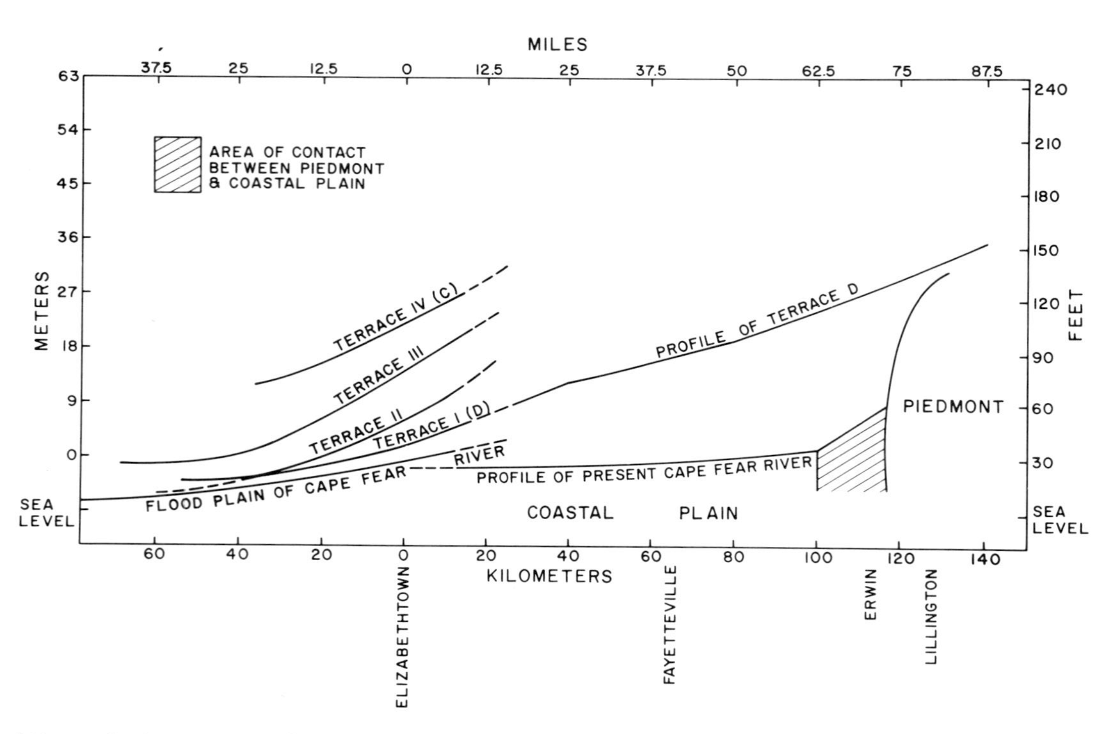


Figure 3. Profiles of the present Cape Fear River and of Terrace D, a late Pleistocene terrace of the Cape Fear River. The profiles of the floodplain of the Cape Fear River and of the Cape Fear River terraces, that are dashed upstream of Elizabethtown, are from Soller (1983).

History and Age of Cape Fear River Terrraces

Because the Cape Fear valley is cut into the lower Pliocene Yorktown Formation, all Cape Fear River terraces must be Pliocene or younger in age. Two $^{14}C$ ages, >40,000 years, on wood from terrace D are the only other data on the ages of the terraces (W-5291 and W-5346, U.S.G.S. $^{14}C$ laboratory, Reston, Va.). Chemical (cation exchange capacity), mineralogical (clay mineralogy), and physical (clay mass) data from soils developed in terrace D indicate an age between 60,000 and 200,000 years. Geomorphic character and weathering profile data suggest a a much greater age for terrace C, but data are not sufficient to determine even a range of ages. Interpretations of the age of the Cape Fear River terraces is presented in Table I. The maximum time span is from about 4 Myr through the late Pleistocene.

Regardless of age, the pattern of river migration since deposition of terrace A has been westward (fig.2). Since deposition of terrace-A gravels some time during the Pliocene, the Cape Fear River north of Rockfish Creek has migrated from a southeastward to a southerly course. Soller (1983) and J. P. Owens (U. S. Geological Survey, written commun., 1984) discuss the same west-southwestward migration for the Cape Fear River downstream of Rockfish Creek. They also considered the migration to be continuous throughout the Pliocene and Pleistocene.

Surface expressions of terrace D, along the reach of the Cape Fear River between Lillington and Rockfish Creek, indicate that the river had a modified anastomosing pattern within a straight channel. The inner part of terrace D is exactly reflected in the course of the present river, suggesting that the terrace was systematically abandoned as the river channelled and began to incise. The present Cape Fear is incised 13 to 28 m into this channel. The 13- to 28-m-deep canyon has formed within the last 200,000 years and is possibly much younger.

The gradient of terrace D increases in the downstream direction, which is consistent with the data of Soller (1983) for the youngest terrace and for terraces intermediate in age between D and C that terminate south of Rockfish Creek (Table I, fig.3). Soller (1983) also shows that downstream from Elizabethtown, near the junction of the Cape Fear and Black Rivers, gradients of all the terraces decrease and become subparallel to the profile of the present Cape Fear (fig.3), thereby suggesting a flexure or zone of transition between areas that have been affected by different rates and/or styles of uplift during the late Pleistocene.

## The Present Cape Fear River

The present Cape Fear River, from Lillington to Rockfish Creek, a distance of some 50 km is incised from 13 to 28 m through the sand and gravel deposits of terrace D into the underlying Upper Cretaceous Cape Fear and Black Creek Formations. For most of this length the depth of incision is 16 to 20 m, the width of the river is less than 0.3 km, and there is no mappable flood plain.

From several kilometers above Lillington to about 8 km south of Erwin, where the Cape Fear River is incised into or flowing on Piedmont crystalline rocks, the gradient of the present Cape Fear River ranges from 0.4 to 1.7 m/km (fig.3). For about 32 km downstream from the Cretaceous-crystalline contact in the present river channel, the river gradient is about 0.09 m/km (fig.3). The decrease in gradient from Fayetteville to Elizabethtown may be due to a change in bedrock or to subsurface structure (fig.3). Data are not yet available to make this determination.

## Tributaries to the Cape Fear River

The three eastward flowing tributaries to the Cape Fear River that have their headwaters in the upper Coastal Plain and outer Piedmont on the Cape Fear River-Pee Dee River divide are large and deeply incised. These tributaries, the Upper and Lower Little Rivers and Rockfish Creek, drain the uplifted highland area north and west of Fayetteville (fig. 2). Near the confluence of each tributary with the Cape Fear River, the tributary assumes the character of the river and is contained within a steep-walled, sluice-like canyon with no mappable floodplain.

Tributaries of the Cape Fear River that drain the upper Coastal Plain between the Cape Fear and Neuse Rivers are small, both in size of drainage basin and in stream flow, compared to those that head on the Cape Fear River-Peedee River divide. They flow parallel to the Cape Fear River for much of its length, coalesce to form the Black River, and join the Cape Fear River only 30 km northwest of Wilmington (fig.1). These tributaries, like the Cape Fear River itself, alter their pattern from relatively straight channels to highly meandering channels in the lower Cape Fear River valley. The line that connects the point of pattern change in the river and its tributaries is approximately coincident with the southeastern boundary of the postulated flexure or zone of faulting.

## DISCUSSION

Geomorphic evidence of late Cenozoic tectonism in the upper Cape Fear River valley can be divided into several categories:

1) Evidence for regional uplift of Pliocene and Pleistocene marine strata of the Coastal Plain of northeastern South Carolina and southeastern North Carolina.

2) Geomorphic data indicating that late Cenozoic deformation associated with the Cape Fear arch has been active since the middle Pliocene.

3) Geomorphic evidence from Cape Fear River terraces, and from drainage patterns of tributaries to the Cape Fear, that suggests a northeast-trending zone of flexure or faulting approximately perpendicular to the axis of the Cape Fear arch and in line with some of the minor faults located along the inner edge of the Coastal Plain northeast of Smithfield. Each category is discussed separately in the following sections.

### Evidence for Regional Uplift

Regional uplift of the South Carolina and North Carolina Coastal Plain is indicated by elevated positions of marine terraces and barriers from Wilmington, North Carolina to Charleston, South Carolina (Markewich and others, written commun., 1983) R. E. Weems, U. S. Geological Survey, written commun., 1983) Data indicate that there have been periods during which the uplift rates over parts of the Cape Fear arch have been less that the average for the region, but that the general trend is increasing uplift rates from the early Pliocene to the present. For the late Pleistocene, rates of uplift seem to be two to five times as great as the average rates noted for the entire Pliocene and Pleistocene. Uplift rates appear to be similar on the south limb of the Cape Fear arch and in the Southeast Georgia Embayment. The post-Yorktown, pre-Bear Bluff regional rate of uplift was about 30 m/Myr for the area of the Cape Fear arch. This is within the range of rates of uplift suggested by Cronin (1981) for the Coastal Plain. As mapped by J. P. Owens (U. S. Geological Survey, written commun., 1984), coincident altitudes of the Bear Bluff and Waccamaw shorelines northeast of the Cape Fear River suggest that in this area there was little or no uplift during the postBear Bluff to pre-Waccamaw period. Southwest of the river, the 15 to 20 m difference in the shorelines suggests that uplift during the post-Bear Bluff to pre-Waccamaw period occurred at a rate of 15-20 m/Myr, which is equal to or slightly less than the average rate of uplift for the region during the Pliocene and Pleistocene (Cronin, 1981). Markewich and others (written commun., 1983) show that if the Conway Barrier, at an altitude of 12-15 m, is only 200,000 years in age, then uplift rates on the southern limb of the Cape Fear arch, in the latter part of the late Pleistocene, could have been as high as 90 m/Myr. Similar high rates have been suggested by R. E. Weems (U. S. Geological Survey, written commun., 1984) for the last 350,000 years in the Charleston area of the southeast Georgia Embayment. Depending on the altitudes and/or ages used, estimates of uplift rates during the latest Pleistocene range from 40 to 1000 m/Myr. Geomorphic, isotopic, and pedogenic data indicate rates of 40 to 240 m/Myr, and suggest that rates of uplift varied areally across the arch.

Incisement of the Cape Fear River throughout its length is the response to the increased rate of uplift throughout the late Cenozoic. The canyon-position of the Cape Fear River north of Elizabethtown is the result of downcutting in response to both regional uplift and to uplift associated with a the flexure or zone of faulting that has affected the highlands north and west of Fayetteville.

## Pliocene and Pleistocene Deformation of the Cape Fear Arch

The successive westward and southwestward migration of the river throughout the Pliocene and Pleistocene suggests that the axis of uplift of the arch has been east and northeast of the river and that uplift in this area has been continuous since regression of the Yorktown sea. There is no evidence to suggest that uplift associated with the arch has been episodic or that faulting is involved.

Northeast trending Flexure or Zone of Faulting

Prowell (1983) indicated that several faults reported by Daniels and Gamble (1972) placed schistose rocks of the Piedmont against unconsolidated clayey sands of Pliocene or Pleistocene age. Both normal and reverse faults were mapped and all the faults are located near Smithfield on the tributary divides of the Neuse River. Reinemund (1955, p. 83) reported that several faults along the eastern edge of the Triassic Sanford Basin, which is coincident with the inner edge of the Coastal Plain north and northwest of Fayetteville, cut Upper Cretaceous and Cenozoic strata (These faults do not reach the surface and are not shown on geologic maps that accompany Reinemund's report). Although the part of the Cape Fear arch between the PeeDee and Neuse Rivers, is where the Fall Line shifts from an east-west to a northeast-southwestward direction (fig. 1), there is little data available on either the style or age of faulting in the area. Some data suggest that at least some offset has occurred during the Neogene, and perhaps in the Pliocene. N. L. Frederiksen, (U. S. Geological Survey, written commun., 1981) suggested a Neogene age for gravels at about 80 m altitude in the Anderson Creek quadrangle. There are at present, however, no definitive data to confirm Pleistocene faulting along the edge of the inner Coastal Plain.

The faults along the inner edge of the Coastal Plain, especially those near Smithfield, suggest the presence of structures subparallel to the Fall Line that extend into the upper Coastal Plain. The Lower and Upper Little Rivers and Rockfish Creek drain an extremely dissected highland north and west of Fayetteville that appears to be bounded on the southeast by a northeast-trending flexure and (or) fault that has affected the profiles of the Pleistocene-age terraces of the Cape Fear River. The shallow 2- to 3-m-thick weathering profiles in the highland area are in sharp contrast to the 4- to 8-m-thick profiles that characterize the Pliocene and lower Pleistocene deposits in the upper Cape Fear valley. The shallow profiles of the highland area are similar to the profiles on the late Pleistocene terrace D of the Cape Fear River. This similarity of weathering and soil profile data suggest that the late Pleistocene, 60,000-200,000 yr, was a time of uplift in this region and provides an explanation for the increased sediment load from the Upper and Lower Little Rivers and Rock Fish Creek during the formation of terrace D. The postulated structure separating these two areas is along a line from Rockfish Creek to the mapped faults northeast of Smithfield (fig.1). This northeast trending line or narrow zone apparently has controlled the sedimentation of the fluvial systems in the region since deposition of terrace A (fig.3). The pinchout of Cretaceous strata along the same boundary suggests that this flexure or fault may have controlled the pattern of sedimentation since the late Mesozoic.

Schumm (1982) noted that tilting of alluvial terraces in a valley is the most convincing evidence of deformation. The oldest terrace will be the most deformed if deformation has persisted over a period of time. Soller (1983) showed this relationship to exist for all the terraces of the Cape Fear River northwest of Elizabethtown that he considered Pleistocene in age (fig.3). In the upper Cape Fear Valley, there is an apparent increase in the gradient of terrace D downstream from Fayetteville. Terraces B and C do not show a similar downstream increase in gradient, which suggests that they have not been affected by the postulated flexure and (or) faulting. Original gradients of terraces B and C were, however, much less than those of the younger terraces mapped by Soller (1983) in the lower Cape Fear valley, because terraces B and C were much nearer their associated shorelines (J. P. Owens, U. S. Geological Survey, written commun., 1984). Up-valley steepening of these older terraces would not be noticeable in comparison to their fluvial neighbors.

Incision of the Cape Fear River from Lillington to downstream of Rockfish Creek, and of the Upper and Lower Little Rivers and Rockfish Creek, is in response to an increased rate of regional uplift and to uplift associated with the flexure or zone of faulting.

Data indicate but do not prove that a fault trending northeast from Rockfish Creek to Smithfield may be associated with the larger proposed flexure or zone of faulting.

## CONCLUSIONS

1) In the area of the Cape Fear arch geomorphic evidence can be used to verify long-term regional deformation and to identify zones of differential uplift, flexures, or zones of faulting. Regional drainage patterns appear to be responding to three types of deformation: a) regional uplift of the Coastal Plain of South Carolina and southeastern North Carolina; b) a flexure and (or) zone of faulting across the Cape Fear River valley from about Lillington to the confluence of the Cape Fear and Black Rivers, and the difference in degree of post-Bear Bluff-preWaccamaw uplift across the valley of the Cape Fear River and c) continuous deformation associated with the Cape Fear arch, with uplift along the Cape Fear River-Neuse River divide, controlling the southwestward migration of the Cape Fear River throughout the Pliocene and Pleistocene.

2) Data on gradients of Cape Fear River terraces suggest a flexure across the Cape Fear River valley from somewhere near Fayetteville to confluence of the Black and Cape Fear Rivers. Stratigrahic data and geomorphic data suggest that a line from Rockfish Creek, to between Dunn and Lillington, to Smithfield marks either the position of a fault associated with the flexure, or simply the top of the flexure.

3) Uplift north and west of Fayetteville was coincident with formation and channelization of terrace D, the 60,000-200,000 yr terrace through which the Cape Fear is incised. The deep incision of the Cape Fear River through terrace D and the low-gradient profile of the river are the river's response to the combined regional and differential uplift affecting the Cape Fear arch sometime after deposition of terrace D materials.

4) Presence of the marine Yorktown Formation across the Cape Fear-Neuse River divide from $60\text{-}80^{+}$ m altitude and 140 km from the present shoreline suggest that this part of the Cape Fear arch was not elevated in early Pliocene time.

REFERENCES

Blackwelder, W. B. and Ward, L. W., 1979, Stratigraphic revision of Pliocene deposits of North and South Carolina: South Carolina Geologic Survey Geologic Notes, v. 23, n. 1, p. 33-49.

Cohee, G. V., and others compliers, 1962, Tectonic Map of the United States, exclusive of Alaska and Hawaii: U.S. Geological Survey and American Association of Petroleum Geologists, scale, 1:2,500,000.

Cooke, C. W., 1930, Correlation of coastal terraces: Journal of Geology, v. 38, p. 577-589.

Cronin, T. M., 1981, Rates and possible causes of neotectonic vertical crustal movements of the submerged southeastern United States Atlantic Coastal Plain: Geological Society of America Bulletin, Part 1, v. 92, p. 812-833.

Dall, W. H., 1898, U.S.G.S. Annual Report Part II, p. 388, published in 1897 as 55th Congress, 2nd session, H, DOC 5.

Daniels, R. B., Gamble, E. E., Wheeler, W. H., and Holzhey, C. S., 1972, Some details of the surficial stratigraphy and geomorphology of the Coastal Plain between New Bern and Coats, North Carolina: Raleigh, N. C., Carolina Geological Society Field Trip Guidebook, p. 58-60.

Gohn, G. S., Smith, C. C., Christopher, R. A., and Owens, J. P., 1980, Preliminary stratigraphic cross sections of Atlantic Coastal Plain sediments of the southeastern United States-Cretaceous sediments along the Georgia Coastal Margin: U.S. Geological Survey Miscellaneous Field Studies Map MF-1015-C.

Harris, W. B., Zullo, V. A., and Baum, G. R., 1979, Tectonic effects on Cretaceous, Paleogene, and early Neogene sedimentation, North Carolina, in Baum, G. R., Harris, W. B., and Zullo, V. A., eds., Structural and stratigraphic framework for the Coastal Plain of North Carolina: Carolina Geological Society and Atlantic Coastal Plain Geological Association Field Trip Guidebook, October, 1979, Wrightsville Beach, North Carolina, p. 19-29.

Markewich, H. W. and Soller, D. R., 1983, The Cape Fear River: A late Quaternary drainage (abs.): Southeastern Section, Geological Society of America Abstracts with Programs, v. 15, no. 2, p. 56.

Prowell, D. C., 1983, Index of Faults of Cretaceous and Cenozoic age in the eastern United States: U.S. Geological Survey Miscellaneous Field Studies Map MF-1269, 2 sheets.

Reinemund, J. A., 1955, Geology of the Deep River Coal Field, North Carolina: U. S. Geological Survey Professional Paper 246, 159p.

Richards, H. G., 1950, Geology of the Coastal Plain of North Carolina: Transactions of the American Philosophical Society, n.s., v. 40, pt. 1, 83p.

Schumm, S. A., Watson, C. C., Burnett, A. W., 1982, Phase I: Investigation of neotectonic activity within the lower Mississippi Valley division: Potamology Program (P-1), Report 2, U. S. Army Corp of Engineer District, Vicksburg, P. O. Box 60, Vicksburg, Mississippi, 39180, 158 p.

Sohl, N. F., and Christopher, R. A., 1983, The Black Creek-Peedee Formational contact (Upper Cretaceous) in the Cape Fear region of North Carolina: U.S. Geological Survey Professional Paper 1285, 37p.

Soller, D. R., 1983, The Quaternary history and stratigraphy of the Cape Fear River valley: unpublished dissertation, George Washington University, 192p.

Weems, R. E., and Lemon, E. M., Jr., in press, Geologic Map of the Mount Holly quadrangle, with text: U.S. Geological Survey Geologic quadrangle map 1579.

# 13
# Appalachian piedmont morphogenesis: weathering, erosion, and Cenozoic uplift

*Milan J. Pavich*

ABSTRACT

Current interest in "tectonic geomorphology" has fostered studies of recent faulting and rapid uplift. However, there has been little study of the factors which produce and control uplift rates of relatively stable areas of the earth's crust. Data are needed from such areas in order to determine the continuity and variability of uplift rates. This paper presents a synthesis of data from independent sources to compare rates of uplift, weathering, and erosion for the Appalachian Piedmont. The mean rates of these processes are within a factor of two; a hypothesis drawn from the similarity of rates is that uplift and denudation are continuous and functionally related over tens of millions of years. It is argued that: 1) long-term ongoing processes have determined the landforms and regolith distribution on the Appalachian Piedmont; and 2) isostatic compensation for mass lost by weathering and erosion has been a major component of continued uplift of the Piedmont through the Cenozoic.

Physiographic stability of the Piedmont can be hypothesized based on the probability of the long-term continuity of hydrologic and geochemical processes. Stream downcutting in response to uplift maintains steep hydraulic gradients in small tributary drainage basins. The movement of ground water along the hydraulic gradient drives the ongoing hydrochemical production of a saprolitic regolith in rocks of steeply dipping structure. The rates of denudation and consequent isostatic uplift are controlled by the rates of chemical weathering and of physical removal of the weathered material.

In this hypothetical model a steady state system of weathering, erosion and isostatic uplift has operated for about 70 million years in the Appalachian Piedmont. Data from a few sources show, however, that perturbations or discontinuities in the rates of erosion and uplift have occured during the late Cenozoic. A methodology is proposed for obtaining data from the weathering profiles and rocks of the Piedmont to directly test the areal variation of uplift and erosion rates through the late Cenozoic.

## INTRODUCTION

The Piedmont physiographic province of the mid-Atlantic and southeastern United States was considered for nearly 100 years to be a peneplain or old-age erosion surface that developed according to the Davisian model of landscape evolution (Davis, 1899). During the past twenty-five years numerous researchers have either augmented or disputed the Davisian model. The concept of "dynamic equilibrium" (Hack, 1960) has been included as an integral part of many of these recently developed models on landscape evolution. Recent studies of drainage basins emphasizing low temperature geochemistry have allowed estimates to be made on the rates of weathering or saprolite formation on the Piedmont (Cleaves and others, 1970). Fission track data from the Piedmont, and paleontological data from Tertiary sediments that overly rocks of the outer Piedmont have also added to our knowledge of rates of uplift and denudation of the Piedmont and the adjacent Coastal Plain.

This paper represents a review of the rates of uplift, erosion and weathering from the Appalachian Piedmont, and discusses some of the processes that have had long-term control on the evolution of the Piedmont landscape. It is argued, as by Stephenson and Lambeck (in press), that the slow epeirogenic uplift following the rapid uplift of a tectonic event takes place over a much longer period of time that does the rapid uplift, and that rebound in response to erosional unloading (isostatic compensation) is a main component of slow epeirogenic uplift. Therefore, it is argued that development of the Piedmont is limited by rates of weathering and erosion. As erosion continues through time, sea level surfaces are uplifted as a result of isostatic compensation as well as any additional components of uplift. In this paper it is argued that rapid uplift generated by a tectonic "event" such as a plate collision lasts about 50 Myr and that subsequent slow epeirogenic uplift can last more than 50 Myr. Data from study of the thermal history of metamorphism, the sedimentation rates on the continental shelf, and the regeneration rate of Piedmont regolith are combined to produce a simplified model of the slow uplift phase.

## GEOGRAPHY AND GEOLOGY OF THE APPALACHIAN PIEDMONT

The Piedmont boundaries as defined by Hack (1982) are shown in Figure 1. Topographically the Piedmont is variable: in many places it is literally at the foot of the mountains, but in other places it is an upland plateau separated from any mountain. The Piedmont is bounded in places by distinct eastern and western structural zones. The Piedmont consists predominately of preCambrian and lower Paleozoic metasedimentary and metavolcanic rocks of varying metamorphic grades. Grabens filled with essentially unmetamorphosed Mesozoic continental clastic sediments comprise a relatively small area.

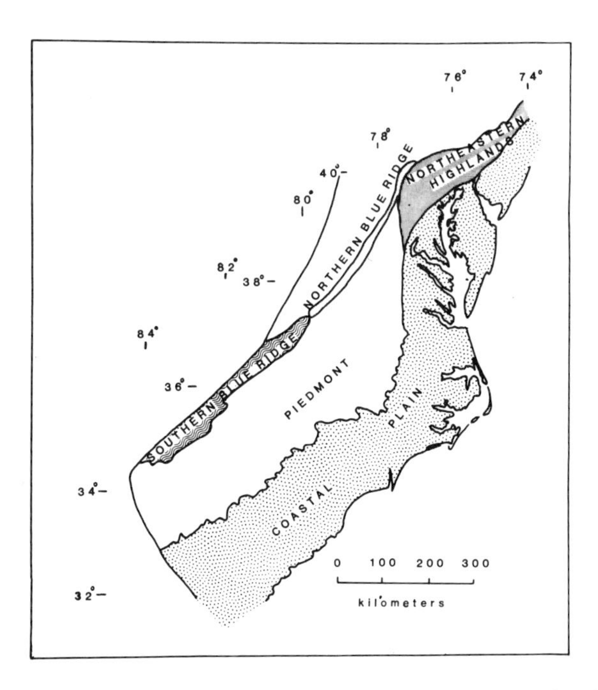


Figure 1 Map of the southeastern United States showing the Appalachian sub-provinces draining to the Atlantic basin.

Regional metamorphism occured in the Paleozoic. Radiometric ages of deformed and undeformed intrusive rocks indicate that igneous activity was continuous along the length of the Piedmont in the Silurian to Early Devonian. This group of ages, between 435 and 390 Myr (Fullagar, 1971) is older than the youngest intrusives found by Pavlides and others (1982) in Virginia ranging in age from 325 to 265 Myr. They correlate these small plutons regionally with similar age plutons dated by Fullagar and Butler (1979) from the Carolinas and Georgia. Based on these widespread plutons, Pavlides and others (1982) suggest that much of the eastern Piedmont experienced a Pennsylvanian to Permian magmatic event or events. No younger regional thermal events are recognized. Mesozoic basalt dikes and sills were associated with rifting and caused contact metamorphism of country rocks, but produced no regional thermal effects of the magnitude caused by the Paleozoic plutonism.

After the Permian, the Piedmont, as part of the larger Appalachian orogen, experienced rifting and uplift through the early Mesozoic. Unmetamorphosed clastic rocks that vary from coarse fanglomerates to fine-grained claystones fill the fault bounded Triassic and Jurassic basins. Rifting stopped after the Jurassic and the Appalachian system of the eastern United States came under a regime of compressive stress, which has continued to the present ( Prowell and O'Connor, 1978; Zoback and Zoback, 1979)

At least part of the present outer Piedmont was covered in the past by nonmarine clastic sediments of Cretaceous and younger age. Localities include: (1) nearly 60 m of upper Paleocene fluvial sands and clays preserved in a downfaulted block on the north side of Pine Mountain, near Warm Springs, Ga. (Christopher and others, 1980), and (2) 20 m of sands and gravels estimated to be late Miocene in age overlying Piedmont rocks along the Fall Line in Virginia (Froelich, 1976) and North Carolina. Many more localities where marine and nonmarine deposits of Cenozoic age overlie Piedmont metamorphic rocks have been reported in the literature. Therefore, at various times during the Cenozoic, at least the outer part of the Piedmont has been covered by sediments that attained thicknesses greater than 50 m. Hack (1982) suggested that essentially all of the sediment blanket had been removed from the Piedmont of Alabama, Georgia, and South Carolina in the post-Paleocene part of the early Cenozoic, but that the Piedmont of Virginia and Maryland had a sediment cover until sometime in the late Cenozoic. The presence of marine Eocene limestone near Raleigh to an altitude of 150 m indicates that the northeast part of the present Cape Fear arch was at sea level in the early Tertiary. Oscillations of the basins and arches of the eastern seaboard are reflected in the late Mesozoic and Cenozoic sediments of the Coastal Plain have been documented by Owens (1970). Adjacent parts of the Piedmont have possibly also experienced vertical displacements due to migrating basins and arches as indicated by minor fault displacements (Mixon and Newell, 1977; Prowell and O'Connor, 1978).

## UPLIFT RATES

Estimation of uplift rates for the Appalachian Piedmont is at present a difficult task and one of varying reliability. The validity of methods such as level-line surveys (Brown, 1978) as indicators of geologically significant uplift is questionable. The thermal metamorphic history of the rocks now exposed on Piedmont can be used to determine long-term post-metamorphic uplift rates. Mineral stabilities and argon closure temperatures in radiometrically dated rocks can be used, assuming constant thermal gradients, to calculate depth-time relationships.

Figure 2 shows simplified histories of the rates of uplift of the inner Piedmont of Georgia and Virginia following late Paleozoic metamorphism. The line connecting the circles is constructed from data presented by Dallmeyer (1978) and Whitney and others (1976) for the Inner Piedmont in Georgia, and the triangles connect the data presented by Durrant (1978) from the eastern Piedmont of Virginia. The numbers in the symbols represent: 1) the maximum depth of metamorphism determined on the basis of phase relationships of high-grade minerals; this depth is used to estimate the pressure and temperature conditions; 2) time and temperature of argon retention by hornblende; 3) time and temperature of argon retention by biotite; and 4) depth and time of intrusion of the Stone Mountain granite in Georgia.

These curves of uplift vs. time give average rates of Paleozoic uplift of 200m/Myr. It is apparent that the trends of both Virginia and Georgia data are similar. In general, the data of Dallmeyer (1978), Whitney and others (1976) and Durrant (1978) suggest that the rate of regional uplift has systematically decreased from approximately 200 m/Myr during the late Paleozoic to about 25 m/Myr for the late Cenozoic.

Hack (1979) presented geomorphic evidence for late Cenozoic uplift in the Appalachians. Cronin (1981) and Newell and Rader (1982) showed that uplift in post late-Miocene time has affected the inner Coastal Plain. It is difficult to determine a minimum rate of Cenozoic uplift. Dallmeyer (1978) estimated only 0.5 km post-Triassic erosion in Georgia, indicating an average uplift rate of 2.5 m/Myr since the Jurassic. This minimum figure is probably unrealistically low for Georgia; certainly for the Piedmont as a whole. For example, the maturity of hydrocarbons in the Culpeper basin in Virginia requires removal of 2.5 km of overlying rocks since Jurassic time ( W.L. Newell, personal communication, 1983). The true average uplift rate appears to lie somewhere between 10 and 40 m/Myr.

The fission track data presented by Zimmerman (1979) for late Mesozoic and Cenozoic uplift of the Piedmont fit that range of rates. In North and South Carolina, his data show that rocks now exposed at the surface passed through the 100$^{\circ}$C isotherm 150 Myr ago near the Fall Zone, and 75 Myr ago near the Brevard Zone along the western margin of the Piedmont. Assuming an average thermal gradient of 33.3 $^{\circ}$C/km over the past 150 Myr, these figures translate to a range of uplift rates of 20 m/Myr (3000 m/150 Myr) to 40 m/Myr (3000 m/75 Myr). The rectangle on figure 2 brackets those ages. Fission track studies of zircon, sphene, and apatite from the Virginia Piedmont by Durrant (1978) also give an average Cenozoic uplift rate of about 20 m/Myr. Although significantly less than the Paleozoic rate, this is fast enough to require the erosion of about 3 km of Piedmont rock during the Cenozoic. Recent analyses of sediment volumes on the Atlantic continental margin show that there is a crude mass balance between the material eroded off the Appalachians and that trapped on the continental shelf.

## CENOZOIC EROSION RATE OF THE PIEDMONT

Cenozoic uplift of the Appalachians could have proceeded without erosion. The fission track data, however, indicate 2 km to 3 km of Cenozoic uplift. Since the Piedmont is not 3 km above base level, it is impossible that erosion

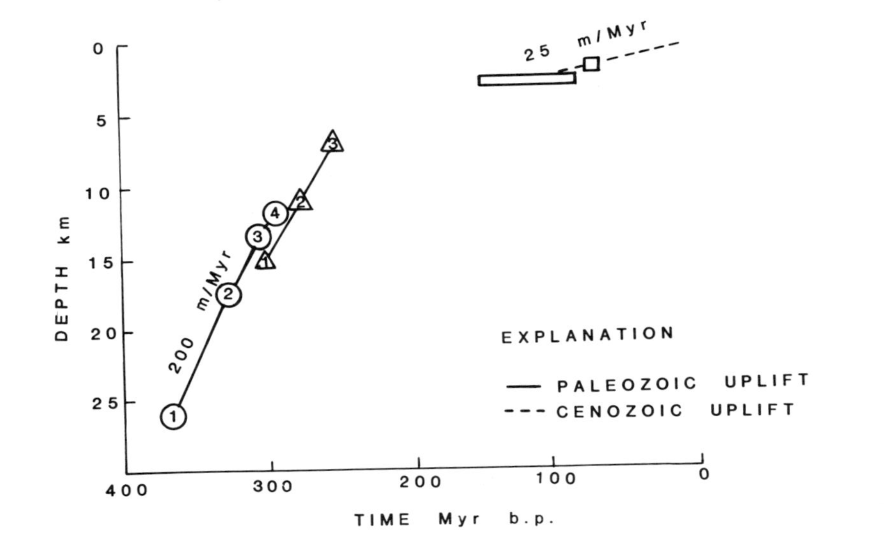


Figure 2 Uplift curve for the Piedmont in the late Paleozoic and Cenozoic. Numbered symbols represent: 1 - regional metamorphic climax; 2 - Argon retention by hornblende; 3 - Argon retention by biotite; 4 - Intrusion of Stone Mountain Granite in Georgia. Sources of data: ○ Dallmeyer (1978); ④ Whitney and others (1976); △ Durrant (1978); ▭ Zimmerman (1979); □ Mathews (1975).

of the Piedmont has not taken place. How much erosion, and the rate of erosion can be estimated from sediment volumes and recent studies of sediment transport rates.

Piedmont erosion is demonstrated by the lack of accumulation of thick transported colluvial and alluvial regolith on the Piedmont (Newell and others, 1980). The Piedmont sits well above present base level, and its major drainages have steep gradients cut into rocky channels at the eastern Fall Line margin (Hack, 1982). The cause of stream incision and erosion is likely due to epeirogenic uplift relative to a eustatically fluctuating base level. Independent of the methods of determining the average Cenozoic uplift rate of the Piedmont, we can estimate the Cenozoic erosion rate of the Piedmont. This estimate is based on an analysis of sediment volume data by Mathews (1975). He derived a figure of 1.6 $km^3/km^2$ of Cenozoic erosion (in 60 Myr) of the southern Appalachians from isopach maps of sediments in the the continental margin sediment prism. Removing this sediment uniformly from the area of the Piedmont in 60 Myr would require an average erosion rate of 27m/Myr (1.6 km/60 Myr). This rate fits the brackets provided by the uplift rate calculated from fission track data. The square on Figure 2 is plotted assuming that sediment volume is equal to the volume of material eroded. The correlation between long-term uplift rate and erosion rate supports the idea that the uplift and erosion of the Piedmont were balanced through the Cenozoic. Although the sedimentation rate data allow for many possible interpretations, it can be argued that the average Cenozoic sedimentation rates are consistent with constant production and stripping of regolith.

Modern Erosion Rate

Hack (1979) presented a compilation of surface lowering rates calculated from total measured chemical and mechanical loads of major rivers. The rates range from 4.6 to 48 m/Myr for basins draining to the Atlantic. The lowest rates are from carefully measured small watersheds. These estimates are conservative and can be considered minimum rates of modern denudation. The highest rates are from large rivers. They are probaly higher than the rates of long-term denudation because no attempt was made to factor out contributions of precipitation, anthropogenic erosion documented by Trimble (1974), and anthropogenic pollution. Since much of the soil disturbed by man-induced erosion has not been removed from the land (Meade, 1982), modern erosion rates may cautiously be used to estimate longer-term rates. The average denudation rate is probably somewhere between the maximum given by the large drainage systems and the minimums given by the smaller drainage systems. Thus, the modern erosion rate data approximate the longer-term rates estimated from sedimentation rates. This suggests that modern denudation processes, other than accelerated soil erosion, are the same as the processes that have acted through the Cenozoic. Analysis of Piedmont regolith demonstrates how soil and rock structure may control denudation rates.

## RATE OF PIEDMONT REGOLITH PRODUCTION

Maps and hypsometric curves show that the Piedmont is the largest area undergoing denudation draining to the Atlantic Ocean. The bedrock map of the Appalachians presented by Hack (1982) shows that large areas are underlain by tightly folded metamorphic rocks northeast and southwest of the Carolina Slate Belt. The thick untransported regolith on tightly folded metasediments is saprolite that averages about 15 to 20 m in thickness beneath the uplands and thins toward the valleys. Regolith or saprolite thickness varies predictably and is a function of position and rock type (Cleaves, 1974; Cleaves and Costa, 1979; Froelich and Heironimus, 1977; Hack, 1982).

Hack (1982) pointed out examples of high relief in Piedmont areas underlain by quartzites and metavolcanic rocks. Mineral stability can explain the resistance of the quartzites to weathering. For the metavolcanic rocks of the Carolina Slate Belt, a more likely explanation is the predominance of fine-grained textures and the shallow angle at which rock structures intersect the ground surface. These rock characteristics impede water infiltration into the slate belt rocks. In these rocks and in massive coarse-grained rocks such as Stone Mountain Granite, resistance to weathering may be largely structural due to impedence of ground water penetration. By contrast, most of the metasedimentary rocks which comprise most of the Piedmont have steeply dipping foliations and fractures. These preferentially transmit water vertically, and zonation of mineral reactions in the regolith reflects the vertical water movement. The regolith can be subdivided into three zones.

The regolith zonation beneath uplands is shown schematically on figure 3. The regolith is predominately rock weathered in situ known as saprolite, the term originally used by Becker (1895). In applying the term saprolite, the following characteristics are generally implied: it is isovolumetric with the underlying bedrock, as indicated by the retention of texture and fabric of the parent material, and it exhibits gradational chemical and mineralogical changes of composition going from the parent rock to the geomorphic surface. The vertically foliated rocks of the Piedmont which contain labile minerals ( e.g. plagioclase, biotite, and pyrite) and resistant minerals (quartz and muscovite) are conducive to formation of a thick structured regolith.

Rock weathering continues through time because of the processes which control water movement and the release of solutes by weathering reactions. The chemical data from the soil, saprolite and the solutions draining basins studied by Cleaves and others (1970) and Pavich (in press) indicate that: a) most of the weathering takes place in the lower saprolite and weathered rock zone, and b) the rate of weathering in a basin will roughly be a function of the mineralogy and recharge to the weathered rock zone, which is controlled by permeability, porosity, and soil partitioning of precipitation.

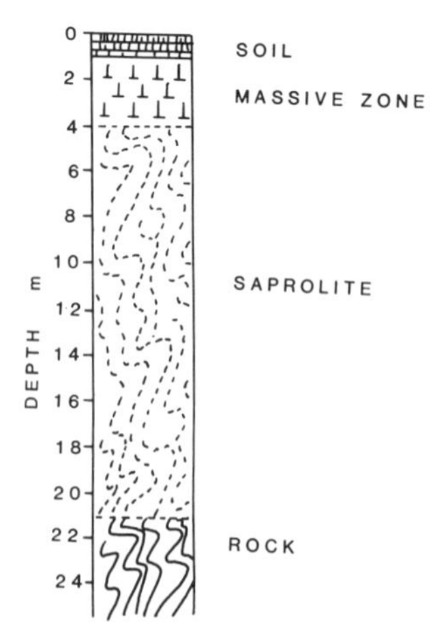


Figure 3 Profile of typical upland regolith of metasedimentary rocks of the Appalachian Piedmont.

The mechanisms responsible for continuous weathering and erosion are complexly related to rock structure and the partitioning of water in the regolith. The elements of the Piedmont regolith which exercise control are: a) the soil which functions to partition the precipitation thereby controlling the volume of water which infiltrates the saprolite and reacts with the labile minerals of the parent rock, b) the rock structure which facilitates the vertical movement of ground water, and c) the mineralogy of the parent rock which provides phases which react with dilute waters at kinetically different rates, slowly reacting phases provide a mechanically stable skeletal framework.

## Rate of weathering and saprolite formation

Table I presents data from various sources, generally given in terms of chemical denudation or surficial lowering by chemical denudation. Most of the dissolved solids in Piedmont streams are $Ca^{+2}$, $Na^{+1}$, $HCO_3^{-1}$, and $H_4SiO_4$. These are derived predominately from weathering of plagioclase at the weathering front. Unlike estimates of surficial lowering rates based on total mechanical and dissolved loads, the Piedmont chemical denudation rates are conservatively calculated from the base flow chemical loads.

Rates are strikingly similar, except for Cleaves and others (1970) in Baltimore County, Md. The rate of 7m/Myr is a more reasonable minimum, and the mean rate for southeastern Piedmont may exceed 10m/Myr. Areal variability of rock weathering rate is not known, and it may range from 2m/Myr to 40m/Myr estimated by Gilluly and others (1951).

At these rates a typical saprolite profile, such as one developed on the gneiss and schist of the Piedmont, can develop in one to two million years. Therefore, the saprolite could have begun development a few million years ago from a fresh bedrock surface which was previously stripped, or the saprolite has been developing continuously over a much longer time period. In the latter case, the rate at which the saprolite develops is equalled by the rate at which it erodes. Other models are possible. Any combination of non-linear erosion and weathering rates is possible as long as erosion rate does not greatly exceed weathering rate for a long time.

Table I Chemical Denudation Rates

| | Surficial Lowering Rate (m/Myr) | Reference |
|---|---|---|
| North American Average | 10 | Garrels and Mackenzie (1970) |
| Northeastern U.S. | 16 | Gilluly (1964) |
| Southeastern U.S. | 40 | Gilluly, Waters, and Woodford (1951) |
| Pond Branch, Maryland | 2 | Cleaves and others (1970) |
| Davis Run, Virginia | 7 | Pavich (in press) |
| Coretta, North Carolina | 40 | Velbel (1984) |
| Equatorial Climate | 20 | Corbel (1959) |

DISCUSSION

The similarity in the range of rates of uplift, denudation, and saprolite production for the Appalachian Piedmont is at least suggestive of a balance among weathering, erosion and uplift rates. Undoubtedly there are components of isostatic compensation, deformation (basin subsidence and domal arching due to loading), and compressive stress due to sea floor spreading involved in uplift of the region. Data from the Atlantic Coastal Plain and adjacent offshore basins indicate that about 3 km of material was removed from the Piedmont-Blue Ridge Province during the Cenozoic. An isostatic response to this mass loss should have occurred. If the isostatic response to mass loss continues, then long term epeirogenic uplift need not produce a significant change in elevation. Instead, a low elevation, low relief plateau like the Piedmont could be maintained as a continuous dynamic system rather than as a static system undergoing neither uplift or denudation. This is represented schematically on figure 4.

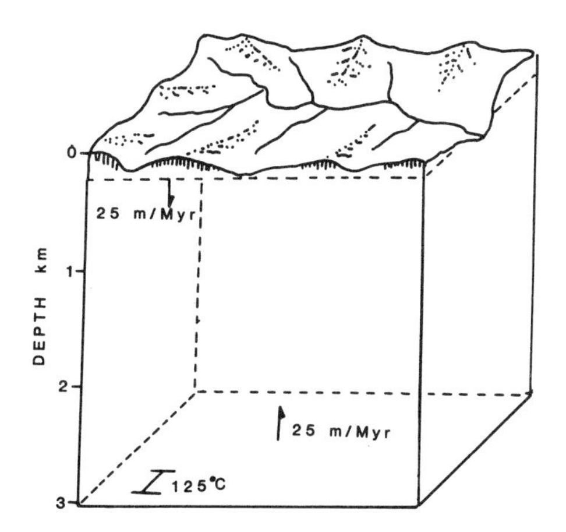


Figure 4 Schematic model of steady state Piedmont denudation and uplift in the Cenozoic.

In this conceptual model, erosion of weathered regolith from the geomorphic surface is balanced by the flux of mass through an imaginary plane representing the 25°C isotherm. This model fits Hack's (1960) concept of dynamic equilibrium under a humid climate. Variation of weathering, erosion, and uplift rates on the Piedmont, and elsewhere in the Appalachians, should not be ruled out. Evidence for timing and rates of accelerated uplift is presented by Markewich (1984) and Newell (1984) elsewhere in this volume. After discussing a model explaining the possible functional relationships of weathering, erosion, and uplift for the Piedmont, a discussion of methods available to assess the spatial variability of late Cenozoic weathering, erosion and uplift is presented.

Steady State Model

Although the data are scant, there is evidence to support the hypothesis of a long term balance of Piedmont weathering, erosion, and uplift rates. This raises the question of whether the driving force for the system is tectonic uplift or, conversely, whether the driving force is denudation and the uplift is an isostatic response. A complete answer cannot be given at this time, but the elements of a model can be presented. There is good evidence that isostatic compensation is an active response to orogeny, ice loading and unloading (Walcott, 1970), sediment loading on continental shelves (Heller and others, 1982), and dissolution of carbonates (Opdyke, 1984). Thus, the crust and mantle respond to gains and losses of mass at the surface of the crust. The volume of sediment accumulated on the Atlantic coastal margin suggests that during the Cenozoic about 3 km of regolith was removed. The uplift due to isostatic compensation can be calculated and compares with the total uplift evidenced by the fission track data. A schematic picture of the relationship between mass removal and isostatic uplift is shown on figure 5. Based on calculations of simple isostasy, this block diagram shows that about 80 percent of the surface lowering produced by denudation is compensated for by uplift. Starting at 1000m above baselevel, and removing material at a constant rate of 25 m/Myr, in 100 Myr the surface would be decreased by 500 m in elevation. Thus the net change in elevation is minor compared to

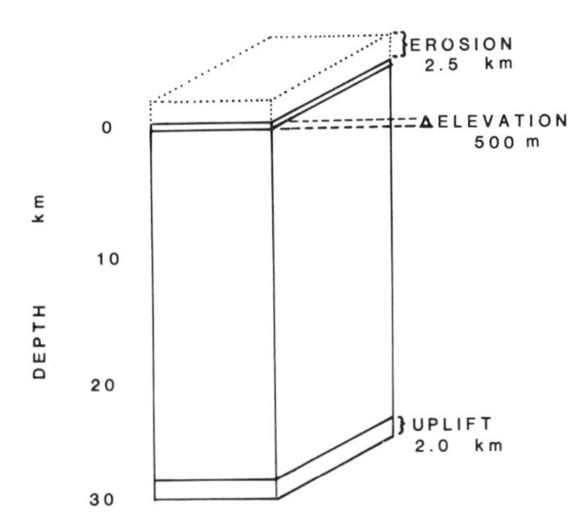


Figure 5 Model of denudation and isostatic uplift of the Piedmont over 100 Myr at 25m/Myr.

the flux of material at the upper and lower surfaces. Applied to the Piedmont, this model shows that the mass removal and uplift indicated by fission track data could be accomplished with relatively little elevation change. The uplift may be a response to the continuous removal of regolith produced from rock by the saprolitization process described previously. Although tectonically accelerated uplift could have occurred at any time during the Cenozoic, and may have occured since the late Miocene (Denny, 1982), it need not be invoked to explain the the long-term rate balance. This model does not require that uplift rates have been the same at all times and places during the Cenozoic history of the Piedmont, but the accelerated rates of uplift produced during times of tectonic activity may have been of short enough duration and of limited areal extent so that the overall balance between uplift, weathering, and erosion was not affected.

The suggestion by Denny (1982) of a late Tertiary acceleration of uplift may be well founded on stratigraphic grounds, but the evidence presented here can be interpreted as placing restraints on the magnitude and duration of that uplift event. If we assume that the uplift occured over a period of 1 million years, an uplift rate on the order of 100m/Myr seems unlikely since an uplift rate of 200m/Myr was last acheived during the thermally driven Paleozoic uplift of the Appalachians. In fact, even the geologically recent tectonic uplift documented for the Adirondaks (Isachsen, 1975), has not had sustained uplift rates of greater than 40m/Myr (Miller and Lakatos, 1983). Thus, evaluation of Denny's (1982) evidence for accelerated uplift is not simple or unequivocal. It is not clear if 40m/Myr is significantly greater than the long-term Cenozoic uplift rate of the Appalachian Piedmont.

Causes and duration of accelerated uplifts are not known. There is evidence of compressive stress in the Appalachians from the Cretaceous to the present (Prowell and O'Connor, 1978; Zoback and Zoback, 1979). It is possible that this compressive stress has produced accelerated local uplift at various places and times. However, the role of that stress in generating the long term or short term uplift of the Appalachains is poorly understood at present. Moreover, accelerated erosion might lag behind the uplift by some unknown amount of time and, therefore, not provide for simple interpretation of the timing and rate of uplift.

Methods of testing spatial variability of uplift rate on the Piedmont

The model as presented can not be used to argue that rates of Piedmont processes are absolutely constant, or that we should not look for evidence of geologically rapid or recent uplift of the Piedmont. Rather, I would argue that the Piedmont is amenable to systematic analysis of landforms and regolith by methods which should reveal evidence for local tectonic events. If rates of uplift and erosion are equated by isostatic compensation for erosion, rather than by a less constant tectonic mechanism, then it is possible that we can measure a few geomorphic and geochemical parameters on the Piedmont which will provide us estimates of the rates of all the significant long term processes. If any of the endogenic or exogenic rates exceed certain limits, then the system will not be in equilibrium.

Hack (1982) examined areas in which that equilibrium has been disturbed by endogenic or exogenic processes. This is not a simple exercise on the Piedmont because many variables may affect the thickness of regolith and its geochemical zonation. First of all, it is not clear whether recent vertical movement would cause thinning or thickening of saprolite. Uplift might increase upland erosion rates, thereby thinning saprolite. Conversely, the rapid response of streams to uplift might increase relief, and the resultant increase of groundwater hydraulic gradients might promote thickening of regolith. Fortunately, techniques are now available by which comparisons of geochemistry and age can be made, thus the question of the rate of denudation can be better defined for areas of apparently continuous erosion.

In addition to standard physical and chemical analyses of soil and saprolite, weathering and erosion rate information can be obtained from isotopic studies such as those of Moriera-Nordemann (1978) of the U/Th disequilibrium in weathered rocks, and that by Pavich and others (1984) of the accumulation of cosmogenic $^{10}Be$ in Piedmont regolith. A fundamental limitation to comparing rates of rock weathering and erosion is the variability of rock resistance. Future studies should attempt to sample and analyze a "standard" rock of comparable structure and mineralogy in different areas of the Piedmont, and from other crystalline rock terrains. Keeping rock resistance constant would simplify comparing other variables. Thermal histories of rocks using argon closure temperatures and fission track studies should also be conducted in conjunction with studies of regolith properties in selected areas.

## CONCLUSIONS

1. Isostatic compensation, or rebound due to erosional unloading, has been a controlling factor on the evolution of the Piedmont landscape during the Cenozoic.
2. A model for development of the Piedmont landscape can be constructed using the concept of continuous isostatic uplift in response to weathering and erosion.
3. The length of time during which any part of the Piedmont has been exposed at the surface is dependent on how recently overlying sediments have been eroded. Hack (1982) suggested that the metasediments and the metavolcanic rocks of the Piedmont in Georgia have not had a significant sediment cover since the early Cenozoic; in Maryland there appears to have been a substantial sediment blanket during at least part of the late

Cenozoic. If correct, then the balanced system of saprolite development, erosion and associated uplift has been active longer in the southeastern Piedmont than in the Piedmont of the mid-Atlantic states. Because the system is balanced, the thickness of saprolite throughout the region is about the same.

4. Because accelerated uplift has been of short duration and limited areal extent, it has caused only perturbations and not long term changes in the regional rates.

5. Spatial variability of Piedmont uplift and denudation rates can be tested by independent methods (e.g. fission track and saprolite isotopic age) and their comparison can be used to identify areas of relatively high rates of long-term and short-term uplift.

## REFERENCES

Becker, G. F., 1895, A reconnaissance of the gold fields of the southern Appalachians, U.S. Geological Survey, 16th Annual Report, p. 289-290.

Berry, J. L., 1977, Chemical weathering and geomorphological processes at Cowetta, North Carolina, abs., Geological Society of America Abstracts with Programs, v. 9, no. 2, p. 120.

Brown, L. D., 1978, Recent vertical crustal movement along the east coast of the United States, Tectonophysics, 44, p. 205-231.

Christopher, R. A., Prowell, D. C., Reinhardt, Juergen, and Markewich, H. W., 1980, The stratigraphic and structural significance of Paleocene pollen from Warm Springs, Georgia, Palynology, v. 4, p. 105-124.

Cleaves, E. T., 1974, Estimated thickness of overburden in Towson Quadrangle, Geologic and Environmental Atlas, Map no. 3 , Maryland Geological Survey.

Cleaves, E. T., Godfrey ,A. E., and Bricker, O. P., 1970, Geochemical balance of a small watershed and its geomorphic implications, Geological Society of America Bulletin, v. 81, p. 3015-3032.

Cleaves, E. T. and Costa, J. E., 1979, Equilibrium, cyclicity and problems of scale- Maryland's Piedmont Landscape, Maryland Geological Survey Information Circular 29.

Corbel, J., 1959, Vitesse de l'erosion Z. Geomorph, v. 3, p. 1-28.

Cronin, T. M., 1981, Rates and possible causes of neotectonic vertical crustal movements of the submerged southeastern U.S. Atlantic Coastal Plain: Geological Society of America Bulletin, part I, 92, p. 812.

Dallmeyer, R. D., 1978, $^{40}Ar/^{39}Ar$ incremental-release ages of hornblende and biotite across the Georgia Inner Piedmont; their bearing on late Paleozoic-early Mesozoic tectonothermal history: American Journal of Science, v. 278, p. 124-149.

Davis, W. M., 1899, The peneplain, repreinted in "Geographical Essays", D. W. Johnson, ed., Dover Publications, 1954, 777 p.

Denny, C. S., 1982, Geomorphology of New England: U.S. Geological Survey Professional Paper 1208, 18 p.

Durrant, J. M., 1978, Structural and metamorphic history of the Eastern Virginia Piedmont province near Richmond, Va., unpubl. Masters Thesis, The Ohio State University, 116 p.

Froelich, A. J., 1976, A geologic and environmental guide to northern Fairfax County, Va.: Geological Society of America Field Trip Guidebook, Short Course, no. 3, 28 p.

Froelich, A. J., and Heironimus, T., 1977, Thickness of overburden map of Fairfax County, Va., U.S. Geological Survey Open-File Report 77-797.

Fullagar, P. D., 1971, Age and origin of plutonic intrusions in the Piedmont of the southeastern Appalachians: Geological Society of America Bulletin, v. 82, p. 2845-2862.

Fullagar, P. D. and Butler, J. R., 1979, 325 to 265 m.y. old granitic plutons in the Piedmont of the southeastern Appalachians, American Journal of Science, v. 279, no. 2, p. 161-185.

Garrels, R. M. and Mackenzie, F. T., 1970, Evolution of Sedimentary Rocks, Norton and Company, New York, 397p.

Gilluly, J., Waters, A. C., and Woodford, A. O., 1951, Principles of geology: San Francisco, W. H. Freeman and Company, 631 p.

Gilluly, J., 1964, Atlantic sediments, erosion rates, and the evolution of the continental shelf, Geological Society of America Bulletin, v. 75, p. 483-492.

Hack, J. T., 1960, Interpretation of erosional topography in humid climates temperate regions: American Journal of Science, v. 258A, p. 80-97.

Hack, J. T., 1979, Rock control and tectonism, their importance in shaping the Appalachian Highlands: U.S. Geological Survey Professional Paper 1126-B, 17 p.

Hack, J. T., 1982, Physiographic dimensions and differential uplift in the Piedmont and Blue Ridge: U.S. Geological Survey Professional Paper 1265, 49 p.

Heller, P. L., Wentworth, C. M., and Poag, C. W., 1982, Episodic post-rift subsidence of the United States continental margin: Geological Society of America Bulletin, v. 93, p. 379-390.

Isachsen, Y. W., 1975, Possible evidence for contemporary doming of the Adirondak Mountains, New York, and suggested implications for regional tectonics and seismicity, Tectonophysics, v. 29, p. 169-181.

Judson, Sheldon, and Ritter, D. F., 1964, Rates of regional denudation in the United States, Journal of Geophysical Res., v. 69, p. 3395-3401.

Markewich, H. W., 1984, Geomorphic evidence for Pliocene-Pleistocene uplift in the area of the Cape Fear Arch, North Carolina, in Morisawa, M. and Hack, J. T., eds., Tectonic geomorphology, proceedings volume of the Fifteenth Annual Geomorphology Symposia Series, held at Binghamton, New York, September 28-29, 1984, Allen and Unwin, International Series, London.

Mathews, W. H., 1975, Cenozoic erosion and erosion surfaces of eastern North America: American Journal of Science, 275, no. 7, p. 818-824.

Meade, R. H., 1982, Sources, sinks, and storage of river sediment in the Atlantic drainage of the United States: Journal of Geology, v. 90, no. 3, p. 235-252.

Miller, D. S., and Lakatos, Stephen, 1983, Uplift rate of Adirondak anorthosite measured by fission-track analysis of apatite: Geology, v. 11, p. 284-286.

Mixon, R. M., and Newell, W. L., 1977, Stafford fault system - structures documenting Cretaceous and Tertiary deformation along the Fall Line in northeastern Virginia: Geology, v. 5, no. 7, p. 437-440.
Moreira-Nordemann, L. M., 1980, Use of $U^{234}/U^{238}$ disequilibrium in measuring chemical weathering rate of rocks: Geochem. Cosmochem. Acta, v. 44, p. 130-108.
Newell, W. L., 1984, Architecture of the Rappahannock Estuary-Neotectonics in Virginia, in Morisawa, M. and Hack, J. T., eds., Tectonic geomorphology, proceedings volume of the Fifteenth Annual Geomorphology Symposia Series, held at Binghamton, New York, September 28-29, 1984, Allen and Unwin, International Series, London.
Newell, W. L., Pavich, M. J., Prowell, D. C., and Markewich, H. W., 1980, Surficial deposits, weathering processes, and the evolution of an Inner Coastal Plain landscape, Augusta, Georgia, "Excursions in Southeastern Geology", V-II, R. W. Frey, ed., Field Trip Guidebook for Geological Society of America, 1980 Annual Meeting, Atlanta, Georgia.
Newell, W. L., and Rader, E. K., 1982, Tectonic control of cyclic sedimentation in the Chesapeake group of Virginia and Maryland, in Central Appalachian Geology, P. T. Lyttle, ed., NE-SE GSA Field Trip Guidebook, p. 1-28.
Opdyke, N. D., Spangler, D. P., Smith, L. D., Jones, D. S., and Lindquist, R. C., 1984, Origin of the epeirogenic uplift of Plio-Pleistocene beach ridges in Florida and development of the Florida Karst: Geology, v. 12, p. 226-228.
Owens, J. P., 1970, Post-Triassic movements in the central and southern Appalachians as recorded by sediments of the Atlantic Coastal Plain, in Studies of Appalachian Geology, Central and Southern, Interscience, New York, N.Y., p.417-427.,
Pavich, M. J.,1974, A study of saprolite buried beneath the Atlantic Coastal Plain in South Carolina, unpub. PhD. thesis, Johns Hopkins University, 133 p.
Pavich, M.J., in press, Processes and rates of saprolite production and erosion on a foliated granitic rock of the Virginia Piedmont, in Colman, S.M., ed., Rates of Chemical Weathering of Rocks and Minerals, Academic Press, New York.
Pavich, M. J., Brown, Louis, Valette-Silver, J. N., Klein, Jeffrey, and Middleton, Roy, 1984, Minimum storage time of untransported Appalachian Piedmont regolith calculated from the $^{10}Be$ inventory, abstract, EOS, v. 65, no. 16, p. 217.
Pavlides, L., Stern, T. W. Arth, J. G., Muth, K. G., and Newell, M. F., 1982, Middle and Upper Paleozoic granitic rocks in the Piedmont near Fredericksburg, Va.; Geochronology, U.S. Geological Survey Professional Paper 1231-B, 9 p.
Prowell, D. C., and O'Connor, B. J., 1978, Belair Fault Zone - evidence of Tertiary fault displacement in eastern Georgia: Geology, v. 6, p. 681-684.
Stephenson, R. S. and Lambeck, Kurt, in press, Erosion-Isostatic rebound models for uplift: an application to southeastern Australia, submitted to Journal of Geophysical Res.
Trimble, S. W., 1974, Man-induced soil erosion on the southern Piedmont 1700-1900, Ankeny, Iowa, Soil Conservation Society of America, 180 p.

Velbel, M. A., 1984, Geochemical Mass Balances and Weathering Rates in Forested Catchments of the Southern Blue Ridge, EOS, v. 65, no. 16, p. 211.

Walcott, R. I., 1970, Isostatic response to loading of the crust in Canada, Canadian Journal of Earth Sciences, v. 7, no. 716, p. 2-13.

Whitney, J. A., Jones, L. M., and Walker, R. L., 1976, Age and origin of the Stone Mountain granite, Lithonia district, Georgia: Geological Society of America Bulletin 87, no. 7, p. 1067-1077.

Zimmerman, R. A., 1979, Apatite fission track age evidence of post-Triassic uplift in the Central and Southern Appalachians, Geological Society of America Abstracts with Programs, v. 11, no. 4, p. 219.

Zoback, M. L., and Zoback, Mark, 1979, State of stress in the conterminous United States, Journal of Geophysical Research, 85, no. B11, p. 6113-6156.

# 14
# Architecture of the Rappahannock estuary – neotectonics in Virginia

*Wayne L. Newell*

ABSTRACT

Upper Mesozoic and Cenozoic Coastal Plain deposits along the west side of Chesapeake Bay in Virginia include clastic fluvial and near-shore marine facies of many transgressive-regressive sequences. In the central Rappahannock area, the Coastal Plain deposits cover complex metamorphic terranes and early Mesozoic rift basins. Faults originating beneath the Coastal Plain have offset Upper Mesozoic and Cenozoic units and controlled the distribution and thickness of their facies. Continued structural control influenced the erosion of pre-existing Coastal Plain deposits and the emplacement of Pliocene and Pleistocene estuarine deposits. The distribution of surficial deposits associated with estuarine terraces, the geomorphology of eroding uplands adjacent to the terraces, and the underlying structural framework suggest that subsurface deformation has been active and cumulative during the repeated excavation and filling of the Rappahannock valley.

## INTRODUCTION

The Tidewater Virginia area, located west of Chesapeake Bay in the mid-Atlantic Region (Figure 1), presents an interesting complex of late Mesozoic and Cenozoic deposits and erosional surfaces. Here, subtle neotectonic effects can be evaluated within a well-studied stratigraphic and structural framework. Across this portion of the Coastal Plain, regionally migrating depositional basins have been filled with transgressive and regressive sequences of marine and non-marine clastic deposits. The sequences are separated by local to regional unconformities. The sediments were derived

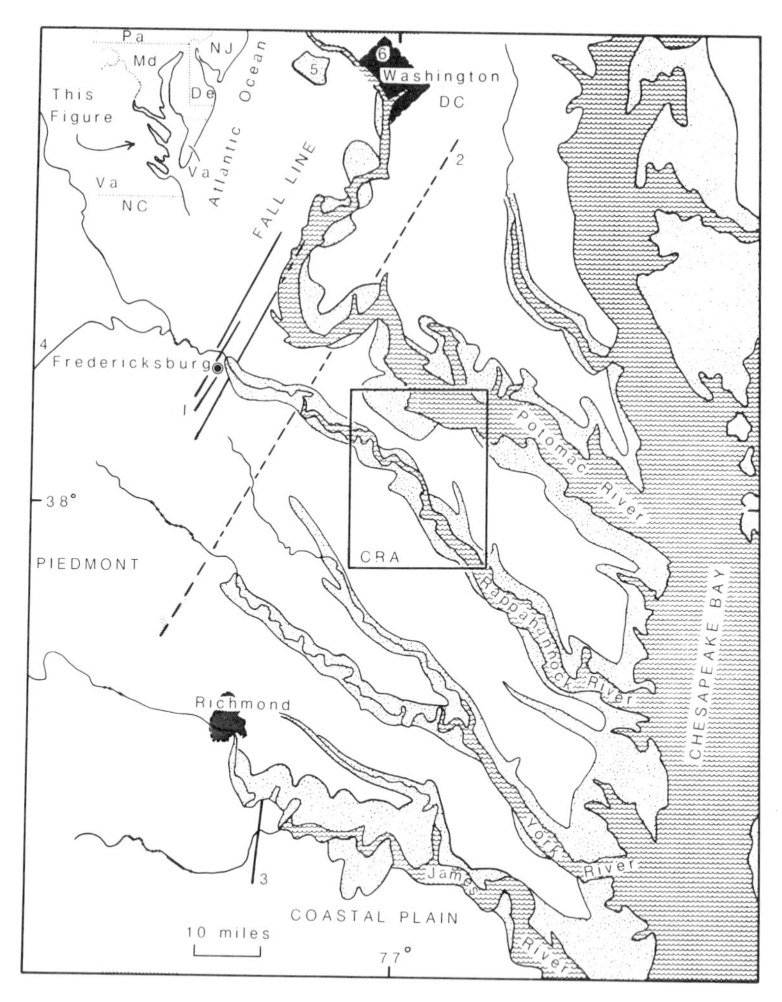


Figure 1 Index map to central Rappahannock area (CRA). The map shows the Quaternary terraces of Chesapeake Bay region estuaries (stippled pattern); deposits of the Coastal Plain uplands ranging from Early Cretaceous to late Pliocene; and rolling Piedmont hills underlain by Paleozoic schists, gneisses, and Mesozoic rift-basin deposits. Known high-angle reverse faults, which are rooted in crystalline rocks and offset Cretaceous and younger deposits, include (1) Stafford fault zone (Mixon and Newell, 1977), (2) Brandywine fault zone (Jacobeen, 1972), (3) Dutch Gap fault (Dischinger, 1979), (4) Mountain Run fault zone (Pavlides, et al., 1983), (5) Tysons Corner fault (Prowell, 1983), and (6) National Zoo fault (Darton, 1950).

from the Piedmont, Blue Ridge, and Appalachians to the west, and from pre-existing Coastal Plain deposits. Regional variations of thickness and facies of the sequences record the lateral shifts of depocenters and the uplift of former basins into arches (Owens, 1970). During the late Pliocene and the Pleistocene, the uplifted deposits of the Virginia Coastal Plain have been overprinted by a pattern of deeply entrenched and filled valleys which document the eustatic rise and fall of sea level accompanying glacial and interglacial stages.

The Coastal Plain strata overlie Paleozoic metamorphic complexes and early Mesozoic rift basins (Figure 2) partly exposed in the Virginia Piedmont west of the Fall Line (Figure 1). Generally, Piedmont schists and gneisses are intensely weathered and thick saprolite underlies uplands and gentle slopes. Fresh rock is exposed in stream beds and steep walled gorges. Locally, weathered alluvium and colluvium cap hill tops underlain by saprolite indicating that weathering, downwasting, and topographic inversion have proceeded for a long time. Surficial geology suggests that on the adjacent Piedmont source area relief of a few hundred feet has been maintained while uplift and erosion have kept pace filling the Coastal Plain depocenters (see Pavich and Markewich, 1984, this volume).

Across the Piedmont, Fall Line, and inner Coastal Plain, a fabric of high-angle reverse faults offsets Cretaceous and Tertiary stratigraphic units and even younger surficial deposits of uncertain age. The faults locally juxtapose crystalline rocks against the Cretaceous and younger sediments. Throughout the history of these persistent but small-scale structures, deformation has been cumulative and interspersed with depositional events (Mixon and Newell, 1977, 1982). These structures have shaped basins, sedimentation patterns, and the geometry of deposits in ways that have yet to be fully documented and explained.

The Rappahannock estuary stretches approximately 100 miles from Fredericksburg, Virginia, to Chesapeake Bay. Commonly the valley is a few miles wide and contains flights of Quaternary terraces bounded by steep upland bluffs. A stratigraphic record ranging from the Early Cretaceous to the Holocene is exposed along various parts of the valley. In the valley of the Rappahannock estuary, the subtle effects of subsurface faults, originating below the Coastal Plain strata, have been documented by geologic mapping with attention to details of facies, contouring of unconformities, surficial deposits, and geomorphic processes.

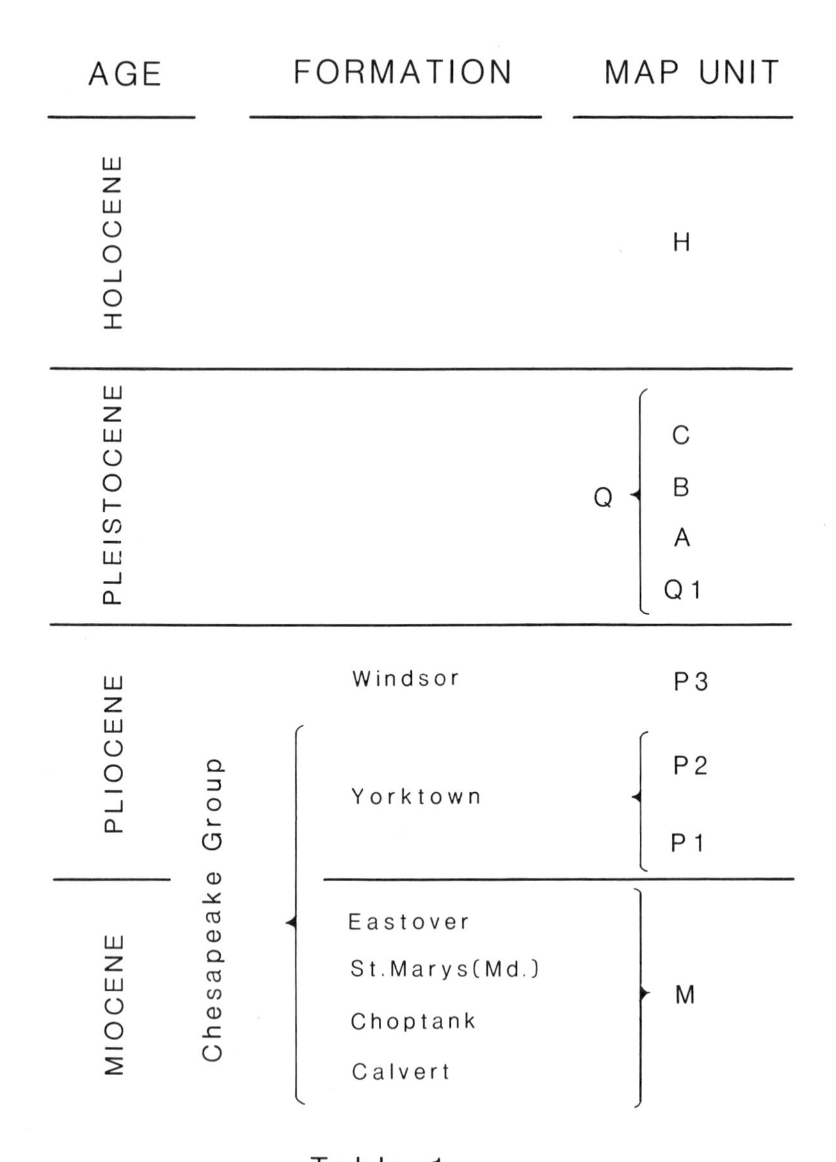

| AGE | FORMATION | | MAP UNIT | |
|---|---|---|---|---|
| HOLOCENE | | | | H |
| PLEISTOCENE | | | Q | C |
| | | | | B |
| | | | | A |
| | | | | Q1 |
| PLIOCENE | | Windsor | | P3 |
| | Chesapeake Group | Yorktown | | P2 |
| | | | | P1 |
| MIOCENE | | Eastover | M | |
| | | St.Marys(Md.) | | |
| | | Choptank | | |
| | | Calvert | | |

Table 1

STRATIGRAPHIC FRAMEWORK

Within the central Rappahannock River area, pre-Miocene sedimentary units are almost entirely in the subsurface. They crop out up dip toward the Fall Line on the west. The modern river valley is entrenched almost totally in the Chesapeake Group, a 200 to 300 foot thick assemblage of Miocene and Pliocene clastic sequences of shelf and near-shore marine deposits (Table I).

The Chesapeake Group includes, locally (from base to top), the Calvert, Choptank, St. Marys (of Maryland), Eastover, and Yorktown Formations (see Ward and Blackwelder, 1980). Originally, each formation included an array of deep-water and shallow-water (regressive) facies recording the rise and fall of sea level. Generally, the shallow-water facies of pre-existing formations were stripped during succeeding transgressions. The Pliocene units (P1, P2, P3, of Figure 3), capping the uplands, most completely preserve the shallow, near-shore deposits. Of the older units, only the marine-shelf facies remain in the stratigraphic record (Unit M of Figure 3). Although the marine-shelf facies of the Chesapeake Group are

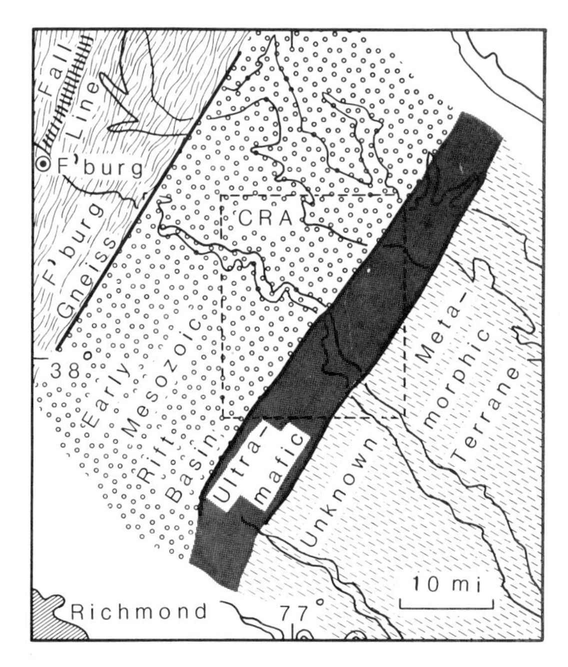


Figure 2 Sub-Coastal Plain terranes under the Rappahannock estuary interpreted from gravity maps (Johnson, 1977), magnetic maps (Zietz, et al., 1977), and bore hole logs (Hubbard, et al., 1978).

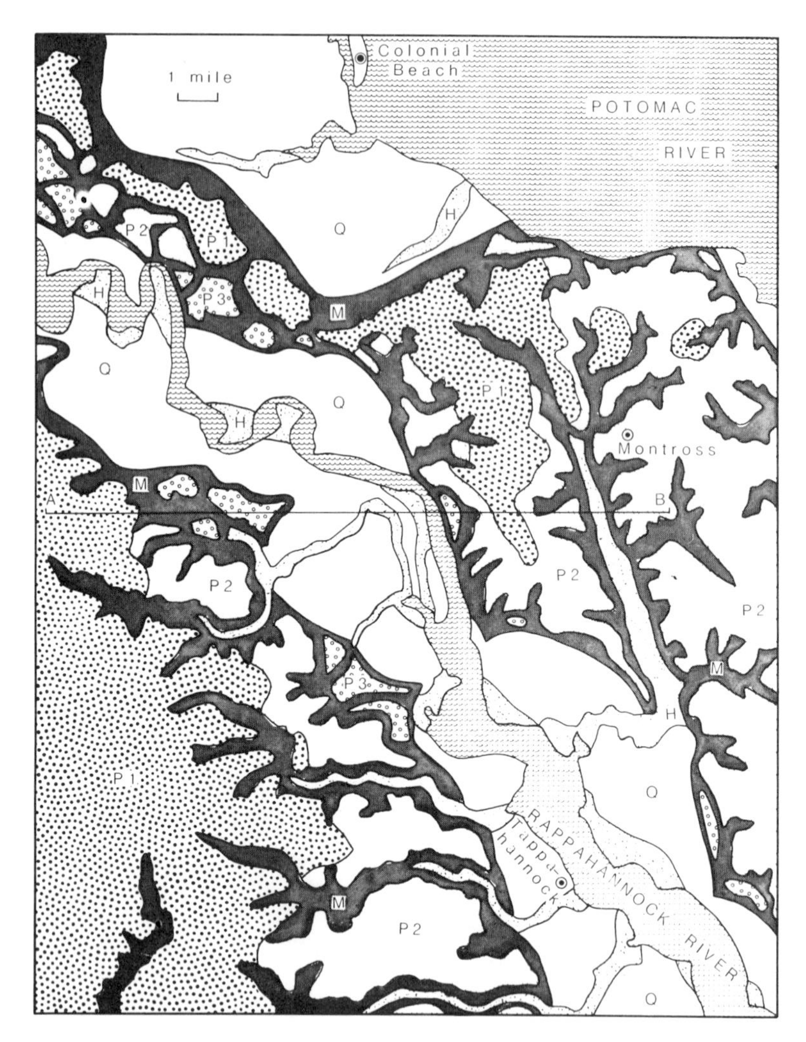


Figure 3 Generalized geologic map of the central Rappahannock River area. Reaches of the Rappahannock River and Potomac River are flanked by Holocene deposits (H) and flights of terraces (generally 80 ft or less in altitude) underlain by Quaternary estuarine deposits (Q). Three upland units record: an early Pliocene lower delta plain (P1, surface greater than 150 ft); the middle Pliocene protoestuary (P2, surface at 140 to 150 ft); and late Pliocene fluvial meanders (P3, surface at 100 to 110 ft). Miocene transgressive-regressive marine deposits (M) underlie both the Pliocene and Quaternary deposits.

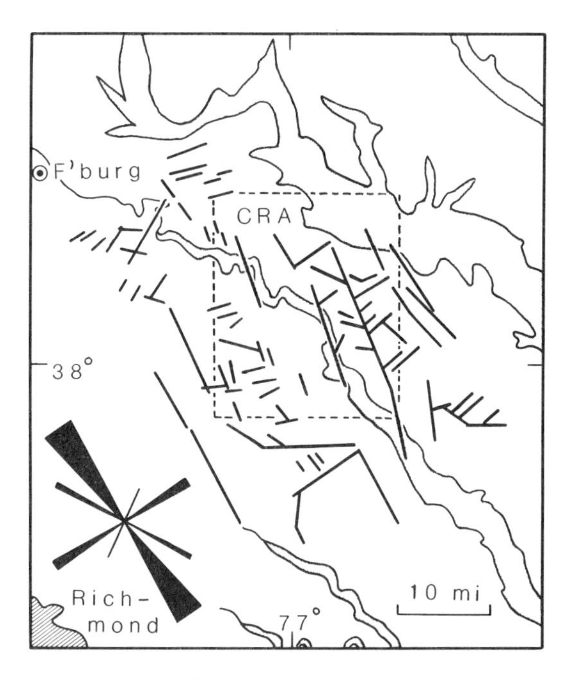


Figure 4 Linear topographic features and rose diagram of tectonic joints in Chesapeake Group sediments of the central Rappahannock area (from Newell and Rader, 1982).

typically fine grained, well sorted, and compact with locally abundant faunal assemblages, their basal unconformities can be distinguished and mapped, revealing subtle structural variations (Newell and Rader, 1982).

Across the mid-Atlantic Region the distribution of Chesapeake Group Formations, their thicknesses, and their facies record the migration of a depositional basin from eastern Maryland in the Miocene to eastern North Carolina in the Pliocene. The pattern of uplift to the north and depression to the south has prevailed through the Pleistocene. Marine Pleistocene units occur east of Chesapeake Bay and thicken southward into eastern North Carolina. The uplifted inner Coastal Plain west of the Chesapeake Bay has been a region of entrenchment and deposition of fluvial to estuarine cycles during the late Pliocene and Pleistocene.

## STRUCTURAL FRAMEWORK

Faults observed to the west of the central Rappahannock River area (Figure 1) generally strike NE-SW and dip steeply either NW or SE. Maximum displacement on any one fault plane commonly is a little over 100 feet. The deformation has been cumulative at a rate of about 3 feet per million years. The faults have reused ancient fracture systems which juxtapose Precambrian and Paleozoic metamorphic terranes. The Paleozoic strike-slip fault systems exposed at the surface today are ductile mylonite zones formed at great depth (L. Pavlides, USGS, verbal commun.). During early Mesozoic rifting, the ductile mylonite zones were exploited by high-angle normal faults and listric faults; cataclastic textures were created at moderate depths as extensional deformation proceeded (A. J. Froelich, USGS, verbal commun.). Late Mesozoic through Cenozoic compressional tectonics produced high-angle reverse faults in the same zones occupied by Paleozoic and early Mesozoic faults. Exposures of the Cenozoic reverse faults indicate that deformation approached the surface. Crushed saprolite, slickensided clay gouge, and oriented clasts occur in the fault zones. The fault planes bifurcate into splays and flatten out near the surface in saprolite and(or) unconsolidated deposits.

The structural framework and tectonic setting of the Cenozoic high-angle reverse faults is not yet well enough understood to explain their significance at the continental scale (i.e., plate tectonic "models"). Nevertheless, Cenozoic high-angle reverse faults have been observed across the inner Coastal Plain, the Piedmont, and the east flank of the Blue Ridge from Georgia to Virginia and possibly farther north. Zoback and Zoback (1980) have documented a present day NW-SE compressional stress field across the same region. A synthesis by Wentworth and Mergner-Keefer (1983) suggests that this stress field has been driving deformation in the same region since the Early Cretaceous.

The major southwest-northeast-trending sub-Coastal Plain boundaries shown in Figure 2 are good possibilities for similarly rooted, compressively induced faults at depth that may extend into the Cretaceous and lower Cenozoic strata of the Coastal Plain. Abrupt changes in thickness and facies of near-surface units (determined by drilling) indicates that cumulative

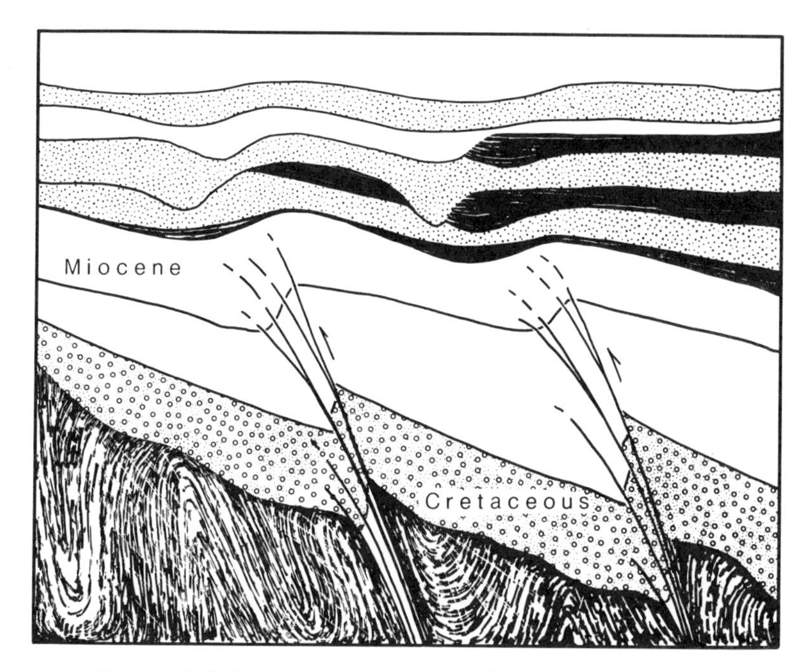


Figure 5 Style of deformation interpreted from outcrop, drill hole, and geophysical data in the Virginia Coastal Plain. High-angle reverse faults originate below the Coastal Plain, offset Cretaceous strata, and die out within the Miocene near the surface. Subsiding troughs and rising arches overlie the fault zones. (Figure not to scale.)

differential movement across the boundaries of different sub-Coastal Plain terranes has probably occurred. Conjugate second- and third-order structures are also likely as witnessed by the joint fabric and corresponding topographic linears (Figure 4) and abrupt thickening or thinning of units across these linear trends. Minor faults offsetting the base of the Chesapeake Group have been observed. In the Central Rappahannock River Area, with the exception of these minor offsets, fault movement near the surface is not apparent.

Within the upper Chesapeake Group, stacked troughs and arches overlie subsurface structural zones (Figure 5). The trough and arch axes are roughly parallel to present day topographic linears and the strike of tectonic joints in the lower formations of the Chesapeake Group. Brittle deformation dies out within the Chesapeake Group near the surface where the erosional and depositional features are localized by the subsiding troughs and rising arches. Contours of unconformities within the Chesapeake Group (Newell and Rader, 1982) indicate progressive oversteepening across the deep structures.

## THE PROTOESTUARY

During the early Pliocene, the Rappahannock River deposited sand and gravel (fluvial and intertidal facies (Pl) of Figures 3 and 6) over an upper and lower delta plain that extended from the Fall Line into the Atlantic Ocean as far as the present day Central Rappahannock River Area. In the middle Pliocene, sea level dropped and the river entrenched its valley through the delta sediments. Sea level subsequently rose, flooding part of the fluvial valley and creating the protoestuary of the Rappahannock River (see Figure 6). In the Central Rappahannock River Area, deposits of the protoestuary are laterally coextensive with fluvial deposits up valley to the Fall Line, and with near-shore marine deposits capping uplands to the south and east. The trend of the protoestuary generally coincides with the eroded, structurally controlled configuration of the base of Pliocene sediments (the contoured horizon of Figure 6).

During the late Pliocene, sea level fell below the middle Pliocene estuarine deposits (P2). Once again, the Rappahannock entrenched its valley through the materials just deposited and into deeper, unconsolidated Pliocene and Miocene sediments. The new valley was largely confined within the bounds established by the protoestuary. Remnants of the alluvium (P3, Figure 3) deposited on the floor of the new valley indicate that the river carved tight meander loops through the loose sand and gravel of the upper Miocene and Pliocene substrate (Figure 6). The Central Rappahannock Valley fluvial deposits (P3) grade down valley into an estuarine sequence of deposits which in turn merge with near-shore marine sediments of yet another transgressive-regressive sequence deposited across southeastern Virginia and eastern North Carolina.

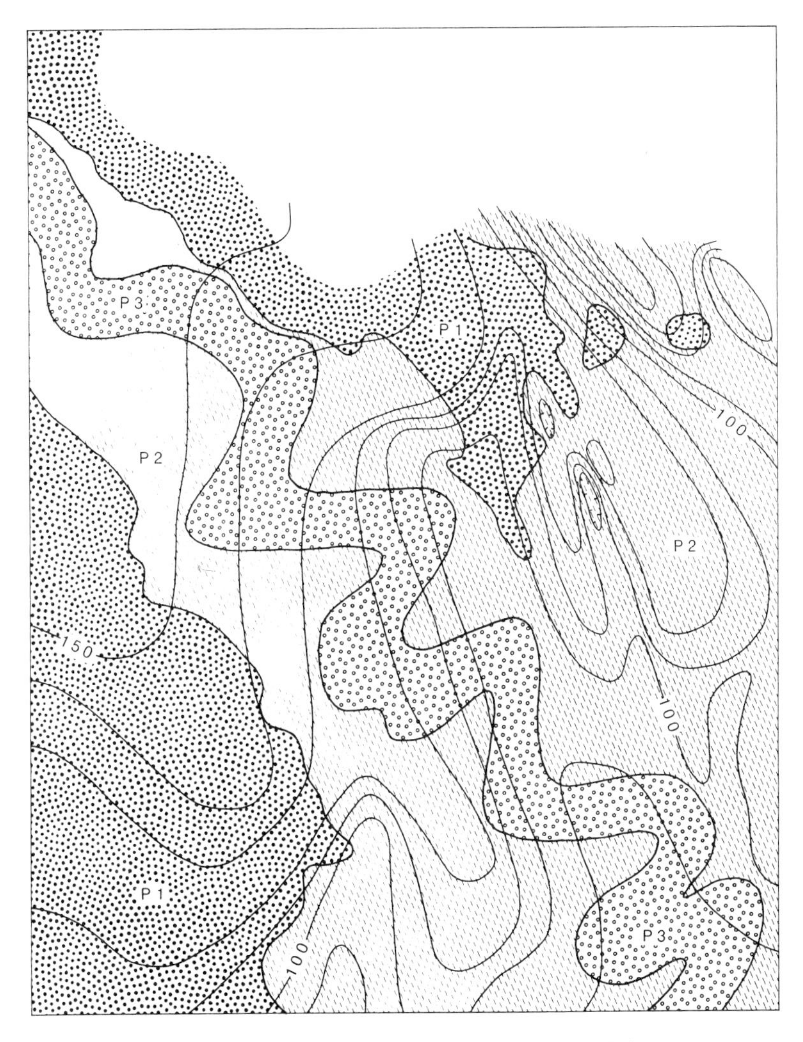


Figure 6 Middle Pliocene protoestuary (P2) and late Pliocene entrenced, sinuous, fluvial channel deposits (P3) reconstructed from data of Figure 3. Contours (10-ft interval) show the base of the Pliocene transgressive deposits (outcropping as P1, Figure 3).

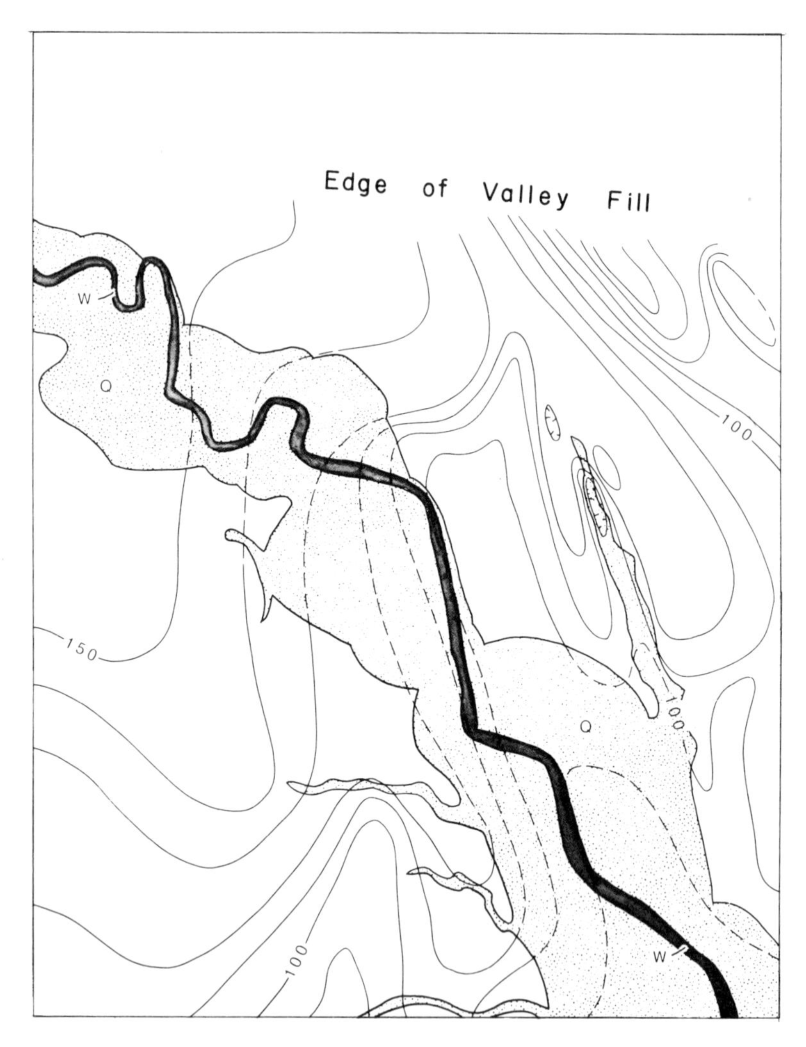


Figure 7 Valley fill of Quaternary deposits (Q). The linear reaches of the fluvial channel (W) were carved during the last sea level low stand (Wisconsinan). Contours (10-ft interval) show the base of the Pliocene transgressive deposits.

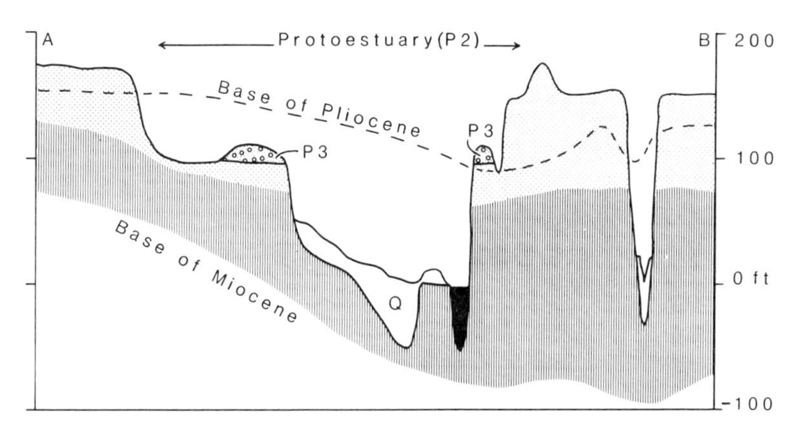


Figure 8 Valley cross-section between points A and B of Figure 3. The late Pliocene meandering channel, defined by deposit P3 is entrenched in friable sand and gravel (dots) of the Pliocene part and the upper Miocene part of the Chesapeake Group. The Quaternary estuary is entrenched into the compact, jointed, fine-grained sediments (vertical lines) of the middle Miocene part of the Chesapeake Group.

## THE ENTRENCHED ESTUARY

As the glacial cycles of the Pleistocene unfolded, and sea level fluctuated as much as hundreds of feet, the Rappahannock valley became entrenched below the level of the late Pliocene sinuous meandering channel deposits shown in Figure 6. During down cutting phases, the river, with steeper gradient and increased erosive power, encountered the compacted, well-sorted, fine-grained marine deposits of the lower formations of the Chesapeake Group. A well-developed joint system (Figure 4) is the major physical inhomogeneity of these strata. The fractures were exploited by the river and in turn left their mark upon the river valley as the main channel developed long, linear reaches parallel to the orientation of the fractures. Rapidly entrenching tributaries also developed linear valleys as the present day drainage system became established (see Figures 4 and 7) (Jacobson and Newell, 1982).

Figure 7 shows that the valley of today although entrenched 100 to 150 ft below the elevation of the protoestuary is still largely confined within the structural bounds set long ago in the Pliocene. The comparison of the late Pliocene channel (Figure 6) and the last entrenched channel (Figure 7) shows the obvious control of the fracture pattern in aligning reaches of the river.

The valley bottom of Figure 7 is a stair-stepped aggregate of terraces varying in altitude from about 80 ft to a few feet above sea level. The terraces are underlain by depositional sequences of fluvial and estuarine facies which filled the entrenched valleys during eustatic sea level transgressions. Some are only 10 ft thick, others approach 100 ft in thickness. A typical cross section (Figure 8) across the center of the map area (Figures 3, 6, and 7) shows the entrenchment of the river through Pliocene and upper Miocene cover into compact, jointed, Miocene marine shelf sediments.

A cycle of valley cutting and sediment infilling occurs in response to the eustatic lowering and raising of sea level. As demonstrated in Figure 9, a typical cycle can be separated into at least four component stages, permitting the comparison of similar erosional events and deposits from a sequence of terraces. The four stages shown in Figure 9 include: (A) sea level low stand - the river erodes to a new base level and the valley bottom is covered by alluvium; (B) transgression - sedimentation keeps pace with rising sea level and intertidal facies include brackish fauna and fresh water peat; (C) sea level high stand - the valley is filled and becomes a confined lower delta plain with intertidal and supratidal deposits; (D) sea level falls and downcutting is initiated wherever the main channel is situated. The cycle is closed when a new transgression begins. (The present estuary is in an advanced stage of transgression.)

Passive control of channel locations by the fracture pattern occurs primarily during vigorous downcutting events (Figure 9, stage D). Conversely, in the central Rappahannock area, aggradational deposits preserved from periods of sea-level high stands (Figure 9, stage C) show differential uplift on underlying structures. During sea-level high stands, surficial deposits from uplifting areas accumulate, and the river channel migrates laterally in the down-structure direction until it is poised for a new down cutting event (Figure 9, stage D). The map and cross section of Figures 10A, B, and C illustrate details that support this scenario, which has been played out many times in the Rappahannock Valley. Channel I (defined by power augering) preceeds deposition of the sediments underlying terrace B. Channel II (defined by bathymetry) was cut during the last low stand and is now being filled during the Holocene transgression. The parallel reaches of the deep channels indicate that the river systematically migrated down

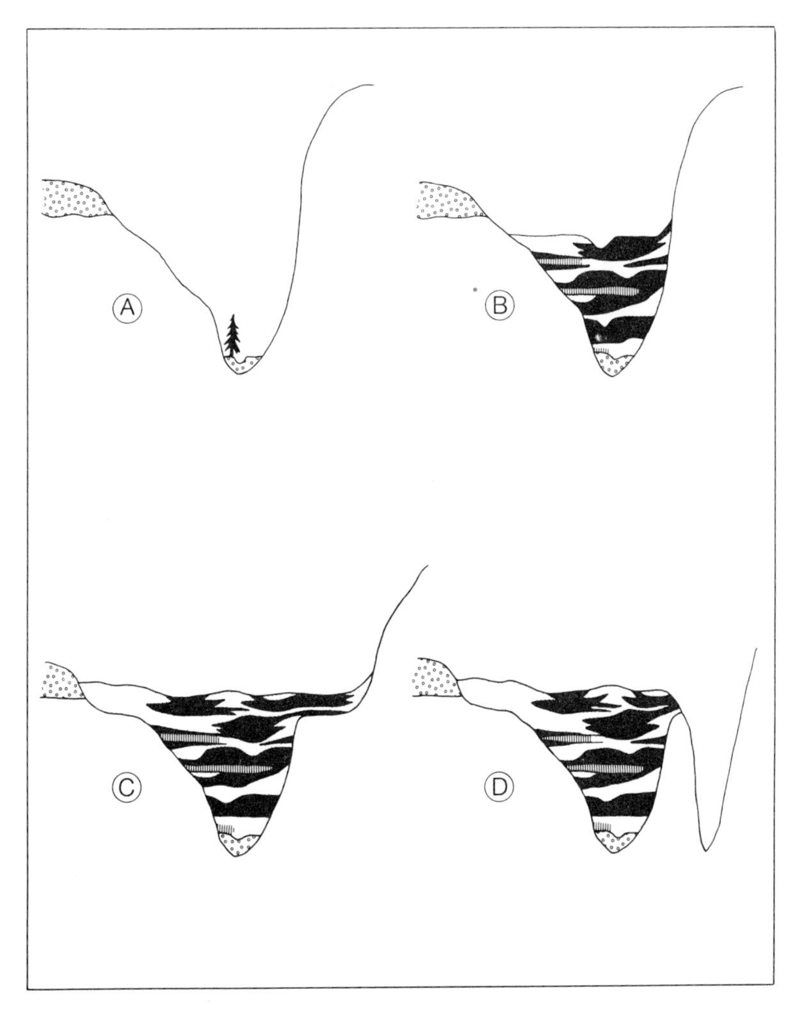


Figure 9 The Quaternary fluvial to estuarine erosional and depositional cycle includes four stages: sea level low stand (A), transgression (B), sea level high stand (C), and regression (D) as interpreted from outcrops and bore holes through terraces in the central Rappahannock River area.

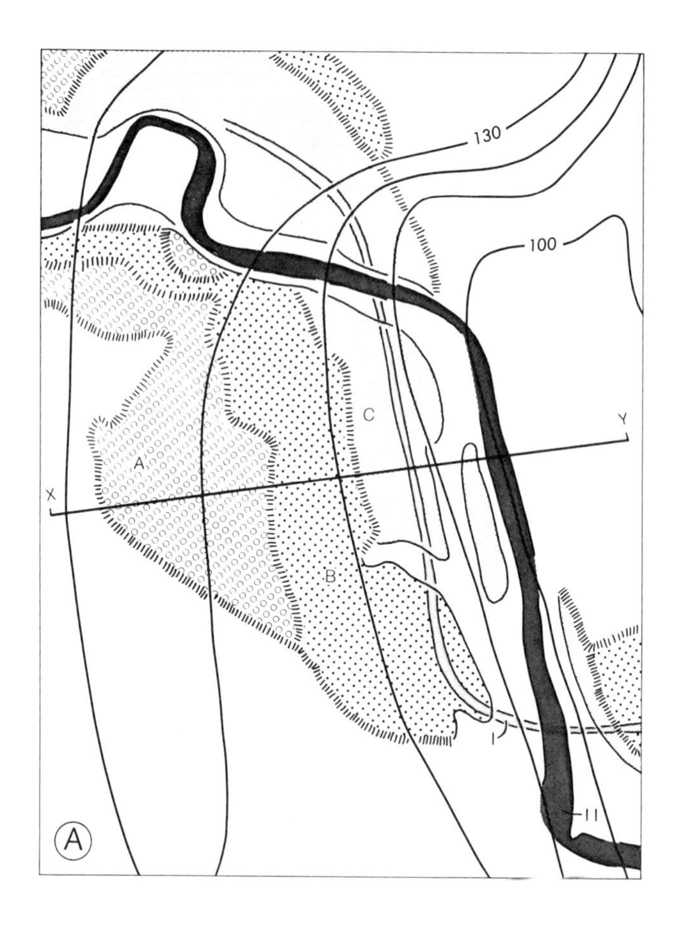
130
100
C
Y
A
X
B
I
II
A

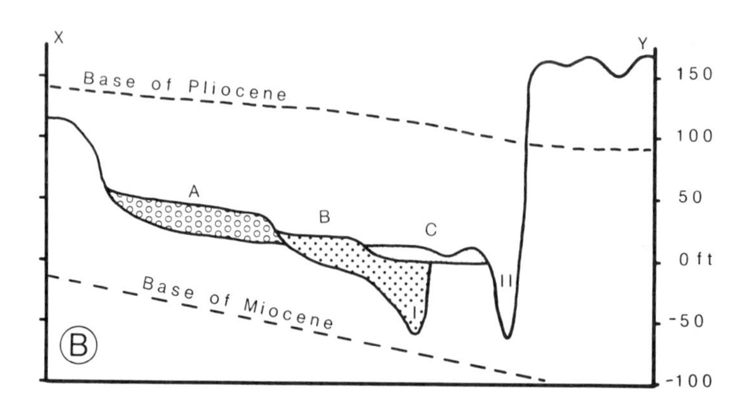
X
Y
Base of Pliocene
A
B
C
I
II
Base of Miocene
B
150
100
50
0 ft
-50
-100

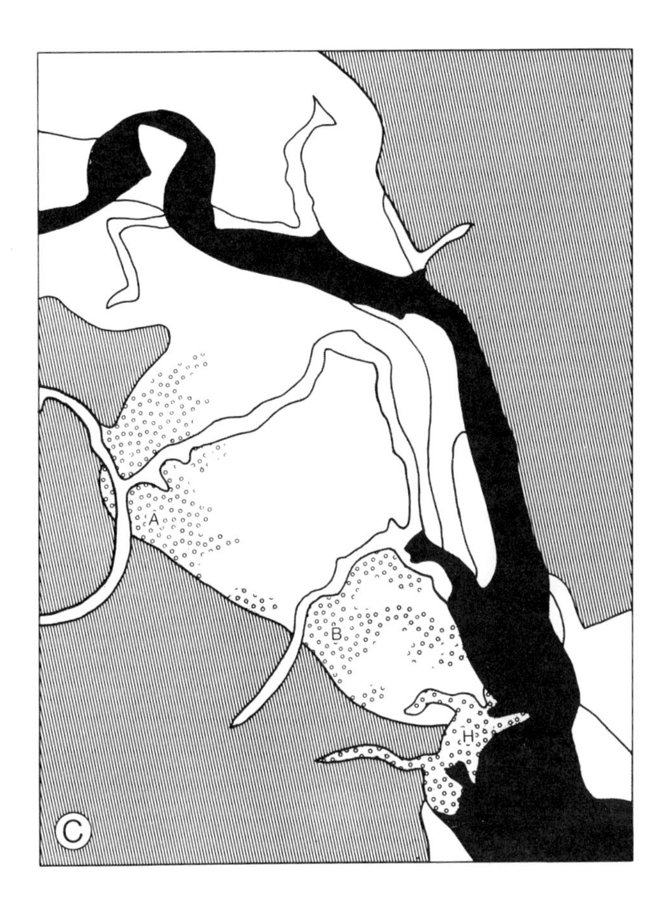


Figure 10 Down-structure migration of the Estuary:

A) A series of terrace deposits (A,B,C) step down the gradient shown by 10-ft contours on the base of the Pliocene. Each deposit has been truncated and inset with a younger, lower unit. A and B are fluvial to estuarine. C is fluvial. Deep fluvial Channels I and II were cut during sea level low stands. (Hachures are eroded scarps.)

B) Cross section along X--Y through terrace deposits of figure 10A. Pliocene and Miocene unconformities show the structural trend.

C) Surficial deposits (alluvial fans and deltas) prograding onto terraces A and B from the uplands. H is a Holocene analog now being deposited during the present transgression.

dip through jointed homogeneous shelf sediments (Miocene) underlying the upland slopes and valley bottoms. The tops of terraces A, B, and C similarly step down dip. There are no hard, dipping beds to control this migration in a passive manner. On Figure 10C surficial deposits labeled A, B, and H are derived from the uplands west of the valley. The source area is the structural and topographic high bordering the trough occupied by the Rappahannock. The deposits include alluvial fans, deltas, and spits which interfinger with and prograde out onto the estuarine terrace deposits. (A and B correspond to terraces A and B of Figure 10A and B. Deposit H is a Holocene analog now accumulating.) These deposits occur only along the flanks of the structural high. Elsewhere, steep valley margins underlain by structural lows are not bordered by such deposits. As can be deduced from the cross sections (Figures 8 and 10B), the dip of the base of Miocene strata is at least twice as steep as the base of the Pliocene strata. The over-steepened Miocene strata indicate cumulative deformation. The geomorphology and surficial geology suggest that this mild warping, possibly connected to sub-Coastal Plain faulting, has continued through the Quaternary.

## DISCUSSION

The altitude of the surface of each estuarine terrace deposit can be partitioned as the sum of an eustatically controlled sea-level high stand (S) plus a cumulative uplift component (U), which is a function of tectonic processes. Each component is a function of two different independent sets of processes. The eustatic component is dependent on global circulation and insolation parameters which effect the rate of change of ice cap volumes. Once a terrace is deposited, the eustatic component of its altitude is fixed for a point in time. The tectonic component is dependent on continental scale rates of deformation coupled with isostatic uplift rates. Through time, the tectonic component is cumulative, assuming constant processes. Rates of faulting (Wentworth and Mergner-Keffer, 1983) and denudation (Pavich, Markewich, 1984, this vol.) indicate this is not a bad assumtion.

By plotting (log log) a super envelope of terrace altitudes (S+U) against their ages (Figure 11) the subtle, cumulative effects of tectonic uplift can be assessed. Major breaks in slope indicate major changes in process. Figure 11 shows three domains: A period of tens of thousands of years where the

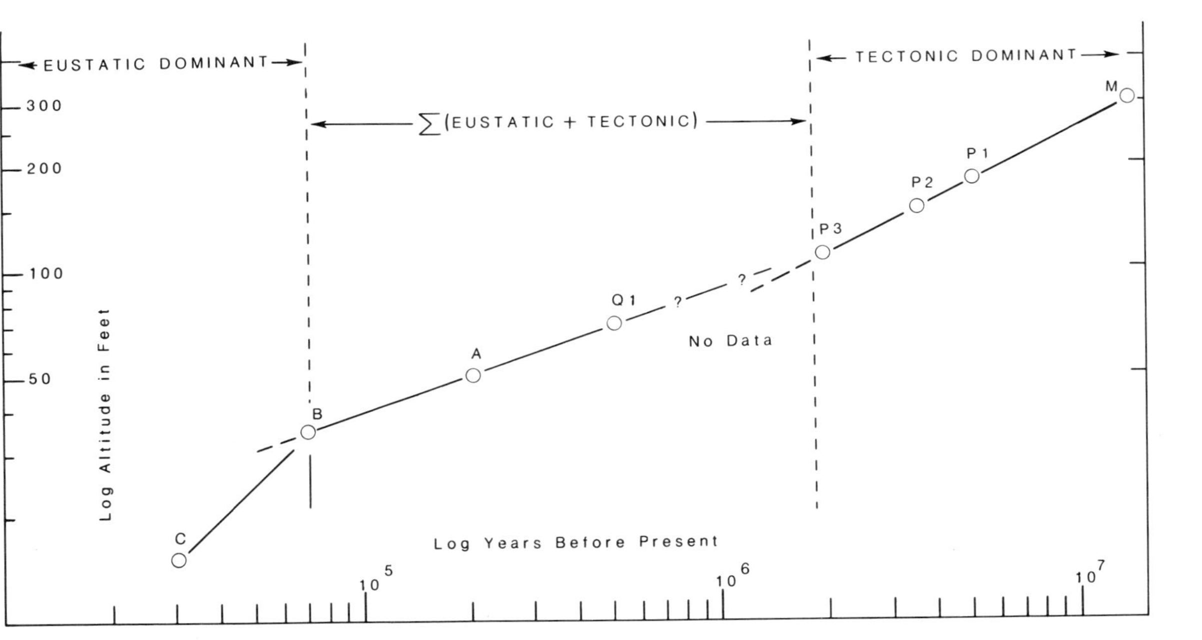


Figure 11 Terrace altitude plotted against age of underlying deposits. The top of each terrace is interpreted to have been at or near a sea level high stand during the time of deposition. A, B, and C are terraces of Figure 10A. Terrace Q1 is older, higher, and occurs within the map area of Figure 3. P1, P2, and P3 are the Pliocene units of Figure 3. M is the highest marginal marine Miocene deposit occuring west of the Fall Line at Fredericksburg (Mixon and Newell, 1982). Ages are extrapolated from dated materials of equivalent map units in the Chesapeake Bay area and coastal area to the south (Mixon, in press).

terrace altitude is dominantly eustatic; a period of 100,000 to 1 million years where the tectonic component of terrace altitude systematically increases; and a period between 1 and 14 million years where the altitudes of the surfaces once graded to sea level are dominated by tectonic uplift. Beyond 1 million years, for this area, the exact determination of sea level is overshadowed by erosional and depositional processes within the same order of magnitude as eustatic sea level changes.

Thus, assuming an appropriate rate of deformation, tectonic uplift outstrips the eustatic components of elevation during periods which exceed a million years. Conversely, the tectonic component diminishes and becomes indeterminant as the Present is approached. The ages of the terrace deposits are interpreted from field evidence and laboratory evidence. However, long range correlations are involved and absolute ages and rates should not be interpreted from this discussion which is intended to put the relative effects of two processes in perspective — slow and steady uplift wins the race.

## CONCLUSIONS

In Virginia, Cenozoic faults have been found cutting thin, surficial deposits on the Piedmont and cutting Coastal Plain strata along the Fall Line. However, it is unlikely that surface-cutting faults will be found east of the Fall Line where hundreds to thousands of feet of unconsolidated sediments overlie fault zones in the sub-Coastal Plain. High-angle, compressional faulting at depth in the Coastal Plain strata apparently is translated into subtly warped arches and troughs near the surface. An inventory of geomorphic features and surficial deposits can be interpreted as the products of cumulative Quaternary deformation if the structural framework has been defined and non-tectonic processes can be factored out.

## REFERENCES

Blackwelder, B. W., and Ward, L. W., 1976, Stratigraphy of the Chesapeake Group of Maryland and Virginia: Geological Society of America, Northeastern-Southeastern Meeting, Arlington, Virginia, 1976, Guidebook for Field Trip 7b, 55 p.

Darton, N. H., 1950, Configuration of the bedrock surface of the District of Columbia and vicinity, U.S. Geological Survey Professional Paper 217, 42 p.

Dischinger, J. B., Jr., 1979, Late Mesozoic and Cenozoic stratigraphic and structural framework near Hopewell, Virginia: Unpublished M.S. thesis, University of North Carolina, 84 p.

Hubbard, D. A., Jr., Rader, E. K., and Berquist, C. R., 1978, Basement wells in the Coastal Plain of Virginia, Virginia Minerals, v. 24, no. 2, p. 16-18.

Jacobeen, F. H., Jr., 1972, Seismic evidence for high angle reverse faulting in the Coastal Plain of Prince George and Charles Counties, Maryland: Maryland Geological Survey Information Circular 13, 21 p.

Jacobson, R. L., and Newell, W. L., 1982, Joint control of Valley form and process in unconsolidated Coastal Plain sediments, Westmoreland, Co., Virginia, Abstracts with Programs, combined NE and SE section meeting, Geological Society of America, Washington, D.C., p. 28.

Johnson, S. S., 1977, Gravity map of simple Bouguer anomaly, Virginia: Virginia Division of Minerals Resources, scale 1:500,000.

Mixon, R. B., in press, Stratigraphic and geomorphic framework of uppermost enozoic deposits in the southern Delmarva Peninsula, Virginia and Maryland, U.S. Geological Survey Professional Paper 1067-G.

Mixon, R. B., and Newell, W. L., 1977, Stafford fault system: Structures documenting Cretaceous and Tertiary deformation along the Fall Line in northeastern Virginia: Geology, v. 5, p. 437-440.

____1982, Mesozoic and Cenozoic compressional faulting along the Atlantic Coastal Plain margin, Virginia, in Lyttle, P. T., ed., Central Appalachian geology, p. 29-54, guidebook for Northeast-Southeast Meeting, Geological Society of America, Washington, D.C., Field Trip No. 2.

Newell, W. L., and Rader, E. K., 1982, Tectonic Control of Cyclic Sedimentation in the Chesapeake Group of Virginia and Maryland, in Lyttle, P. T., ed., Central Appalachian Geology, p. 1-27, Guidebook for Northeast-Southeast Meeting, Geological Society of America, Washington, D.C., Field Trip No. 1.

Owens, J. P., 1970, Post-Traissic tectonic movements in central and southern Appalachians as recorded by sediments of the Atlantic Coastal Plain, in Studies of Appalachian geology: central and southern: New York, Interscience Publishers, p. 417-427.

Pavich, M. J., and Markewich, H. W., 1984, Appalachian Piedmont Morphogenesis: Weathering, Erosion, ad Cenozoic Uplift, in Morisawa, M., and Hack, J. T., eds., Tectonic geomorphology, proceedings volume of the Fifteenth Annual Geomorphology Symposia Series, held at Binghamton, New York, Sep. 28-29, 1984, Allen and Unwin, International Series, London.

Pavlides, L., Bobyarchick, A. R., Newell, W. L., and Pavich, M. J., 1983, Late Cenozoic faulting along the Mountain Run Fault Zone, Central Virginia Piedmont, Abstracts with Programs, SE Section Meeting, Geological Society of America, Tallahassee, Florida, p. 55.

Prowell, D. C., 1983, Index of faults of Cretaceous and Cenozoic Age in the eastern United States, U.S. Geological Survey, Miscellaneous Field Studies Map MF-1269.

Ward, L. W., and Blackwelder, B. W., 1980, Stratigraphic revision of upper Miocene and lower Pliocene beds of the Chesapeake Group-Middle Atlantic Coastal Plain: U.S. Geological Survey Bulletin 1482-D, 61 p., 5 pls.

Wentworth, Carl M., and Mergner-Keefer, M., 1983, Regenerate Faults of Small Cenozoic offset-Probable Earthquake sources in the southeastern United States, in Gohn, G. S., Studies Related to the Charleston, South Carolina, Earthquake of 1886 - Tectonics and Seismicity, U.S. Geological Survey Professional Paper 1313, 51-20.

Zietz, I. Calver, J. L., Johnson, S. S., and Kirby, J. R., 1977b, Aeromagnetic map of Virginia: U.S. Geological Survey Geophysical Investigations Map GP-916.

Zoback, M. L., and Zoback, M., 1980, State of stress in the Conterminous United States: Journal of Geophysical Research, v. 85, no. B11, p. 6113-6156.

# PART III

# 15
# Marine terraces and active faults in Japan with special reference to co-seismic events

*Yoko Ota*

ABSTRACT

The Japanese Islands, as active island arcs along subduction zones, are dominated by various tectonic landforms as direct products of Quaternary tectonic movement. Marine terraces and active faults are well developed and have been intensively studied in terms of origin and process of formation. They are considered the key for understanding Quaternary tectonics or seismicity and provide fundamental data for earthquake prediction. Many recent studies of marine terraces and active faults have been made with special reference to coseismic events.

Marine terraces can be used as well-defined key surfaces to detect tectonic deformation in uplifting areas. Based on the height distribution of the former shoreline of the last interglacial terrace of about 125KA, the Japanese Islands are divided into four tectonic regions, different in pattern and rate of deformation. Type D deformation, on the Pacific coast of central and southwestern Japan, is characterized by the greatest rate of uplift, attaining 1.5m/KA and by a landward tilting which usually shows progressive deformation throughout the late Quaternary and is interpreted as a result of accumulation of activity of subsidary reverse faults associated with subduction of the Phillippine Sea plate.

The marine terrace of the Muroto Peninsula of southern Shikoku is the best example of an accumulated result of coseismic deformation, because progressive landward tilting of the terrace is concordant in pattern with coseismic uplift of the 1946 earthquake (M=8.1). The Holocene marine terrace of the southern tip of the Boso Peninsula is subdivided into four steps, the lowest of which emerged at the time of the 1703 earthquake (M=8.2). Three other terraces can be regarded as a result of coseismic uplift, on the basis of similarity of width of terraces and height of terrace risers to those of the 1703 terrace. It is thought that at least three large earthquakes similar to that in 1703 occurred since about 6,000 yBP, resulting in this series of terraces. Coseismic uplift and tilting is observed in terraces located on the Japan Sea coast, but recurrence intervals of the earthquakes seem to be much longer than on the Pacific side.

Topography displaced by active faults is one of the major tectonic landforms in Japan and has been extensively studied. "Active Faults in Japan", showing the distribution of active faults on a scale of 1:200,000 with detailed inventories for each fault including the distribution, regional characteristics, nature, amount of displacement and degree of activity of each marked the final step of the "age of discovery of active faults". However, there are still many problems to be solved. Trenching of active faults is one of the main ways to get information on time of the latest event and recurrence of faulting, necessary for long term earthquake prediction. Exact natures of fault zones are also disclosed by trenching. About thirty trenches across ten fault systems were excavated in the past several years.

One is the trenching of the Tanna fault in the Izu Peninsula, along which the latest seismic displacement accompanied the Kita-Izu earthquake (M=7.3) of 1930. Eight events, excluding the 1930 earthquake, were recognized in the deposits after deposition of "Ah" volcanic ash, dated at 6,300 yBP. Recurrence intervals of the earthquakes are estimated at 700-1,000 years, which fits fairly well with the assumption that faulting took place about every 1,000 years, based on the comparison of surface displacement in 1930 with total amount of offset of streams dissecting middle Quaternary volcanoes.

Trenching across the Ito-Shizu Tectonic Line, which marks the western margin of Fossa Magna, provides a second example. Historical surface breaks were not recorded along the fault. Two trenches excavated in 1983 on the western margin of two tectonic bulges disclosed notable fault zones at both localities. Four events are recognized since about 8,000 years at one of the trenches. In three of the events the eastern side of the fault zone is upthrown, consistent with the sense of faulting that was topographically estimated. The latest event is inferred to be about 1,200 yBP which probably corresponds to the historical earthquake of 841 A.D.

## INTRODUCTION

The Japanese Islands as active island arcs along subduction zones are dominated by various tectonic landforms which are direct products of Quaternary tectonic movements (Ota, 1980). Comparison of present relief and amount of Quaternary vertical displacement revealed that more than one half, or two thirds of the present relief was attributed to Quaternary vertical deformation (The Research Group for Quaternary Tectonic Map, 1973). Among tectonic landforms marine terraces and active faults are well developed in the Japanese Islands and have been intensively studied in terms of origin and processes of formation. Furthermore, they are considered to be the key for understanding of Quaternary tectonics or Quaternary seismicity and provide fundamental data for long term earthquake prediction. Many recent studies of marine terraces and active faults have been carried out with special reference to coseismic events, as explained in this paper.

## DEFORMATION OF MARINE TERRACES ASSOCIATED WITH HISTORICAL GREAT EARTHQUAKES

Coseismic uplift and tilt of coastal areas of central and southwestern Japan, along the Pacific Ocean. Many great earthquakes occurred in and around the Japanese coast since the beginning of the 1770's. Of these, eleven earthquakes were accompanied by coseismic uplift and tilt of coastal areas (Table 1 and Fig. 1). In particular, the coastal area of central and southwestern Japan, located close to the Sagami, Suruga and Nankai Trough (the northern margin of the Phillippine Sea Plate), has experienced notable coseismic deformation associated with great earthquakes, larger than 7.9 in magnitude. Late Quaternary

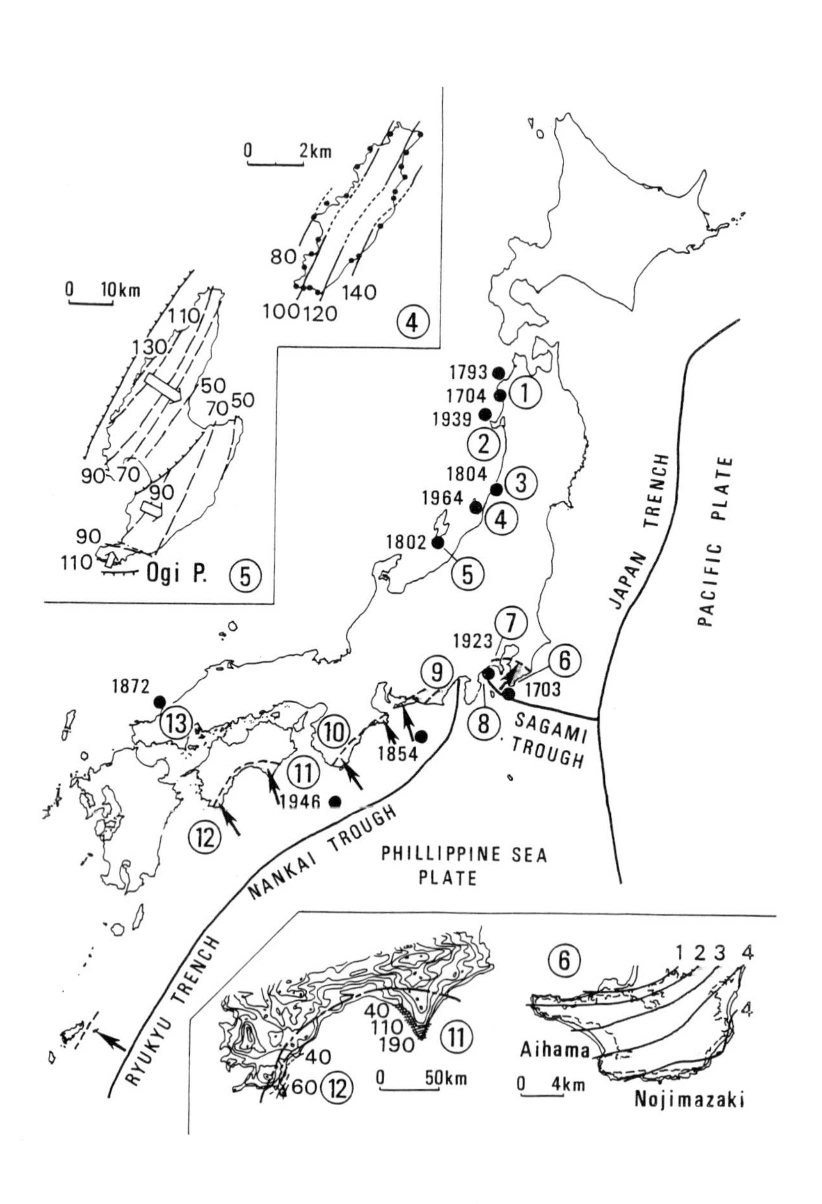
0 2km
80
100 120
140
4
0 10km
110
130
50
70 50
90 70
90
90
110
Ogi P.
5
1793
1704
1
1939
2
1804
3
1964
4
1802
5
JAPAN TRENCH
PACIFIC PLATE
7
1923
6
9
1703
1872
8
SAGAMI TROUGH
13
10
1854
11
1946
12
NANKAI TROUGH
PHILLIPPINE SEA PLATE
RYUKYU TRENCH
6
1 2 3 4
4
Aihama
40
110
190
11
40
60 12
0 50km
0 4km
Nojimazaki

Figure 1. Map showing distribution fo historical earthquakes accompanied by coseismic uplift and tilt. Solid circles and numbers: epicenter and date of earthquakes; Arrows and dashed line: landward tilt of terraces and its inland limit (hinge line); Numbers encircled correspond to area numbers of Table 1.

Inset (area 4): Coseismic tilt of Awashima caused by the 1964 Niigata earthquake (Nakamura et al., 1964).

Inset (area 5): Three tilted blocks of Sado Island on the basis of the height of the last interglacial terrace (Ota et al., 1976).

Inset (area 6): Landward tilt of the southern Boso Peninsula associated with the 1703 earthquake. Former shoreline of N1 is shown in a dashed line and that of N4 in a thin solid line (after Matsurda et al., 1978).

Inset (areas 10 and 11): Landward tilt of the last interglacial terrace (Ota, 1975). General topography is shown in thin solid lines (100m contour interval) height is shown in meters.

Table 1. RATE OF UPLIFT ESTIMATED FROM HEIGHT OF FORMER SHORELINES AND COSEISMIC DEFORMATION IN HISTORICAL TIME

| Area | Height of the marine terrace and rate of uplift | | | | Historical earthquake which caused crustal deformation of coastal area | | |
|---|---|---|---|---|---|---|---|
| | Holocene | | Last inter-glacial | | | | |
| | I | II | I | II | year | M | Max. uplift (m) |
| 1 Nishi-tsugaru | 10* | 1.3 | 100 | 0.7 | 1793 | 6.9 | 2.5 |
| | | | | | 1704 | 6.9 | 2.0 |
| 2 Oga Pen. | | | 90 | 0.7 | 1939 | 7.0 | 0.5 |
| 3 Kisakata | | | | | 1804 | 7.1 | 1.8 |
| 4 Awashima I. | 10 | 1.3 | 60 | 0.5 | 1964 | 7.5 | 1.6 |
| 5 Sado I. | 9* | 1.2 | 120 | 0.9 | 1802 | 6.6 | 2.0 |
| 6 Boso Pen. | 25* | 3.8 | | | 1703 | 8.2 | 4.4 |
| | | | | | 1923 | 7.9 | 1.5 |
| 7 Oiso | 30* | 4.6 | 160 | 1.2 | 1923 | 7.9 | 1.5 |
| 8 Hatsushima I. | 10* | 1.3 | ≦60 | ≦0.5 | 1923 | 7.9 | 1.5 |
| 9 Omae-zaki | 10 | 1.3 | ≧90 | ≧0.7 | 1854 | 8.4 | 1.0 |
| 10 Kii Pen. | | | 60 | 0.5 | 1946 | 8.1 | 1.0 |
| 11 Muroto Pen. | 13* | 1.8 | 195 | 1.5 | 1946 | 8.1 | 1.3 |
| 12 Ashizuri Pen. | | | 60 | 0.5 | 1946 | 8.1 | 0.4 |
| 13 Hamada | | | | | 1872 | 7.1 | 1.2 |

I: Maximum height (m), II: Maximum rate of average uplift (m/1,000y)
*: Holocene terrace is subdivided into several steps.

deformation of these areas, as deduced from the height of former shorelines of marine terraces, is characterized by a high rate of uplift and a progressive landward tilt, which is concordant in pattern with historical coseismic tilt (Ota, 1975; Yonekura, 1975; Ota and Yoshikawa, 1978). For example, M1 terrace of the Muroto Peninsula, correlated with the last interglacial age, is about 200m high at the southern tip of the peninsula and decreases northwestward (landward) to about 40m high (inset 11, Fig. 1). Results of relevelling of bench marks over two different time spans indicate that the landward tilt of the terrace has a positive relationship to the deformation pattern between 1924 and 1947, which represents coseismic deformation due to the Nankai earthquake of 1946, and has a negative relation to that of the interseismic period between 1895 and 1929. As the amount of coseismic landward tilt is much larger than that of the interseismic southward tilt, the former has accumulated during the late Quaternary resulting in progressive landward tilt of the terraces (Yoshikawa _et al._, 1964).

However, general topography of the coastal area decreases southward as shown by contours of inset 11, Fig. 1. It is understood that such a landward tilt started after formation of the oldest marine terrace of this area, which was probably formed during high sea level of the penultimate, or one stage older, interglacial period. This landward tilt is different from the tectonic movement which has continued over a longer period, producing the growth of mountains. A hinge line is located to mark the boundary between these two different tectonic areas (Yoshikawa, 1968).

Four well-developed Holocene marine terraces, N1-N4, flanking the southern Boso Peninsula (inset 6, Fig. 1, Matsuda _et al._, 1978), provide data for an estimation of the recurrence interval of major earthquakes. The oldest (highest) N1 terrace, c.6200-6900 yBP, represents the culmination of post-glacial sea level rise. The lowest, N4, the Genroku terrace, was uplifted with a northward tilt in association with the Genroku earthquake of AD 1703 which occurred at the plate boundary. Coseismic uplift is clearly recorded in old documents and maps. For example, at Aihama, a map drawn in 1654, indicates a location of the former shoreline, just before the 1703 earthquake. Point Nojima-zaki, a stack in the sea before the earthquake, was connected to the mainland with a rocky terrace about 5m high by coseismic uplift. The Holocene marine terrace shows a progressive landward tilt, i.e., a gradient of $1.4\times10^{-3}$ for N1, $0.9\times10^{-3}$ for N2, $0.7\times10^{-3}$ for N3 and $0.3\times10^{-3}$ for N4 (Ota, 1982).

Coseismic origin for three older terraces can be deduced, by using similarity of height of terrace risers and width of terrace surfaces to those of the Genroku terrace, in addition to the progressive landward tilt mentioned above. Thus, it is inferred that at least three great earthquakes could have occurred prior to the 1703 event, c.6200 yBP, 4400 yBP and 2900 yBP, respectively (Nakata _et al_., 1979).

_Coseismic uplift of two islands located in the Japan Sea, off central Honshu_. Coseismic uplift and tilt are also known in coastal areas of northern and central Honshu along the Japan Sea, although the magnitude of earthquakes in this area is less than 7.5, smaller than those of the Pacific side (Fig. 1 and Table 1). Except for area 13, where no prominent marine terraces were observed, the deformation pattern of marine terraces in each area is consistent with that of crustal deformation due to historical earthquakes which might be caused by the activity of relatively short submarine reverse faults.

Awashima Island (area 4) is the nearest land to the epicenter of the Niigata earthquake of June 16, 1964. Observations 70 hours after the shock indicated that the coseismic uplift and tilt were 0.8-1.6m and 55.7", respectively in the direction of $298.6^{\circ}$, suggesting that the island tilted as single rigid body (inset 4, Fig. 1, Nakamura _et al_., 1964). Marine terraces of different ages and Miocene sedimentary rocks of the island also tilt northwestwards, consistent with the pattern of coseismic tilt in 1964, but progressive in amount. Therefore, such a tilt is considered to be an accumulated result of coseismic events associated with great earthquakes, repeated throughout approximately the last one million years, with a recurrence interval of about every thousand years, on the basis of data summarized in Table 2A.

Sado Island (area 5), 35km off central Honshu, is fringed by a series of Pleistocene and Holocene marine terraces. The island is composed of three tilted blocks based on the deformation pattern of these terraces (inset 5, Fig. 1). The southern Ogi Peninsula was uplifted at the time of the Ogi earthquake of December 9, 1802. The coseismic uplift was 2m at the south, decreasing northwards. The gradient of the northward tilt of the Holocene marine terrace, probably of c.6000 yBP, is nearly the same as that of the 1802 tilt, implying that the latter was the first event after formation of the Holocene terrace. However, as Pleistocene terraces are characterized by a progressive northward tilt, an accumulation of coseismic tilt can be deduced. The average earthquake recurrence interval is estimated by comparing amount of tilt of Pleistocene terraces with that of the 1802 tilt (Table 2B, Ota _et al_., 1976).

The coseismic uplift and tilt at the time of the 1802 earthquake were significantly larger than those of Awashima mentioned above, although the magnitude of the former is smaller than the latter. This fact and the limited areal deformation by the 1802 earthquake suggest that the epicenter of this earthquake was located very close to the Ogi Peninsula, probably within a few kilometers of the south coast. It is possible that the formation of two other tilted blocks was caused by reverse faults shown in the inset 5, Fig. 1. No historical earthquakes, however, were known along these faults.

## RECURRENCE INTERVALS OF GREAT EARTHQUAKES DEDUCED FROM TRENCHING DATA ACROSS ACTIVE FAULTS

_Active faults and trenching studies in Japan_. Topographies dislocated by active faults are one of the prominent tectonic landforms in Japan. Most major boundaries of topographic units are delineated by active faults. The study of active faults in Japan started with an appearance of the surface break along the Neodani Fault, central Japan, associated with the 1891 Nobi earthquake. The understanding of an earthquake as a sudden movement along a fault has been accepted and the study of active faults was included as an item in the first Japanese earthquake prediction program in 1962. Geological and geomorphological

Table 2. PROGRESSIVE COSEISMIC TILT OF TWO ISLANDS, AWASHIMA (A) AND SADO ISLAND (B), LOCATED IN THE JAPAN SEA, OFF CENTRAL JAPAN

A. AWASHIMA ISLAND

(Nakamura et al.,1964 partly revised in 1978)

| | Strike | Dip | Rate of tilt sec./year |
|---|---|---|---|
| Coseismic tilt of the 1964 earthquake | N28.6°E±2.6° | 55.7"NW±5.0" | - |
| -25m terrace (c.10-20x1000y.) | 30° | 15'-30' | 0.05-0.2 |
| 50-70m terrace (c.125 x1000y.) | 25° | 2° | 0.06 |
| Miocene strata | 28° in average | 15° in average | - |

B. SADO ISLAND (Ota et al.,1976)

| | | A | B | C | D | E |
|---|---|---|---|---|---|---|
| Name of terrace | | Estimated age (yBP) | Tilting: Direction of maximum slope | Tilting: Dip angle | | |
| 1802 terrace | | 173 | N11.9' E±19.85' | 1.59'±0.17' | | |
| Holocene terrace | | 6,000 | N20.5' E±16.98' | 2.25' ±0.31' | (1) | (>6000) |
| Plei-sto-cene t. | IV terrace | 80,000 | N19.0' W±24.83' | 14.85' ±0.32' | 9 | 8600 |
| | III terrace | 120,000 | N7.5' W±72.16' | 38.02' ±12.92' | 24 | 5020 |
| | II terrace | ? | N3.0' E±72.21' | 57.10' ±9.53' | 36 | |

D: Numbers of earthquake since formation of the terrace (C/1.59 )

E: Average interval of earthquakes, yrs. (A/D)

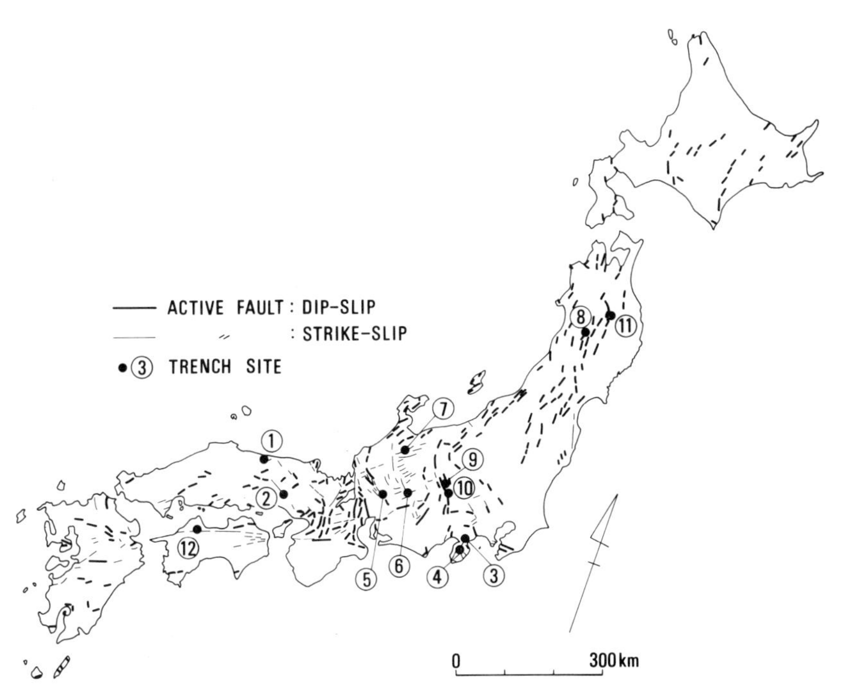


Figure 2. Map showing distribution of active faults and trench sites in Japan. Simplified from The Research Group for Active Faults, 1980.

studies on active faults revealed that the pattern of displacement along an earthquake fault is the same as that of the active fault which has been active through recent geological times.

Active faults over the whole Japanese Islands were mapped with a uniform standard, and the result was published (The Research Group for Active Faults, 1980). It consists of sheet maps (1:200,000) of active faults on land and of submarine areas, with detailed inventories on each mapped fault and a text. A compiled map of the faults was also prepared on a scale of 2,000,000, summarized in Fig. 2. This work marks the final step in the "Age of discovery of active faults in Japan". A number of problems, however, remain to be solved. One of them is to establish the date and amount of previous events, and recurrence intervals of the faulting. These data are important for estimating the date of the next faulting, accompanied by great earthquakes. Although the Japanese Islands have a relatively long written history, only nineteen surface breaks caused by historical earthquakes are known since 1847. In order to obtain more information, trenching studies of active faults started in 1979 in Japan. About thirty trenches across ten major fault systems were excavated (Fig. 2 and Table 3). Two examples of the trenching are discussed here.

Trenching of the Tanna faults. The Tanna fault (N-S trend) consists of a conjugate fault system with NW-SE trending faults in the northern part of the Izu Peninsula (Fig. 3). The latest surface break along the Tanna fault was accompanied by the Kitaizu earthquake (M=7.3) of 1930. The maximum amount of the surface break was 3m in strike slip (left-lateral) and 2m in dip slip (Matsuda, 1972). Trenching was carried out in 1980 and 1982 on the alluvial plain of the Tanna Basin. Holocene deposits of the trench site are composed of mainly silt, peat and volcanic sand and gravel with a thin layer of dated airfall tephras (Akahoya ash, c.6300 yBP; Kawagodaira pumice, c.2900 yBP; Zunasawa scoria, c.2500 yBP and Kozushima ash, AD 838) (Fig. 4). Nine events including the last 1930 earthquake were recognized after the deposition of Akahoya ash (Table 3, The Tanna Fault Trenching Research Group, 1983). For recognition of previous events, special attention was paid to the observation of prism beds

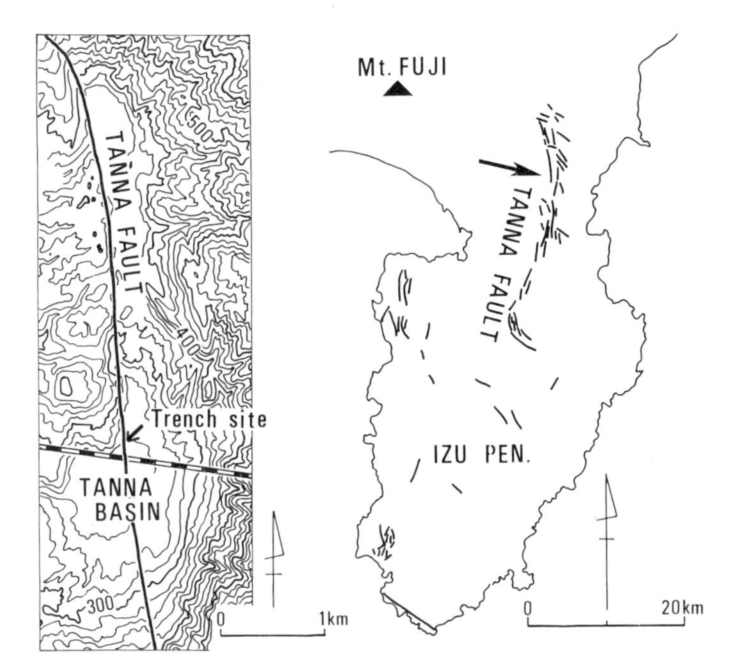


Figure 3. Map showing active faults in the Izu Peninsula and the trench site across the Tanna Fault (after The Tanna Fault Trenching Research Group, 1983).

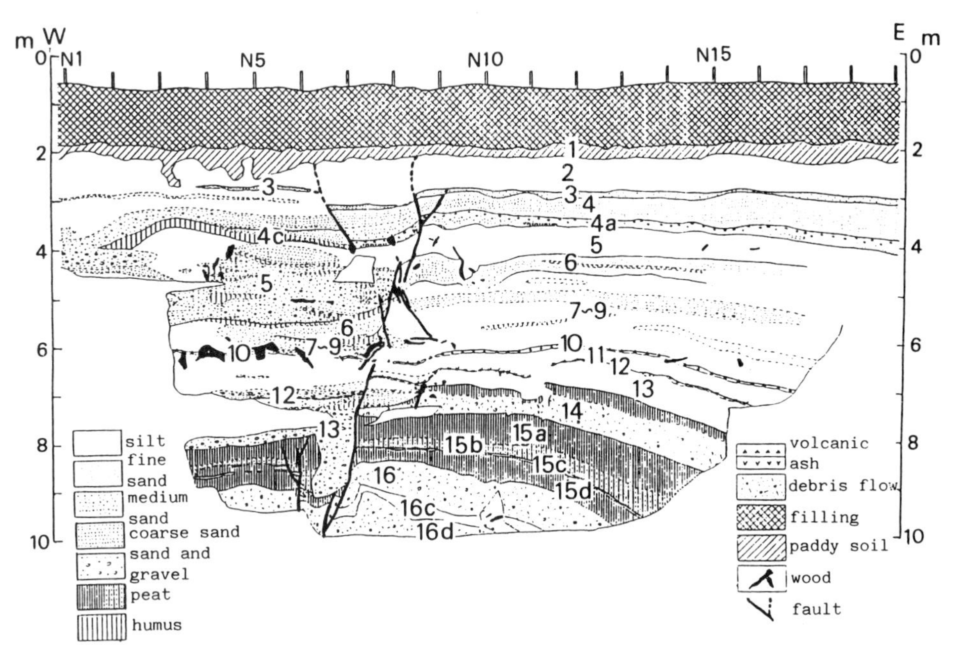


Figure 4. Sketch of the north wall on the 1982 trench across the Tanna Fault (The Tanna Fault Trenching Research Group, 1983).

Table 3. SUMMARY OF TRENCHING OF ACTIVE FAULTS IN JAPAN

| Loc. no. | Date of trenching | Fault name | Size of trench 1 | 2(m) a | b | Age of event (yBP) | Reference |
|---|---|---|---|---|---|---|---|
| 1 | Dec. 1978 | Shikano | 3 | 15 | 3 | AD 1943<br>1490-7970 | Tsukuda et al., 1979 |
| 2 | Mar. 1979 | Yamasaki | 4 | 25 | 3 | 750-1170<br>(AD 868?)<br>1800-5560 | Okada et al., 1980 |
| 3 | Oct. 1980<br>Feb. 1982 | Tanna (Myoga) | 2 | 20 | 7 | AD 1930<br>400-600<br>900(AD 841?)<br>1100-2500<br>c.2900<br>c.3300<br>c.4000<br>4700-5300<br>5300-6000 | The Tanna Fault Trenching Research Group, 1983 |
| 4 | Dec. 1980 | Tanna (Ukihashi) | 1 | 20 | 3 | < 2900<br>2900-4670 | Yamazaki et al., 1981 |
| 5 | Jul. 1981 | Nobi | 2 | 26 | 5 | AD 1891<br>19200-20900<br>27000-30100<br>c.32300 | Disaster Prevention Research Institute, Kyoto Univ., 1983 |
| 6 | Oct. 1981 | Atera | 1 | 25 | 3 | < 5500<br>c.6500<br>c.9500<br>c.12300 | Geological Survey of Japan, 1982 |
| 7 | Jul. 1982 | Atotsugawa | 1 | 23 | 13 | <820<br>(AD 1858?)<br>5200±250<br>7500±800<br>8600±400<br>and 6 events older than 8600yBP. | The Atotsugawa Fault Trenching Research Group, 1983 |
| 8 | Oct. 1982 | Senya | 5 | 20 | 5 | AD 1896<br>c.3000? | The Senya Fault Trenching Research Group, (in press) |
| 9 | Jul. 1983<br>Jul. 1984 | Okaya | 4 | 14 | 3 | 2500-2700<br>c.5000<br>6900-7800<br>c.13500<br>c.19500 | The Okaya Fault Trenching Research Group, 1984 |
| 10 | Oct. 1983 | I.T.L. | 2 | 30 | 7 | 1130-1290<br>(AD 841?)<br>1750-3550<br>4170-6470<br>9330-c.15000 | I.T.L. Trenching Research Group (in preparation) |
| 11 | Nov. 1983 | Urata | 1 | | 4 | < 7000 | Shimokawa & Awata, 1984 |
| 12 | Jan. 1984 | M.T.L. | 1 | 20 | 5 | < AD 450~650<br>AD 450~650-<br>AD 200~450 | Ando et al., (in preparation) |

1:number of trench, 2:a;length, b;depth

burying the former fault scarp or flexure scarp, to deposits filling the open crack along the fault zone and to unconformities between the lower, more deformed bed and the upper, less deformed bed. Recurrence interval of the event over the past c.6000 years is estimated at about 700-1000 years. This value fits fairly well with an assumption that the faulting took place about every 1000 years, obtained from a comparison of amount of surface displacement in 1930 with the total amount of offset of streams (c.1km), dissecting the middle Quaternary volcano (Kuno, 1936).

Trenching of the Itoshizu Tectonic Line (ITL). ITL marks the western boundary of "Fossa Magna", a major tectonic depression between southwestern and northeastern Japan. It runs from Itoigawa on the Japan Sea coast to Shizuoka on the Pacific coast, crossing mountainous central Japan. In contrast to the north-south trend of ITL in the northern and southern part, it strikes northwest-southeast in the central part, where a series of elongated tectonic bulges is developed (Fig. 5). Although topographic expressions of ITL are distinguishable, an exact fault feature was not known, and no surface break was recorded in historical time. At both trench sites, notable fault zones bordered by high angle fault planes were discovered exactly on the topographically estimated fault traces and repeated activities along these faults were recognized. It is estimated that faulting took place at least four times in the last c.15,000 years at trench B, along the fault zone shown in Figs. 7 and 8, interpreted as follows: The western margin of the layer 3 peat bed, dated at 1290±80 yBP at the upper part, is faulted and overlain by undeformed layer 2b, dated at 1130±80 yBP. Therefore, the latest event is regarded to have occurred after deformation of layer 3 and prior to that of layer 2b, that is, about 1200 yBP.

An event of similar age was discovered further north along ITL (Togo et al., in preparation). Therefore, the event of c.1200 yBP seems to cover an extensive area. An old document described damages of fences and houses by a large earthquake in AD 841, epicenter estimated at 138°0'E, 36.2°N, north of these trench sites. It is likely that this earthquake corresponds to the latest event of c.1200 yBP. The penultimate event is estimated

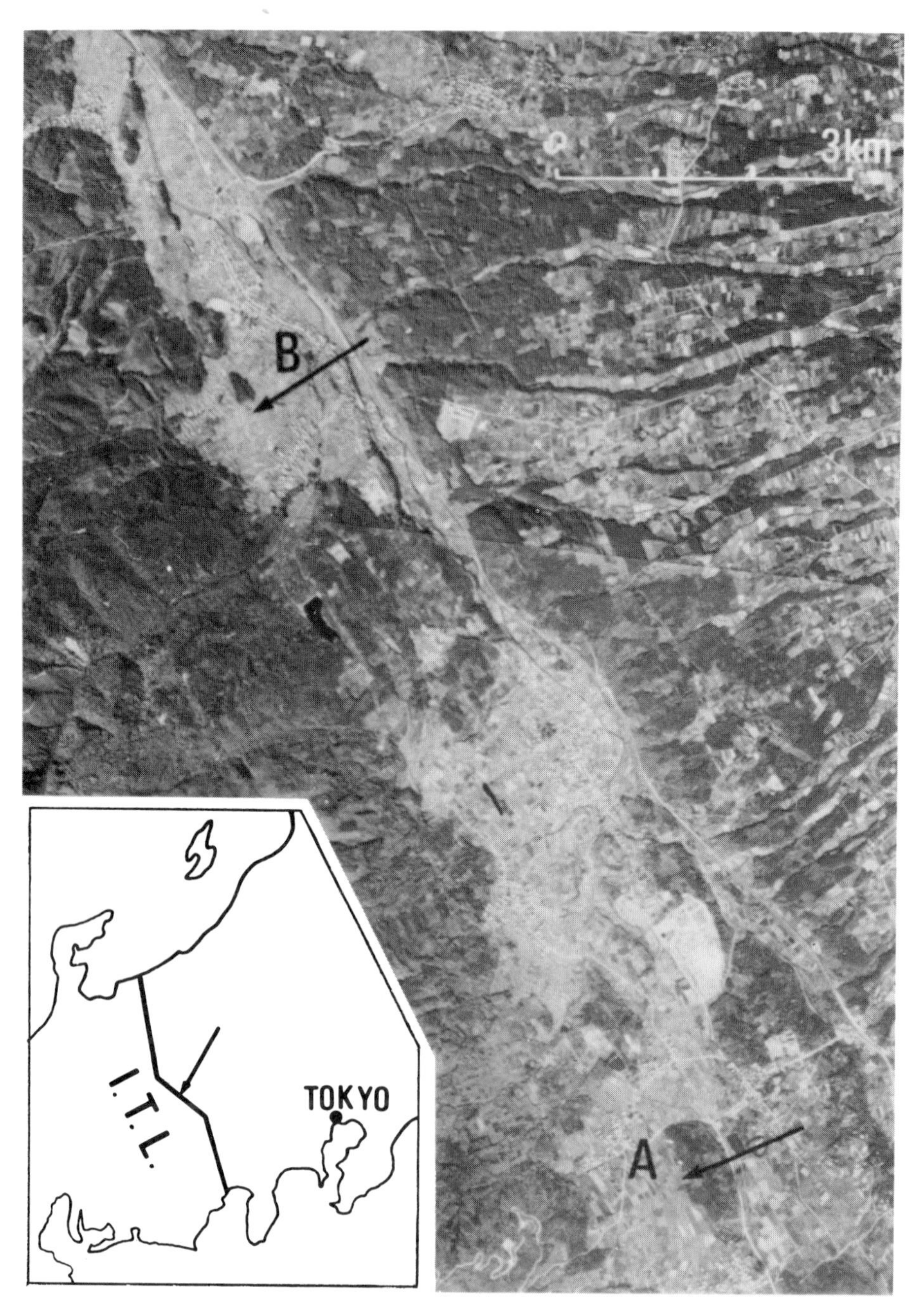


Figure 5. Air-photo showing a series of tectonic bulges along ITL and trench sites.

between 3550 yBP and 1750 yBP, because the faulted peat of layer 3, dated at 1750±85 yBP at its base, fills a fault depression dislocating layer 4a, dated at 3550±95 yBP. An unconformity between a more deformed layer of 4b, dated at 6470±120 yBP and a less deformed layer of 4a, suggest that faulting took place between c.3600 yBP and 6400 yBP. The oldest event is represented by layer H3, which was formed as the surface soil developed on a fault scarp dislocating bed 5b. An estimated age for this event is between c.9000 yBP and c.14000-17000 yBP, judging from C14 ages for H3 and 5b beds. However, multiple faulting might have occurred in this relatively long time span.

The east side of the fault zone was upthrown in three of these four events. The same sense of faulting is observed at trench A. These facts fit with the presence of tectonic bulges to the east of the fault zone in the area shown in Fig. 5.

Characteristic fault features found at trench A and B across ITL are 1) very steep, nearly vertical fault planes, 2) arrangement of gravels and drag structures parallel to the strike of the fault, 3) uncomparable facies of deposits at both sides of the main fault zone, and 4) steeper northern slope of the bulge than that on the southern side. These features strongly suggest a left lateral displacement along the central part of ITL, although no systematic offset of streams and ridges was recognized.

## DISCUSSION AND SUMMARY

Terrace data and trenching data across active faults provide fundamental information for understanding the activity of offshore- or inland-type earthquakes, which have been active repeatedly during recent geological times and will be active in the future. Results obtained by recent studies in Japan and problems to be solved are summarized as follows:

1) A progressive landward tilt of marine terraces is a good indicator of coseismic deformation which accompanied historical great earthquakes. Landward tilt is recognized on the Japan Sea coast as well as on the Pacific coast. The areal extent of the landward tilt affected by great earthquakes is greater on the Pacific coast, close to the plate boundary, than on the Japan Sea coast.

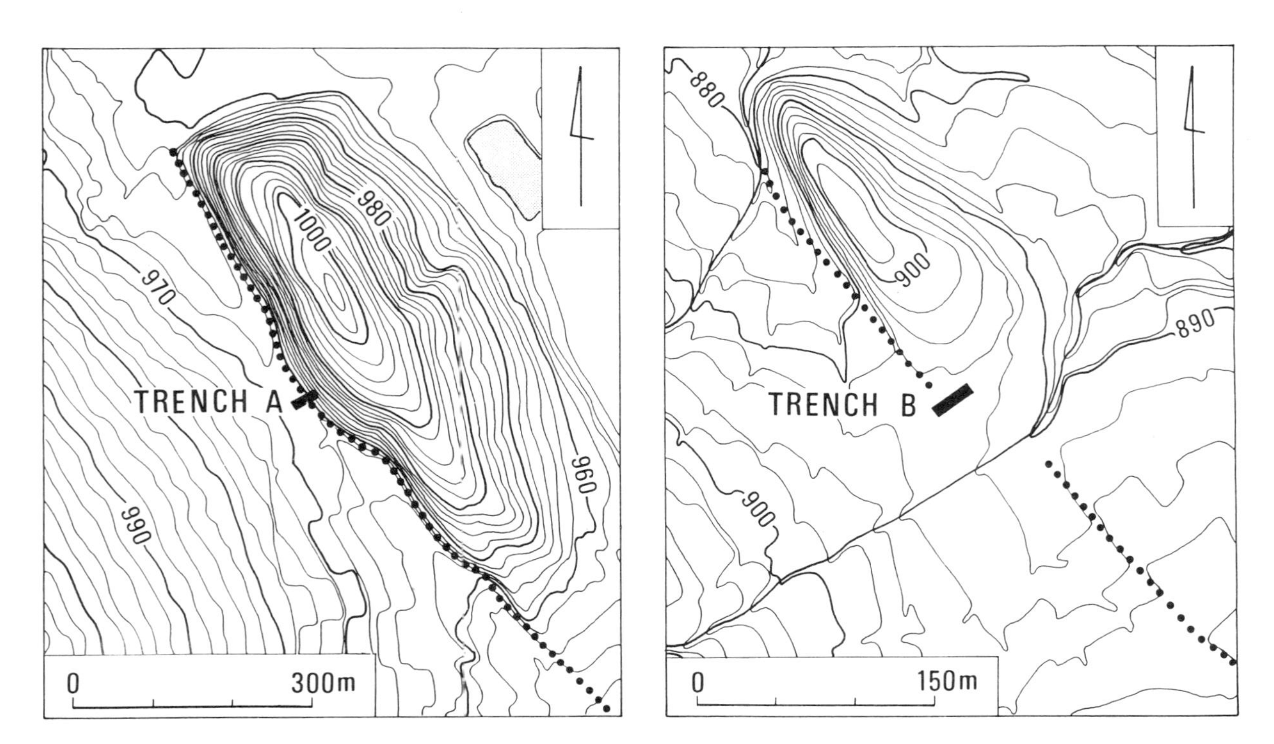


Figure 6. Contour maps of two tectonic bulges and trench sites along ITL.

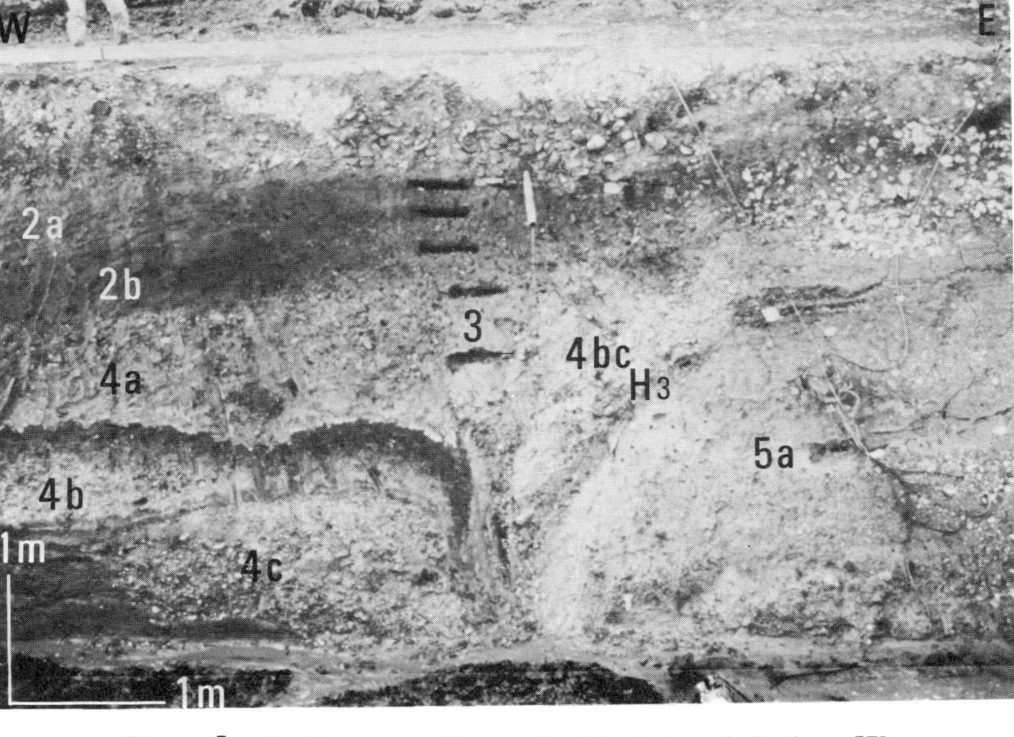


Figure 7. Fault zone on the northwall at trench B along ITL.

Figure 8. Fault zone on the south wall at trench B along ITL.

2) Based on either high rate of uplift, landward tilt or subdivided Holocene terraces, probable coseismic deformation can be identified even in areas where no great earthquakes are known in historical times. Examples are Kikai Island of the north Ryukyu, close to the Ryukyu Trench (Ota, 1982; Nakata et al., 1979), and the northeastern coast of North Island, New Zealand, near the Kermadec Trench (Yoshikawa et al., 1980). The coastal area protruding into the sea between Cape Arago and Cape Bionco, south Oregon, is very likely a coseismic deformation area, judging from the presence of a high rate of uplift and notable landward tilt (Admas, 1984; Heaton, Ota and Okada, in preparation).

3) A comparison of tilt rate enables an estimation of number and recurrence interval of great earthquakes, assuming that the tilt rate has been uniform throughout recent geological times. This assumption itself, however, should be examined. Age and number of subdivided Holocene marine terraces also can be used for reconstruction of previous events.

4) The repeated activities of some major faults in recent geological times was confirmed by the trenching study. Dates of events reconstructed are summarized in Table 3. Some of the previous events are probably correlated with great earthquakes recorded in historical documents, which have no description of surface breaks. For instance, the event of c.1200 yBP at trench B of ITL seems to correspond to the Shinano earthquake of AD 841, the event of c.900 yBP at the Tanna Fault to an earthquake in the Izu area of AD 841, and the event about 750-1170 yBP of the Yamasaki Fault to the Harima earthquake of AD 868 (Okada et al., 1980). Approximate recurrence intervals of faulting range from one to several thousand years. Accuracy of the estimation depends on numbers and quality of datable material.

5) Detailed fault features can be observed by trenching, giving information on the sense of faulting. For example, left lateral displacement is strongly suggested along the ITL by trench data, in spite of the lack of systematic horizontal offset of topography.

6) Amount of displacement and length of a fault formed by a single event are essential data for estimation of magnitude of

earthquakes due to a fault activity. However, those data are still insufficient. It is especially difficult to estimate the amount of strike-slip. The number of trenchs is too few to cover the major fault systems over a long distance or to estimate the length moved by a single event.

7) A number of large earthquakes reconstructed by terrace and fault data may represent the minimum events because of subsequent removal of topographic and geologic evidences. Recurrence interval of events thus may become smaller as more detailed observation progresses.

ACKNOWLEDGEMENTS

I would like to express my sincere thanks to Professor M. Morisawa, State University of New York at Binghamton, for giving me an opportunity to present this review paper and revising the manuscript. I am grateful to the National Science Foundation, U.S.A., for travel funds to attend the symposium on tectonic geomorphology, held at the State University of New York at Binghamton. Thanks also to many colleagues for their collaboration and stimulating discussions on various aspects of tectonic geomorphology, especially to Professor T. Matsuda, Earthquake Research Institute, University of Tokyo. Lastly, I am obliged to Miss M. Miyoshi and Mrs. K. Yokoyama, Yokohama National University, for drafting figures.

REFERENCES

Adams, J., 1984, Active deformation of the Pacific northwest continental margin: Tectonics, v. 3, p. 449-472.

Disaster Prevention Research Institute, Kyoto Univ., 1983, Trenches across the trace of the 1891 Nobi Earthquake Faults: Rep. Coordinating Committee for Earthquake Prediction, v. 29, p. 360-367.**

Geological Survey of Japan, 1982, Exploratory excavation and fault activity of the Atera Fault, central Japan: Rep. Coordinating Committee for Earthquake Prediction, v. 28, p. 299-303.**

Kuno, H., 1936, On the displacement of the Tanna fault since the Pleistocene: Bull. Earthq. Res. Inst., Univ. Tokyo, v. 14, p. 621-631.

Matsuda, T., 1972, Surface faults associated with Kita-Izu Earthquake of 1930 in Izu Peninsula, Japan: Hoshino, M. and Aoki, H. eds., _Izu Peninsula_, Tokai Univ. Press, p. 73-79.*

Matsuda, T., Ota, Y., Ando, M. and Yonekura, N., 1978, Fault Kanto district, Japan, as deduced from coastal terrace data: Geol. Soc. Am. Bull., v. 89, p. 1610-1618.

Nakata, T., Koba, M., Jo, W., Imaizumi, T., Matsumoto, H. and Suganuma, T., 1979, Holocene marine terraces and seismic crustal movement: Sci. Rep. Tohoku Univ., 7th series, v. 29, p. 196-204.

Nakamura, K., Kasahara, K. and Matsuda, T., 1964, Tilting and uplift of an island Awashima, near the epicenter of the Niigata earthquake in 1964: Jour. Geod. Soc. Japan, v. 10, p. 172-179.

Okada, A., Ando, M. and Tsukuda, T., 1980, Trenches across the Yamasaki Fault in Hyogo Prefecture: Rep. Coordinating Committee for Earthquake Prediction, v. 24, p. 190-194.**

Ota, Y., 1975, Late Quaternary vertical movement in Japan estimated from deformed shorelines: Royal Soc. N.Z. Bull., v. 13, p. 231-239.

_______, 1980, Tectonic landforms and Quaternary tectonics in Japan: Geo. Journal, v. 4, p. 111-124.

_______, 1982, Holocene marine terraces of uplifting areas in Japan: Colquhoun, D.J. ed., Holocene Sea Level Fluctuations, Magnitude and Causes, p. 118-134.

_______, Matsuda, T. and Naganuma, K., 1976, Tilted marine terraces of the Ogi Peninsula, Sado Island, central Japan, related to the Ogi earthquake of 1802: Jour. Seism. Soc. Japan, Ser. II, v. 29, p. 55-70.*

_______, and Yoshikawa, T., 1978, Regional characteristics and their geodynamic implications of late Quaternary tectonic movement deduced from deformed former shorelines in Japan: Jour. Phy. Earth, v. 26, Suppl. s.379-s.389.

The Atotsugawa Fault Trenching Research Group, 1983, Trenching study for Atotsugawa Fault: The Earth Monthly, v. 5, p. 335-340.**

The Okaya Fault Trenching Research Group, 1984, Trenching study for the Itoshizu Tectonic Line at Okaya area in 1983: Rep. Coordinating Committee for Earthquake Prediction, v. 32, p. 363-372.**

The Research Group for Active Faults, 1980, Active Faults in Japan Sheet Maps and Inventories, Univ. Tokyo Press, 363p.*

The Research Group for Quaternary Tectonic Map, 1973, Explanatory text of the Quaternary tectonic map of Japan: National Research Center for Disaster Prevention, Sciences and Technology Agency, Tokyo, 166p.

The Tanna Fault Trenching Research Group, 1983, Trenching study for Tanna Fault, Izu, at Myoga, Shizuoka Prefecture, Japan: Bull. Earthq. Res. Inst., v. 58, p. 797-830.*

Tsukuda, T., Ando, M. and Okada, A., 1979, Trenching study for Shikano and Yamasaki Faults: Chiri, v. 24, p. 64-71.**

Yamazaki, H., Kakimi, T., Tsukuda, E. and Awata, T., 1981, Trench survey of the Ukihashi-chuo Fault, Tanna Fault System: Program and Abstracts Seism. Soc. Japan, 1981, no. 1, p. 118.**

Yonekura, N., 1975, Quaternary tectonic movements in the outer arcs of southwest Japan with special reference to seismic crustal deformations: Bull. Dept. Geogr. Univ. Tokyo, v. 7, p. 21-71.

Yoshikawa, T., 1968, Seismic crustal deformation and its relation to Quaternary tectonic movement on the Pacific coast of southwest Japan: The Quat. Res., v. 7, p. 157-170.*

____________, Kaizuka, S. and Ota, Y., 1964, Mode of crustal movement in the late Quaternary on the southeast coast of Shikoku, southwestern Japan: Geogr. Rev. Japan, v. 37, p. 627-648.*

____________, Ota, Y., Yonekura, N., Okada, A. and Iso, N., 1980, Marine terraces and their tectonic deformation on the northeast coast of the North Island, New Zealand: Geogr. Rev. Japan, v. 53, p. 238-262.*

*in Japanese with English abstract
**in Japanese

# 16
# Tectonic geomorphology and its application to earthquake prediction in China

*Mukang Han*

ABSTRACT

The research of the author and his colleagues on tectonic landforms (morphotectonics) in the various earthquake regions of China has resulted in the recognition and analysis of many interesting morphotectonic features and their relation to seismotectonics, and in the successful prediction of earthquake hazard sites (Wang and Han, 1984). The present paper outlines some of the author's approaches and the information obtained from this work.

## GEOMORPHIC EXPRESSIONS OF NEOTECTONIC STRESS FIELDS

The landscapes of China may be subdivided into three general elevation categories regarded by Chinese geomorphologists as morphotectonics of the first order (Fig. 1). The high level is the Qinghai-Xizang (Qinghai-Tibet) plateau in the southwest with a mean elevation of 4,000 m, being bounded on the south by the Himalayan mountain ranges. The middle level bordering the high level on the north and east comprises the northwestern area of China and the Yunnan-Guizhou plateau. Its altitudes are generally lower than 1,000-2,000 m, with interspersed basins of varying sizes. The lowest level includes the southeastern hill area of China, with elevations generally below 500 m, and the vast North and East China plains generally lying below 50 m. The demarcations between the three major levels coincide fairly well with changes of gravity and crustal thickness.

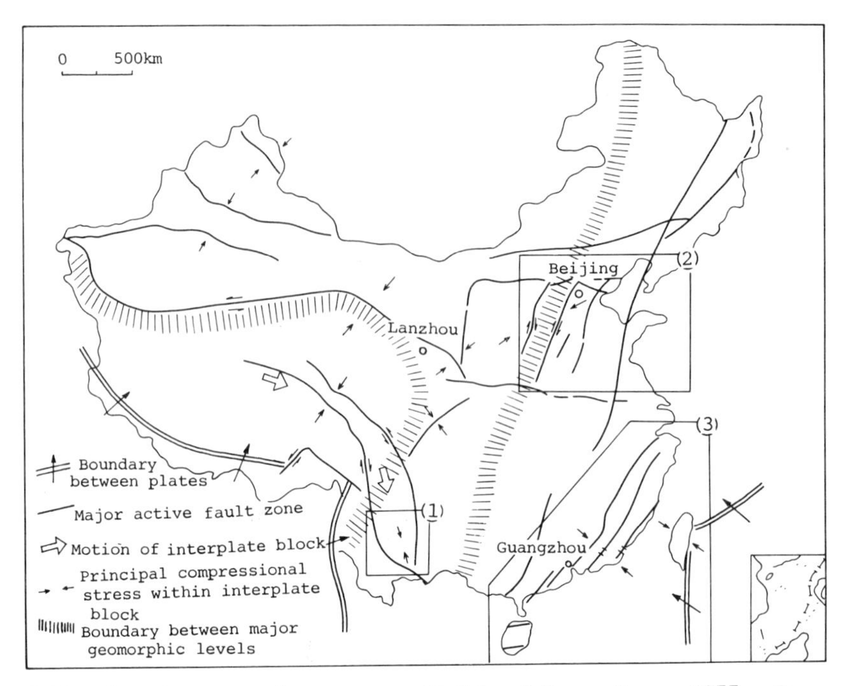


FIGURE 1. Neotectonic stress fields (after Kan, 1977; Deng, 1979), demarcations among three major geomorphic levels, and index maps of morphotectonically studied areas. For Inset 1, see Fig. 2; for Inset 2, see Fig. 4; for Inset 3, see Fig. 7.

These major geomorphic levels are recognized as having been generated by horizontal as well as vertical crustal movements, especially through the effect of neotectonic stress fields presumably related to interactions of the Indian, Eurasian, and Pacific lithospheric plates. Associated morphotectonic patterns and features in various areas are found to be controlled by different principal compressional stress axes within the interplate blocks and are best expressed near to or along the first order morphotectonic demarcations (Fig. 1, 2, 3, and 4).

## Yunnan-Guizhou Plateau

In the western part of Yunnan-Guizhou plateau (1 in Fig.1) close to the demarcation between the high and middle geomorphic levels, uplift and northeastward compression of the Tibetan plateau, caused by subduction of the Indian plate, leads to the uplift, warping, and dislocation of Neogene planation surfaces, and intense folding and faulting of Neogene-early Pleistocene lacustrine and fluvio-lacustrine strata in many intermontane basins (Han and Chai, 1979; Han et al., 1981a).

For instance, on the east side of the NNW-trending elongate Pu'er basin, which is close to the southeastern margin of the Tibetan plateau, the Neogene strata appear to be overturned. Farther to the east, in the narrow N-S striking Yuanmou graben, a series of compressional shear controlled drag folds and faults are formed in the early Pleistocene lacustrine and fluviolacustrine strata close to the major boundary fault (Fig. 2). The fold axes trend at a sharp angle to the major boundary fault and demonstrate a right-lateral motion of the major fault in the earlier stage. However, these folds are cut by a reverse fault that angles into the major fault and shows a left-lateral motion of the major fault in the later stage. In addition, development of river terraces only on the eastern side of basins, and their westward tilting, indicates compressional crustal movements there as well (Han and Chai, 1979).

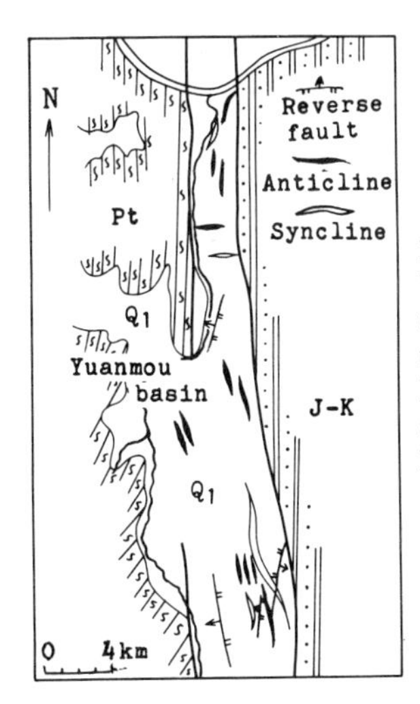


FIGURE 2. Morphotectonics of the Yuanmou basin showing compressional-shear-controlled drag folds and thrusts generated by two neotectonic stress fields of different geological stages. The NW-trending were formed earlier; the NE-trending were formed later. For location, see Fig. 1.

In the southern part of Yunnan-Guizhou plateau, the southeastward movement of a NW-trending rhombic interplate block (Fig. 3) produces an eastern boundary fault, the N-S trending Xiaojiang fault zone, with left-lateral offset. This movement results in the formation of a series of NW-trending en echelon grabens and horsts in the Neogene-early Pleistocene gravels in the extensional shearing downfaulted basin. The western boundary fault of the rhombic block, the famous NW-trending Red River fault zone, is subject to right-lateral motion, leading to the formation of a series of associated morphotectonic features such as offset drainage and displaced alluvial fans.

It is interesting to note that northeast of the Red River fault zone there is another smaller but important NW-trending seismic fault, the Qujiang fault, where a great earthquake (M = 7.7) occurred in 1970. This fault experienced right-lateral compressional shear and pivotal movements resulting not only in offsets of drainage and alluvial fans, but also in the spectacular vertical dislocation of Neogene planation surfaces and Pleistocene river terraces, both across the fault and in its northwestern and southeastern sections. This pivotal movement causes northward tilting of a lake basin and northward displacement of the shoreline of Qilu Lake, leaving four ancient shorelines between 83 m and 6 m above the southern margin of the lake, dated from 12,756 years to about 180 years B.P. (Fig. 3; Han et al., 1981b, 1983a).

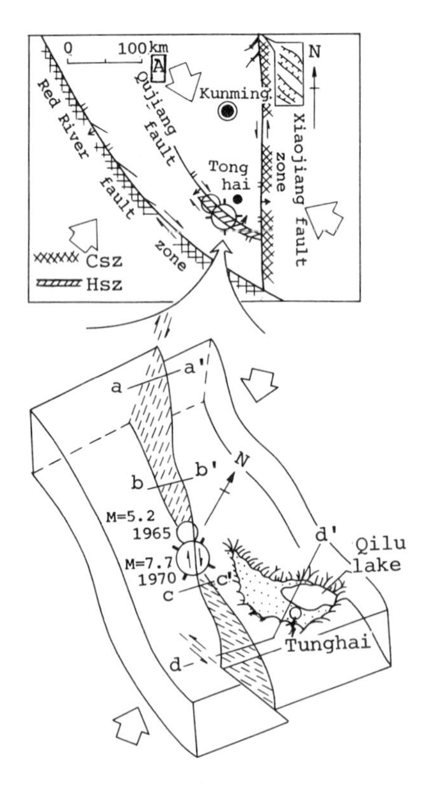


FIGURE 3. Morphotectonics in the southern part of Yunnan-Guizhou plateau (upper; for location, see Fig. 1), and in the 1970 Tonghai earthquake region (lower). Csz = compressional shear zone, Hsz = Historical and small earthquake zone. For cross sections a-a', b-b', c-c', and d-d', see Han et al, 1983a.

Taihangshan Range

A particularly interesting geomorphic expression of neotectonic stress fields is seen along the demarcation between the middle and low geomorphic levels in North China (Fig. 4). Here the effect of northeastward compression of the Qinghai-Tibet Plateau weakens relative to the influence of the westward movement of the Pacific plate. Right-lateral shearing, influenced by the NE-trending principal compressional stress, has created a series of well expressed en echelon morphotectonic features on both sides of the Taihangshan mountain range, which separates the middle and low geomorphic levels in the vicinity of Beijing.

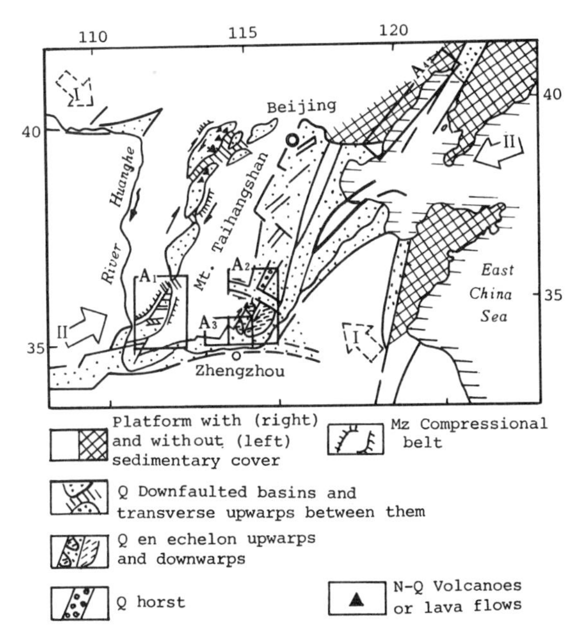


FIGURE 4. Geomorphic expressions of the neotectonic stress fields in North China. For location, see Fig. 1. A = areas studied in detail: A1--Linfen Basin, see Fig. 6; A2--Tangyin graben, see Fig. 5; A3--see Fig.13; A4--see Han et al, 1983c; Arrow I--direction of principal compressional stress during Mesozoic Yanshan Movement; Arrow II--direction of principal compressional stress during Neogene-Quaternary Himalayan Movement.

On the western side of the Taihangshan is the famous Shanxi graben system, consisting of a series of NNE- and NE-trending downfaulted, en echelon, extensional basins of Quaternary age. They are filled with Plio-Pleistocene lacustrine and fluvial deposits and alternate with transverse highlands, which are actually dome-like upwarped Pliocene planation surfaces with overlaying gravels and fluviolacustrine deposits.

On the eastern side of the Taihangshan are two en echelon pairs of NW-striking Quaternary upwarps and downwarps within the NNE-trending Tangyin graben (Fig. 5) in the southern section of the piedmont fault zone (Han, 1983; Han et al., 1983b). The upwarps consist of Pliocene lacustrine and fluviolacustrine deposits, and the downwarps are filled with Quaternary alluvium.

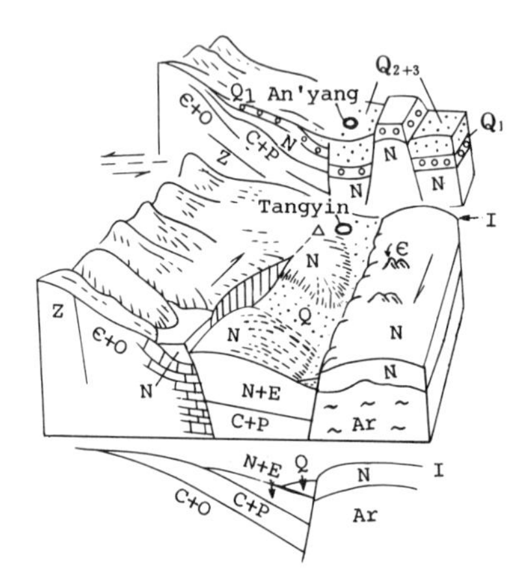


FIGURE 5. Block diagram of the Tangyin graben showing the earthquake risk location (open triangle) predicted by the author and verified by a subsequent seismic event. For location, see Fig. 4. N--Neogene lacustrine and fluviolacustrine strata; $Q_1$--early Pleistocene; $Q_{2+3}$--middle and late Pleistocene alluvial deposits.

These en echelon morphotectonics are still active at present as shown by the deformation and displacement of river terraces (Fig. 6) and Plio-Pleistocene strata, changes in elevation of the boundary between Holocene floodplain and channel deposits, changes in the elevation of archeological materials, and variations in drainage density, which is conspicuously greater on the eastern side of the Taihangshan Range. Repeated levelling confirms ongoing deformation (Er and Li, 1976; Han, 1983; Han et al., 1983b).

A compressional belt with the same strike as that of the extensional boundary faults is evident on the mountain side of some basins in the Shanxi graben system. This takes the form of low-angle thrust faults and folds overturned toward the mountains (Er and Li, 1976; Wang, 1979). This compressional belt is developed in the Archaeozoic-upper Paleozoic formations and was produced by the Mesozoic Yangshan orogeny in a stress field of left-lateral shear under the action of a NW-SE principal compres-

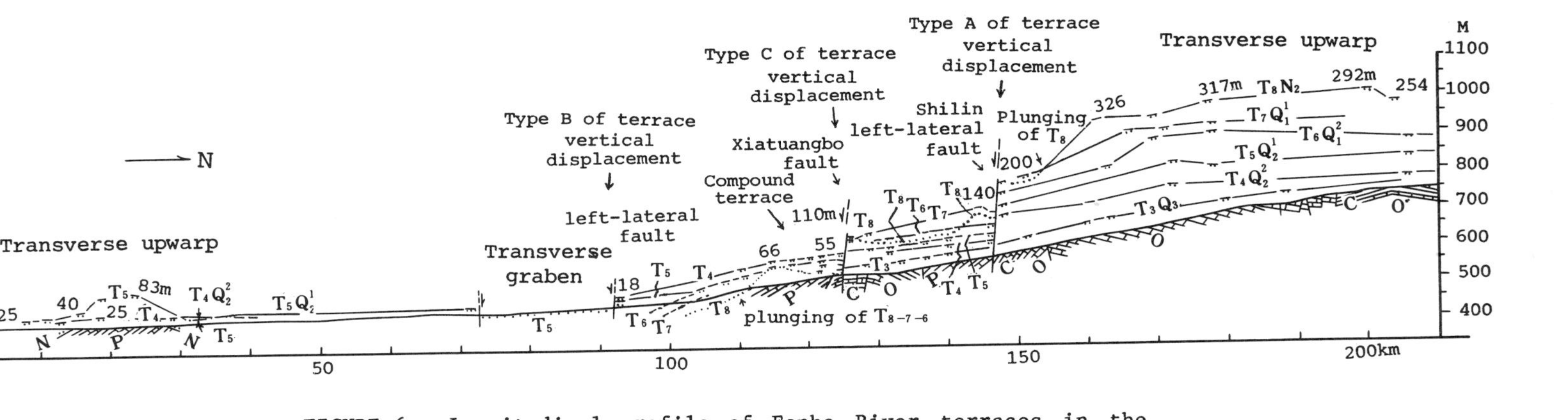


FIGURE 6. Longitudinal profile of Fenhe River terraces in the Linfen basin of the Shanxi graben system. For location, see Fig. 4. O--Ordovician; P--Permian; C--Carboniferous; N--Neogene fluvial and lacustrine deposits; $T8N_2$--terrace with Neogene deposits; $T7Q_1^1$--terrace with deposits of the early stage of the early Pleistocene; $T6Q_1^2$--terrace with deposits of the later stage of the early Pleistocene; $T5Q_2^1$--terrace with deposits of the early stage of the middle Pleistocene; $T4Q_2^2$--terrace with deposits of the later stage of the middle Pleistocene; $T3Q_3$--terrace with late Pleistocene deposits; T2 and T1--Holocene terraces, omitted from the profile. For the types of terrace displacement, see text and Figs. 9, 10, and 11.

sional stress. It follows that between Mesozoic time and the post-Neogene period the neotectonic stress field in North China has changed from a left-lateral shear pattern with a NW-SE principal compressional stress to a right-lateral shear pattern with a NE-SW principal compressional stress.

It is significant that obvious imprints of dissimilar tectonic stress fields of different geological stages with quite different directions of principal compressional stress are present in the morphotectonics of the Shanxi graben system (Er and Li, 1976; Wang, 1979; Han, 1983), the Yuanmou basin (Han and Chai, 1979), and the Yongshen basin (Han et al., 1981a). This shows that in the study of tectonic landforms and geomorphic development close attention must be paid not only to vertical and horizontal crustal movements, but also to the changing patterns of tectonic stress fields through time. Additionally, morphotectonic patterns and features in neotectonic stress fields are found to be influenced by older tectonic structures and the physical properties of the rocks of an area. For instance, in the area consisting of plastic formations, as are observed on the Yunnan-Guizhou plateau, horizontal crustal movements commonly create nearly parallel or arcuate upwarped mountain ranges alternating with elongate downwarped or downfaulted basins. In the more rigid areas, such as the ancient massif of the Southeast China hilly region (Fig. 7), smaller rhombic blocks, cut by NE- and NW-trending faults, are differentially uplifted or tilted up and warped or tilted down or alternated with downfaulted basins. The large uplifted block closely adjacent to the large downfaulted plains are characteristic of the rigid ancient platform without sedimentary cover, such as the Liaodong Peninsula, Shandong Peninsula, and the western coastal region of Liaoning Province. The latter region exhibits striking flights of marine terraces basically parallel and more or less vertically displaced (Han et al., 1983c).

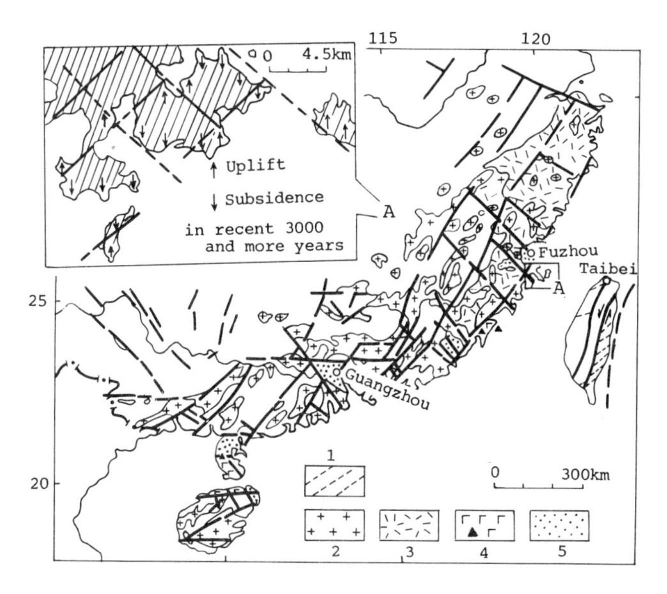


FIGURE 7. Morphotectonics of the southeastern China hilly area (part of data obtained from Bi and Zhou, 1984, and Zhang, 1982), showing the rhombic blocks cut by NE- and NW-trending faults in Fujiang and Guangdong provinces and left-lateral-shear-controlled drag folds on the eastern and western coasts of Taiwang Island (Wang and Han, 1984). For location, see Fig. 1. 1--Shear-conrolled folds; 2--Mesozoic granite; 3--Mesozoic volcanic rocks; 4--Cenozoic basalt (black triangle shows Quaternary volcano); 5--Meso-Cenozoic basin.

## SOME RESPONSES OF RIVER TERRACES AND ALLUVIAL FANS TO ACTIVE TECTONICS

### 1. Terrace Deformation in Warped Areas.

It has been found that intermittent progressive warping usually lends to corresponding deformation of river terraces traversing the warped area. In the transitional area between up- and down-warped areas can be seen the plunging of older terraces under younger terrace deposits or the present floodplain. Therefore, a point of convergence which corresponds with the axis of crustal flexuring can be defined (Fig. 8). Sometimes, two or more terrace plunge points can be produced due to intermittent expansion or tectonic migration of the upwarp center toward the downwarped area, and the migration rate can be determined if the age of the deformed terraces is known. Additionally, the buried alluvial strata in a previously downwarped area can be uplifted and together with overlying younger alluvium can constitute a terrace of a unique type after subsequent fluvial downcutting. It is proposed that this type of terrace be named a compound terrace. Its

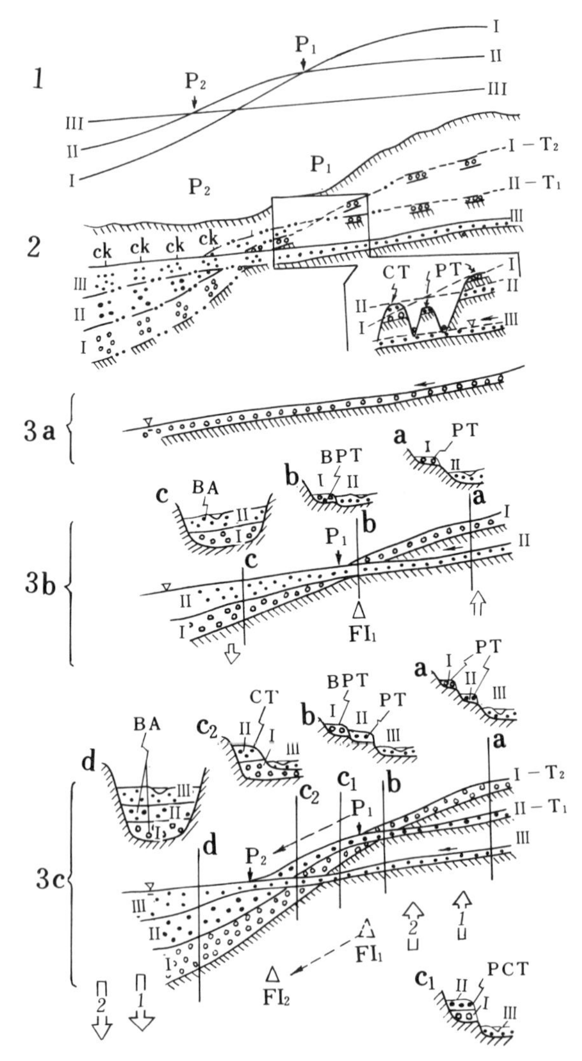


FIGURE 8. Deformation of river terraces in warping areas. A--schematic profile; B--actual profile observed in the field; C1, C2, C3--stages of warping; P--plunging point; F1--flexure axis; PT--terrace with bedrock pedestal; BPT--terrace with buried bedrock pedestal; BA--buried alluvial strata; CT--compound terrace; PCT--compound terrace with bedrock pedestal; I,II,III--different geological ages. Arrow shows direction of tectonic movement.

appearance is that of a terrace of the accumulative type, but it is composed of two or more alluvial formations that differ in age and that correspond to the terraces of different levels in the adjacent upwarped area, but in a reverse order; the lowest and oldest alluvial stratum in the downwarped area corresponds to the highest and oldest terrace in the upwarped area. The appearance of such a terrace implies that this area has experienced two tec-

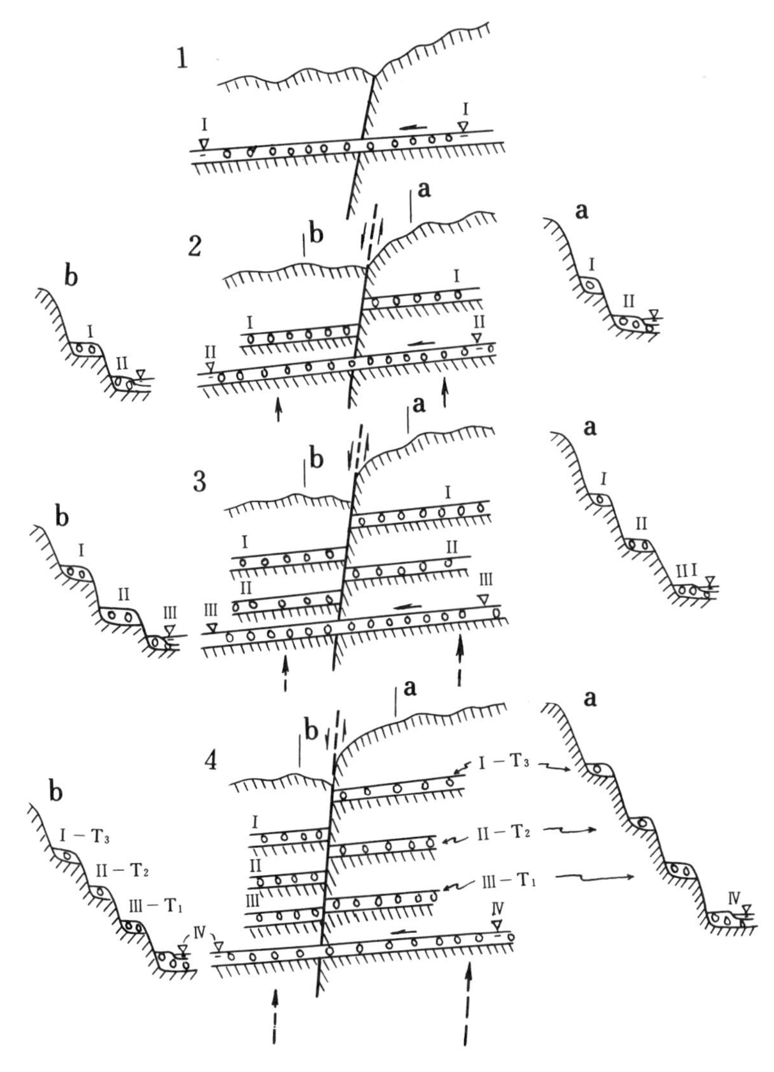


FIGURE 9. Type A of terrace vertical displacement. 1,2,3,4--different stages of displacement; a,b--cross section; I,II,III--different geological ages; T1,T2,T3--terrace order. Arrow shows direction of tectonic movement.

tonic stages: first, intermittent downwarping, then, reverse upwarping. Compound terraces of this type can be seen on both sides of the Taihangshan: in the Linfen basin of the Shanxi graben system on the western side (Fig. 6) and along the piedmont zone on the eastern side (Han and Zhao, 1980).

2. Terrace Vertical Displacement by Faulting.

In the area where a river valley is crossed by an active fault, three types of terrace displacement have been recognized.

Type A: Vertical displacement of terraces related to differential uplift on the two sides of a fault (Fig. 9). In this case, terraces of the same geological age are higher on the upthrown side and lower on downthrown side. Sometimes, a larger number of terraces are present on the upthrown side than on the downthrown side.

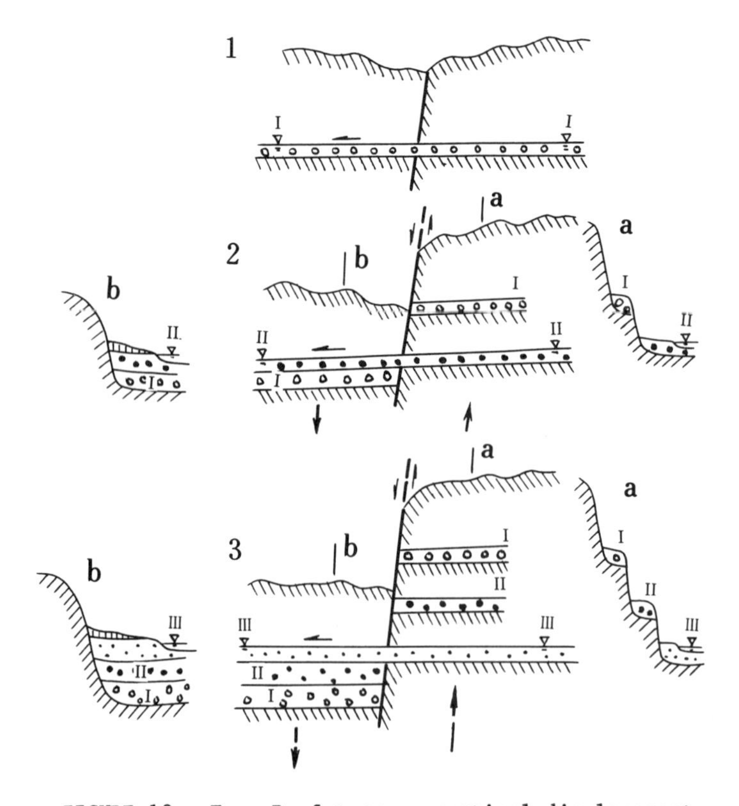


FIGURE 10. Type B of terrace vertical displacement.

Type B: Vertical displacement of terraces resulting from contrasting motion on the two sides of a fault (Fig. 10). In this case, intermittent uplift of the upthrown side has resulted in the formation of many cut terraces and intermittent subsidence of the downthrown side led to formation of a well developed floodplain over a great depth of underlying alluvial strata with different geological ages. There is no terrace on the downthrown side, where each of the buried alluvial strata corresponds to a terrace of the same age on upthrown side in the reverse order, i.e. the lowest and oldest buried stratum coincides in age with the highest and oldest terrace on the upthrown side.

Type C: Vertical displacement of terraces reflecting compound motion (Fig. 11). In this case, terrace displacement experiences two stages. In the first stage, terrace displacement occurs in the manner of contrasting motion as in Type B. On the upthrown side a series of terraces is formed due to intermittent uplift and on the downthrown side several alluvial strata are buried owing to intermittent subsidence. In the second stage, terrace displacement took place in the manner of differential uplift, as in Type A. The upthrown side continued to undergo intermittent uplift resulting in formation of a series of younger and lower terraces, and the downthrown side reversed from subsidence to intermittent uplift, raising the buried alluvial strata and forming a compound terrace as described earlier in the case of terrace deformation in a warped area. Such a tectonic reversal from subsidence to uplift on the downthrown side can be caused also by expansion of the uplifted area or migration of the upwarp center toward the downwarped area.

These three types of terrace displacement are different geomorphic expressions of growth faults in Quaternary time, and the rate of tectonic activity can be estimated if the elevation of each terrace and the thickness or depth of each buried alluvial stratum and its age are known.

Terrace displacement of the above types can be observed on both sides of the Taihangshan and even in the piedmont zone not far from the city of Beijing. Such types of terrace displacement,

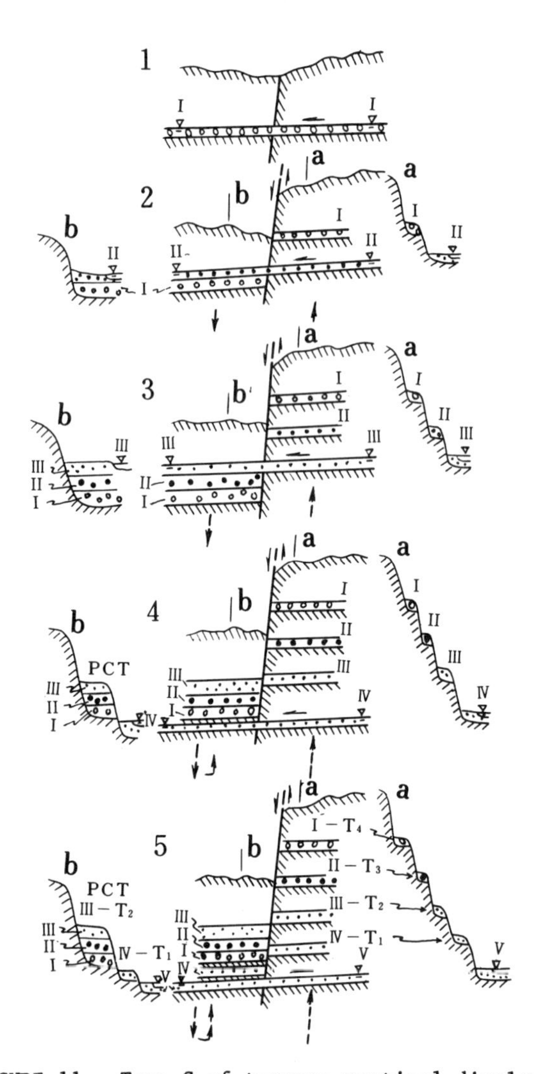


FIGURE 11. Type C of terrace vertical displacement.

especially Types A and B can be seen also in the case of marine terraces (Han et al., 1983c).

3. Response of Alluvial Fans to Active Tectonics.

Four types of alluvial fan structures associated with different patterns of neotectonic movements are recognized (Fig. 12).

Type A: Inset fans (Fig. 12A). In this case, each successively younger alluvial fan is inset into an erosional breach in the previous fan, leading the latter to be a sharply defined fan

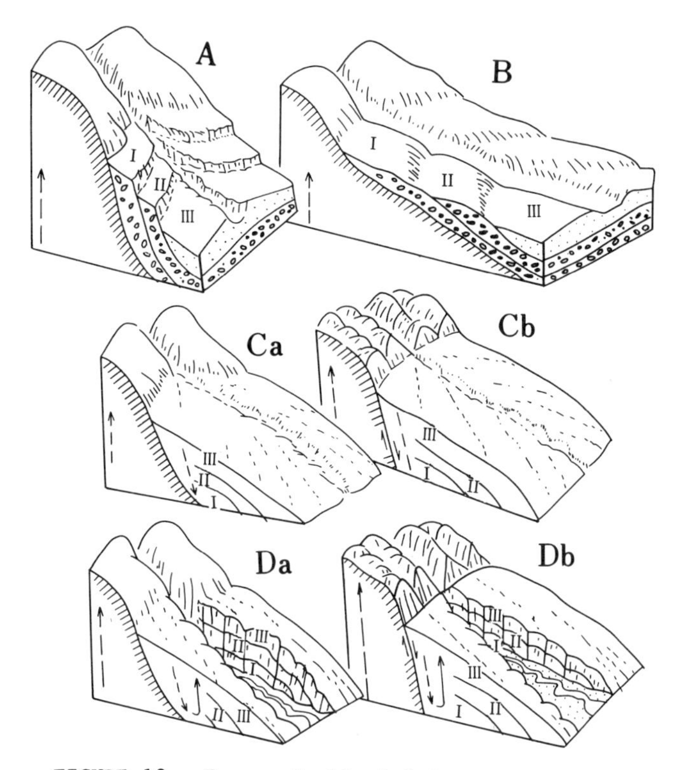


FIGURE 12. Types of alluvial fan occurrence.

terrace. Such nested fans usually occur in the narrow piedmont zone between an intermittently but vigorously upwarped mountain area and a closely situated, strongly downwarped or subsided basin or plain.

Type B: Imbricated fans (Fig. 12B). Here, each younger fan laps onto the middle to lower portion of the next older and higher one. This usually occurs in a broad piedmont zone between a lightly upwarped mountain area and a somewhat distant downwarped or subsided basin or plain.

Type C: Buried fans (Fig. 12C). In this case, each successively older fan is buried by the next younger one. It occurs in the area where the piedmont zone is intermittently downwarped (Fig. 12A) or downfaulted (Fig. 12B).

Type D: Compound fans (Fig. 12D). Formation of this type of fan, like that of a compound river terrace, requires two stages.

In the first stage, intermittent downwarping or downfaulting of the piedmont zone results in the formation of a series of superposed alluvial fans (Type C). Then in the second stage tectonic reversal to upwarping causes the buried fans to be uplifted and incised, and hence to form a compound fan consisting of several generations of fan bodies. This type of fan is also produced by expansion of the uplifted area or tectonic migration of the upwarp center toward the downwarped (Fig. 12Da) or downfaulted (Fig. 12Db) area.

All of these types of fans can be seen on both sides of the Taihangshan and not far from the city of Beijing. Investigation of differences in alluvial fans along a great piedmont zone or periphery of a large intermont basin can reveal the characteristics of Quaternary fault activity and active tectonics.

## APPLICATIONS OF MORPHOTECTONIC RESEARCH TO EARTHQUAKE PREDICTION

According to Chinese experience, morphotectonic research can be used in two fields for earthquake prediction.

The first field is to study the active tectonics and seismotectonics of prehistoric and historic earthquake regions for the purpose of summarizing the morphotectonic patterns and features associated with them, and then to use the obtained knowledge in combination with information from the related sciences such as geophysics, seismology, and geodetic research in identifying the potential earthquake risk region and in providing advice as to where and how to monitor hazardous tectonic activity.

The second field is to provide information and advice in assessing potential seismic intensity for application to construction and building codes, and also in the siting of large projects such as reservoirs and nuclear power plants.

In the first field, the Chinese examples indicate that most earthquakes occur at specific tectonic locations, as where active faults end, curve sharply, or intersect: where the tectonic stress is easy to concentrate and a locked area is likely to be produced. Such tectonically specific locations are usually demonstrated in morphotectonic patterns and features and can be revealed by morphotectonic research (Han, 1982).

In addition, one of the author's studies has shown that the pivotal part of a compressional-shearing pivotal fault is also a tectonically specific location prone to earthquakes. For example, in the southern part of the Yunnan-Guizhou plateau, the 1970 Tonghai earthquake (M = 7.7) was found to occur on a NNW-trending right-lateral compressional-shearing pivotal fault, where the dip of fault plane changes from southwestward to northeastward. Such a mode of Quaternary faulting has been indicated by vertical displacement of Neogene planation surfaces and Pleistocene river terraces in different ways on the two sides of the northwestern and southeastern sections of this fault, along with right-lateral offset of drainage, alluvial fans and an ancient valley (Fig. 2; Han, 1980; Han et al., 1983a).

It is reported that in the western part of Sichuang province the 1973 Luhuo earthquake (M = 7.9) also occurred in the pivotal part of a fault, but it is a NNW-NW-NNW striking left-lateral compressional-shearing pivotal fault, the famous Xianshuihe (Fresh Water River) fault (Huang, 1982).

After intensive morphotectonic research in the southern section of the eastern piedmont fault zone of the Taihangshan--the Tangyin graben--this investigator predicted an earthquake at the terminus of the western boundary fault of the Tangyin graben, which was found to be a hinge fault, passing into an unfaulted flexure. This prediction has been verified by a subsequent earthquake (M = 3.9) at the point where the fault passes into the flexure (Fig. 6; Han and Zhao, 1980).

If there are many tectonically specific locations in a potential earthquake risk region, what we want to know is where the tectonic stress will be concentrated, hence where the most dangerous location will be. For this purpose, in 1982 the author and his graduate, Shilong Zhu, conducted two-dimensional computer modeling of morphotectonics and its relationship with seismicity by applying the finite element method during their morphotectonic research in the southern piedmont area of the Taihangshan (Fig. 13).

It has been found that all the Quaternary active faults in

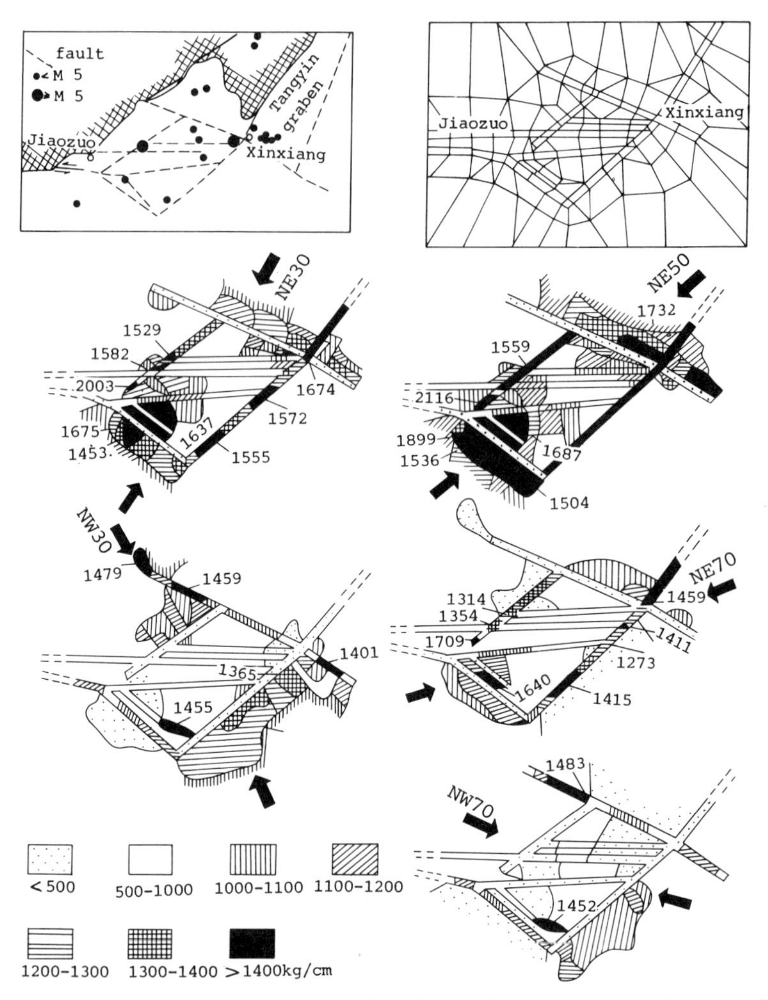


FIGURE 13. Morphotectonics and seismicity in the southern piedmont area of Taihangshan and results of computer modelling using the finite element method. Upper left--the area studied; Upper right--finite element grids for modelled area; Middle and lower right--distributions of the maximal shear stress value on the model under the action of principal compressional shear stresses of varying direction; Lower left--grade of maximum shear stress value.

the study area constitute a buried complex rhombic block traversed by several through-going faults. This rhombic block controls the drainage pattern, the changes in river gradients, and the distribution of Neogene-Pleistocene strata, and is situated in the neotectonic stress field of North China having a NE-trending principal compressional stress. Several historical and recent small and medium earthquakes have occurred within and adjacent to this rhombic block.

Since the morphotectonic, geodetic and seismologic data indicate that horizontal crustal movements have been predominant in this area in Quaternary and Recent time, the horizontal principal stresses in different directions are applied to the morphotectonic model and their effects are calculated by computer. The results of morphotectonic modeling show that under a N50$^{\circ}$E principal compressional stress there will be extreme values of shear stress at the western and eastern corners of the rhombic block (Fig. 13). That means these two locations are most prone to strong earthquakes (Zhu, 1984; Han et al, in preparation). Our future task is to carry out corresponding observations and monitor the area by different techniques for full detection of the seismic hazard.

ACKNOWLEDGEMENTS: The author is very grateful to Professor Theodore M. Oberlander of the Department of Geography, University of California, Berkeley, to Lucille Oberlander, and to Professor Marie Morisawa, Department of Geological Sciences, State University of New York, Binghamton, for their assistance in preparing, reviewing, and editing this paper.

REFERENCES

Bi, Fuzhi and Zhou, Caizhong, 1984, Uplift and subsidence of the coast south of Putian county, Fujian province, in the last 3,000 and more years: Acta Oceanologica Sinica, v. 6, no. 1, pp. 55-60.*

Deng, Qidong, et al., 1979, On the tectonic stress field in China and its relation to plate movement: Seismology and Geology, v. 1, no. 1, pp. 11-22.**

Er, Ke and Li, Zhizhen*, 1976, Characteristics of neotectonic movements in the Linfen basin, Shanxi province, and their relations to seismicities: Geological Sciences and Technology, no. 4, pp. 52-62, Geol. Pub. House. (* Pen names of the present author and his co-author)

Han, Mukang, 1980, A discussion on the location of the 1970 Tonghai earthquake of magnitude 7.7 in Yunnan province: Seismology and Geology, v. 2, no. 2, p. 30.*

---------, 1982, Morphotectonic features of three earthquake regions in China and their relations to earthquakes, in Dongsheng Liu, ed., Quaternary geology and environment of China: China Ocean Press, Beijing, pp. 185-187.

---------, 1983, Geomorphic expressions of the neotectonic stress fields bordering the Taihangshan range of North China: presented at the annual meeting, Association of American Geographers, Denver, Colorado, April, 1983.

Han, Mukang and Chai, Tianjun, 1979, Neotectonic movements in the northern section of the Yuanmou fault zone, Yunnan province, and the earthquake risk involved: Seismology and Geology, v. 1, no.1, pp. 56-65.**

Han, Mukang, and Zhao, Jingzhen, 1980, Seismotectonic characteristics of Tangyin graben, Henan province, and its earthquake risk: Seismology and Geology, v. 2, no. 4, pp. 47-58.**

Han, Mukang et al., 1981a, The seismological features in the Yongshen earthquake region, Western Yunnan: Journal of Seismological Research, v. 4, no. 1, pp. 35-43.**

---------, 1981b, The Holocene migration of the Qilu lake in southern Yunnan and its relation to the neotectonic movements in the 1970 Tonghai earthquake area: Geological Review, v. 27, no. 6, pp. 491-495.

---------, 1983a, Morphotectonic features of the Tunghai earthquake region, Yunnan province: Acta Geographica Sinica, v. 38, no. 1, pp. 41-54.**

---------, 1983b, Geomorphic expressions of the Quaternary tectonic stress field in the southern section of the Eastern piedmont fault zone of Taihangshan mountain: Acta Geographica Sinica, v. 38, no. 4, pp. 348-357.**

---------, 1983c, Marine terraces, neotectonic movements, and degree of earthquake risk in the western coastal region of Laoning province: Acta Oceanologica Sinica, v. 5, no. 6, pp. 743-753.*

---------, (in preparation), Morphotectonics, its relationship with seismicity and result of their computer modelling in the southern piedmont area of the Taihangshan mountain, North China.

Huang, Shengmu, 1972, A preliminary classification of active faults and the earthquake structures in Sichuang province, in Committee on Seismogeology of the Seismological Society of China, ed., The active faults in China: Seismological Press, Beijing, pp. 236-241.*

Kan, Rongji, et al., 1977, A discussion on the recent tectonic stress field and the characteristics of recent tectonic movements in the southwestern area of China: Acta Geophysica Sinica, v. 20, no. 2, pp. 96-109.**

Wang, Nailiang, and Han, Mukang, 1984, Theories, methods, application, and tendency of tectonic geomorphology, in The Geographical Society of China, eds., selection of papers presented at the All China Symposium on tectonic geomorphology, held in Datong, Shanxi province, Aug. 1982: Science Press, pp. 1-09.*

Wang, Yipeng, 1979, Interplate earthquake and Meso-Cenozoic stress field in China: Seismology and Geology, v. 1, no. 3, pp. 1-11.**

Zhang, Hunan, 1982, On activity of the NW-trending fractures along the Fujian-Guangdong region: Seismology and Geology, v. 4, no. 3, pp. 17-25.**

Zhu, Shilong, 1984, Recent tectonic stress state of Qixiang-Xinxiang-Jiaoxuo area along the piedmont belt of Taihang mountain and its relation to seismicity: Seismology and Geology, v. 6, no. 1, pp. 13-20.**

*In Chinese.

**In Chinese with English abstract.

# Morphotectonics working group of the International Geographic Union

*Mario Panizza*

The Working Group started activity as a Sub-commission of the Geomorphological Survey and Mapping Commission of the International Geographic Union. The proposal to form the Sub-commission on Morphotectonics was formulated during the meeting in Baku (USSR) in June 1978. This proposal was accepted, Professor Piotrovski was appointed Chairman and I was appointed Vice-Chairman. The most important aims of the Sub-commission concern methodology rather than discoveries and results. In particular, a program to develop a geomorphological methodology for the identification of neotectonic movements was proposed. The work, in which many scientists took part, continued during the Meeting in Basel (Switzerland) in September, 1978.

Numerous researchers of different nationalities participate informally in this Sub-commission, both to discuss problems of methodology and to study and discuss in field excursions the relations between landforms and tectonic structures. In particular, during the meeting of the Geomorphological Survey and Mapping Commission in Italy, September, 1979, numerous papers were presented on this theme and some excursions were organized on questions of morphotectonics. The texts of the papers and accounts of the proceedings and the excursions of the meeting are collected in the volume: "Proceedings of the 15th Plenary Meeting I.G.U. Commission, Geomorphological Survey and Mapping, Modena - Italy, September 1979".

In the course of the XXIV International Congress of Geography held in Tokyo (Japan) twenty papers on this theme were presented in Section I "Geomorphology and Glaciology". During the sessions there were lively discussions, showing scientific interest in the relations between geomorphology and tectonics.

These activities have shown the desirability and even the necessity for scientists interested in morphotectonics to meet in order to discuss common problems connected with this theme. From these discussions, the initiative to establish a Working Group within the International Geographical Union was started. This Working Group was formally approved in 1981.

In August 1982 the first meeting took place in Rio Claro (Brazil) within the Latin American Regional Conference of the International Geographical Union. Proceedings of the meeting have been published by the International Institute for Aerial Survey and Earth Sciences (Enschede, The Netherlands).

The second meeting took place in Sofia (Bulgaria) in October 1983, with the participation of many researchers coming from eight countries. Fifteen scientific papers were presented and a two-day field trip was held in southwestern Bulgaria. Of particular interest were the problems of global plate tectonics and the genesis of large landforms closely associated with it; tectonic-magmatic activation and its role in late relief formation; neotectonic movements and their decoding by means of morphostructural analysis; recent crustal movements; earthquakes and their correlation with the morphostructural blocks. The Proceedings of the meeting were printed by the Bulgarian Academy of Sciences.

The third meeting was held in Paris (France) in August 1984, during the 25th International Geographical Congress, with the participation of about 50 scientists.

At present the Working Group is constituted of about seventy members (Full Members and Corresponding Members) from about thirty countries on all the continents.

The subjects of the Working Group are:

1) From a general point of view: the W.G. proposes to define the basic concepts regulating the relationships between landforms and tectonism, the latter being considered in both static and dynamic roles.

2) Regarding methodology: the W.G. proposed to prepare a "guidebook" with explanatory notes and sketches giving a list of all the landforms which can be associated with neotectonism.

3) Regarding applied earth sciences: it underlines the growing importance assumed by morphoneotectonics in its practical applications (such as in the construction of large building complexes, dams and nuclear power plants) where seismic risk should be evaluated. In this respect, morphotectonics has a fundamental role together with other disciplines such as geology, geophysics, geochemistry, etc.

In particular now we are working on a Glossary of Morphotectonics. A first draft was presented during the 25th International Geographical Congress in Paris, August 1984. The final version is to be presented at the 26th International Geographical Congress in 1988 in Australia.

The second work is a Guidebook on Morphoneotectonics. An index of the following topics has been proposed:

Methodologic base of Morphoneotectonics;
Methods of Morphoneotectonics research;
Methodology of Morphoneotectonic analyses;
Morphoneotectonic Mapping;
Applied Morphoneotectonic Studies;
Complex and applied morphoneotectonic mapping;
Studies and examples in different countries.

A third new program will be prepared for the next meeting concerning the relationships between plate tectonics and geomorphology. The title will be Global Geomorphology.

In order to deal with the above mentioned programs, special commissions have been established with the purpose of collecting, standardizing and preparing the scientific material proposed by the members of the Working Group to be presented during general assemblies. The present members will, in the end, approve a final document which could be either a final result or a stage for further work.

The next meetings will be held in:

Brno (Czechoslovakia), in 1985;
Manchester (United Kingdom), on the occasion of the First International Conference on Geomorphology, in 1985;

Barcelona (Spain), during the International Geographic Conference of the Mediterranean Countries, in 1986.

Italy, in 1987;

Australia, in 1988.

I wish to invite those taking part in this Symposium, who are interested in the Morphotectonic Working Group work, to collaborate with us and to keep in contact with me for any other detail.